Birkhäuser

Advanced Courses in Mathematics
CRM Barcelona

Centre de Recerca Matemàtica

Managing Editor:
Carles Casacuberta

For further volumes:
http://www.springer.com/series/5038

Roberto Cominetti
Francisco Facchinei
Jean B. Lasserre

Modern Optimization Modelling Techniques

Editors for this volume:
Aris Daniilidis (Universitat Autònoma de Barcelona)
Juan Enrique Martínez-Legaz (Universitat Autònoma de Barcelona)

Birkhäuser

Roberto Cominetti
Departamento de Ingeniería Industrial
Universidad de Chile
Santiago de Chile
Chile

Francisco Facchinei
Department of Computer, Control, and
Management Engineering Antonio Ruberti
Università di Roma "La Sapienza"
Roma
Italy

Jean B. Lasserre
LAAS-CNRS
Toulouse
France

ISBN 978-3-0348-0290-1 ISBN 978-3-0348-0291-8 (eBook)
DOI 10.1007/978-3-0348-0291-8
Springer Basel Heidelberg New York Dordrecht London

Library of Congress Control Number: 2012944421

Mathematics Subject Classification (2010): Primary: 90-XX, 91-XX; Secondary: 12Y05, 14P10, 65K10, 65K15, 90B20, 90C22, 90C26, 90C30, 90C33, 90C40, 91A10, 91A13, 91A26, 91B51

Printed on acid-free paper

Springer Basel AG is part of Springer Science+Business Media (www.birkhauser-science.com)

Contents

Part II: Computation of Generalized Nash Equilibria: Recent Advancements

Francisco Facchinei

Part III: Equilibrium and Learning in Traffic Networks

Roberto Cominetti

Preface

During the period July 20–24, 2009, the research group on Optimization of the Autonomous University of Barcelona organized an advanced course at the CRM, with the aim of promoting research in the area of optimization in all of its components: theory, algorithms, and practical problems. This volume is a unified version of the material presented in the course.

The advanced course was entitled *Optimization: Theory, Methods, and Applications*. The courses and the written material were accordingly divided into these three main parts. The theoretical part of the book is a self-contained course on the general moment problem and its relations with semidefinite programming, presented by Jean B. Lasserre, senior researcher at the CNRS (France), world-leading specialist of the domain and author of a recent research monograph on this topic (Imperial College Press, 2009). The second part is dedicated to the problem of determination of Nash equilibria from an algorithmic viewpoint. This part is presented by Francisco Facchinei, professor at the University of Roma "La Sapienza", established researcher and co-author of an extended monograph on this topic (Springer, two volumes). The third part is a study of congestion models for traffic networks. This part develops modern optimization techniques for finding traffic equilibria based on stochastic optimization and game theory. It has been presented by Roberto Cominetti, professor at the University of Chile, who has been working for several years on congestion models of the traffic of the municipality of Santiago de Chile.

This advanced course was an i-MATH activity (ref. 2009 MIGS-C4-0212), which was also supported by the Spanish Ministry of Science and Innovation (Complementary Actions, ref. MTM2008-04356E). We wish to thank the CRM direction and administrative staff for the logistic support, and our three main lecturers for the excellent course and the quality of the material presented. We also thank our colleagues Emilio Carrizosa (Sevilla), Laureano Escudero (Rey Juan Carlos), Claude Lemaréchal (INRIA Rhône-Alpes), and Justo Puerto (Sevilla), who agreed to deliver invited talks complementary to the courses, as well as the 70 participants of the event. Our special thanks to Sabine Burgdorf (Konstanz), Vianney Perchet (Paris), Philipp Renner (Zürich), Marco Rocco (Bergamo), and Guillaume Vigeral (Paris), who accepted to review carefully several parts of this material.

Bellaterra, February 2011

Aris Daniilidis
Juan Enrique Martínez-Legaz

Part I

Moments and Positive Polynomials for Optimization

Jean B. Lasserre

Introduction

Consider the optimization problem

$$f^* = \inf_{\mathbf{x}} \{f(\mathbf{x}) \,:\, \mathbf{x} \in \mathbf{K}\} \tag{1}$$

for some given measurable function $f: \mathbb{R}^n \to \mathbb{R}$ and a Borel subset $\mathbf{K} \subset \mathbb{R}^n$. Here we insist on the fact that f^* is the *global* minimum on $\mathbf{K}$, as opposed to a *local* minimum. In full generality, problem (1) is very difficult and there is no general purpose method, not even to approximate f^*.

Observe that, when one searches for the global minimum f^*, problem (1) has two equivalent formulations:

$$f^* = \inf_{\mu \in \mathcal{M}(\mathbf{K})_+} \left\{ \int_{\mathbf{K}} f \, d\mu \,:\, \mu(\mathbf{K}) = 1 \right\} \tag{2}$$

and

$$f^* = \sup_{\lambda} \{\lambda \,:\, f(\mathbf{x}) - \lambda \geq 0, \ \forall \mathbf{x} \in \mathbf{K}\}, \tag{3}$$

where $\mathcal{M}(\mathbf{K})_+$ is the space of finite Borel measures on $\mathbf{K}$ (the convex positive cone of the vector space $\mathcal{M}(\mathbf{K})$ of finite signed Borel measures on $\mathbf{K}$).

To see that (2) is indeed equivalent to (1), observe that, for every $\mathbf{x} \in \mathbf{K}$, $f(\mathbf{x}) = \int_{\mathbf{K}} f \, d\mu$ with $\mu \in \mathcal{M}(\mathbf{K})_+$ being the Dirac measure $\delta_{\mathbf{x}}$ at $\mathbf{x}$. On the other hand, if $f^* > -\infty$ then, as $f - f^* \geq 0$ on $\mathbf{K}$, integrating with respect to any probability measure $\mu \in \mathcal{M}(\mathbf{K})_+$ yields $\int_{\mathbf{K}} f \, d\mu \geq f^*$. In addition, if $\mathbf{x}^* \in \mathbf{K}$ is a global minimizer then $\mu = \delta_{\mathbf{x}^*}$ is also a global minimizer of (2).

In fact, the two formulations (2) and (3) are dual of each other in the sense of classical LP-duality if one observes that (2) is the infinite-dimensional LP

$$f^* = \inf_{\mu \in \mathcal{M}(\mathbf{K})} \{\langle f, \mu \rangle \,:\, \langle 1, \mu \rangle = 1; \ \mu \geq 0\}, \tag{4}$$

where $\langle g, \mu \rangle = \int_{\mathbf{K}} g \, d\mu$ for every $\mu \in \mathcal{M}(\mathbf{K})$ and every bounded measurable function g on $\mathbf{K}$ (e.g., assuming that f is bounded measurable on $\mathbf{K}$). Notice that, in (4), when $f \in \mathbb{R}[\mathbf{x}]$ the unknown measure μ only apppears in $\int_{\mathbf{K}} f \, d\mu$ through its moments $y_{\boldsymbol{\alpha}} = \int_{\mathbf{K}} \mathbf{x}^{\boldsymbol{\alpha}} \, d\mu$, $\boldsymbol{\alpha} \in \mathbb{N}^n$.

The optimization problems (3) and (4) are convex and even linear but also infinite-dimensional! In (3) there are uncountably many constraints $f(\mathbf{x}) \geq \lambda$, $\mathbf{x} \in \mathbf{K}$, whereas in (4) the unknown is a signed Borel measure $\mu \in \mathcal{M}(\mathbf{K})$ (and not a finite-dimensional vector as in classical LP). With no other assumption on neither f nor $\mathbf{K}$, one does not know how to solve (3) and (4) because one does not have *tractable characterizations* of Borel measures supported on $\mathbf{K} \subset \mathbb{R}^n$ (for solving (4)), or functions nonnegative on $\mathbf{K}$ (for solving (3)). And so in general (3) and (4) are only a *rephrasing* of (1).

However, if one now considers problem (1) with the restrictions that

(i) $f\colon \mathbb{R}^n \to \mathbb{R}$ is a polynomial (or even a rational function), and

(ii) $\mathbf{K} \subset \mathbb{R}^n$ is a compact basic semi-algebraic set,

then one may solve (or approximate as closely as desired) the linear programs (3)–(4). Indeed, relatively recent results from real algebraic geometry permit to characterize polynomials that are positive on $\mathbf{K}$, which is exactly what we need to solve (3). In addition, it turns out that those characterizations are tractable as they translate into *semidefinite* (or sometimes *linear*) conditions on the coefficients of certain polynomials that appear in some appropriate representation of the polynomial $\mathbf{x} \mapsto f(\mathbf{x}) - \lambda$, positive on $\mathbf{K}$. Moreover, the previous representation results have a nice dual facet which is concerned with sequences of reals $\mathbf{y} = (y_{\boldsymbol{\alpha}})$, $\boldsymbol{\alpha} \in \mathbb{N}^n$, that are the *moments* of a measure μ supported on $\mathbf{K}$, which is exactly what we need to solve the linear program (4) when $f \in \mathbb{R}[\mathbf{x}]$.

Even with the above two restrictions (i)–(ii), problem (1) still encompasses many important optimization problems. In particular, it includes 0/1 optimization problems, modelling $x_i \in \{0,1\}$ via the quadratic polynomial constraint $x_i^2 = x_i$. Therefore, one should always have in mind that one addresses NP-hard problems in general.

In the sequel we first describe results from real algebraic geometry on the representation of polynomials positive on $\mathbf{K}$, and their dual counterparts in functional analysis about sequences $\mathbf{y} = (y_{\boldsymbol{\alpha}})$ that have a finite representing Borel measure μ supported on $\mathbf{K}$. We then show how to use those results to define a hierarchy of *convex relaxations* for problem (1), whose associated monotone sequence of optimal values converges to the global minimum f^* (sometimes with even finite convergence). Depending on the type of representation which we use, those convex relaxations are either *semidefinite* or *linear* programs whose size increases with the rank in the hierarchy.

Recall that this general purpose approach (that one may call the *moment approach*) aims at solving (or approximate as closely as desired) difficult NP-hard problems. On the other hand, one also knows that there is a large class of convex optimization problems that are considered "easy" as they can be solved efficiently by several *ad hoc* methods. Therefore a highly desirable feature of a general purpose approach is to *recognize* easier convex problems and behave accordingly (even if it may not be as efficient as specific methods tailored to the convex case). We show that this is indeed the case for the moment approach based on semidefinite relaxations which uses representation results based on *sums of squares*, whereas it is not the case for the LP-relaxations which use other representation results.

And so it is an important and remarkable feature of the moment approach based on semidefinite relaxations, to *not* distinguish between convex continuous problems (considered easy) and non convex (possibly discrete) problems (considered hard). For instance, a boolean variable x_i is not treated with any particular attention and is just modeled via the quadratic equality constraint $x_i^2 - x_i = 0$, just one among the many other polynomial equality or inequality constraints in the definition of the feasible set $\mathbf{K}$. Running a local minimization algorithm of continuous optimization with such a modeling of a boolean constraint would not be considered wise (not to say more)! Hence this might justify some doubts concerning the efficiency of the moment approach by lack of specialization. Yet, and remarkably, the resulting semidefinite relaxations still provide the strongest relaxation algorithms for hard combinatorial optimization problems and in the same time they also recognize easy convex problems as in this case convergence is even finite (and sometimes at the first semidefinite relaxation)!

Next, we also consider *parametric* optimization problems, that is, optimization problems where the criterion to minimize as well as the constraints that describe the feasible set may depend on some parameters that belong to a given set. Sometimes, in the context of optimization with data uncertainty, some probability distribution on the set of parameters is also available. The ultimate and difficult goal is to compute or at least provide some information and/or approximations on the global optimum and the global minimizers, viewed as *functions* of the parameters. This is a very challenging problem and, in general, one may only obtain local information around some nominal value of the parameter. However, in the context of polynomial optimization, we show that much more can be done. Indeed, we describe what we call the "joint + marginal" approach to parametric optimization and show that the moment methodology is well-suited for providing good approximations (in principle, as closely as desired) when the parametric optimization problem is described via polynomials and basic semi-algebraic sets. Finally, we show how to use these results of parametric optimization to define what we call a "joint + marginal algorithm" for polynomial programs (4). The basic and simple idea is (a) to consider the first coordinate x_1 as a parameter in some interval $\mathbf{Y}$, and (b) using the "joint + marginal methodology", approximate the optimal value function $y \mapsto J(y) = \min\{f(\mathbf{x}) : \mathbf{x} \in \mathbf{K};\, x_1 = y\}$ by a univariate polynomial. In minimizing this univariate polynomial on $\mathbf{Y}$ (an easy task), one obtains an estimate $\tilde{x}_1$, and then one iterates with x_2, etc., until one obtains a point $\tilde{\mathbf{x}}$ which may be used as an initial guess in some standard local minimization algorithm.

Chapter 1

Representation of Positive Polynomials

In one dimension, the ring $\mathbb{R}[x]$ of real polynomials in a single variable has the fundamental property (Theorem 1.6) that every nonnegative polynomial $p \in \mathbb{R}[x]$ is a sum of squares of polynomials, that is,

$$p(x) \geq 0, \quad \forall x \in \mathbb{R} \quad \Longleftrightarrow \quad p(x) = \sum_{i=1}^{k} h_i(x)^2, \quad \forall x \in \mathbb{R},$$

for finitely many polynomials (h_i). In multiple dimensions, however, it is possible for a polynomial to be nonnegative without being a sum of squares. In fact, in his famous address in a Paris meeting of mathematicians in 1900, Hilbert posed important problems for mathematics to be addressed in the 20th century, and in his 17th problem he conjectured that every nonnegative polynomial can always be written as a sum of squares of rational functions. This conjecture was later proved to be correct by Emil Artin in 1926, using the Artin–Schreier theory of real closed fields.

An immediate extension is to consider characterizations of polynomials p that are nonnegative on a basic semi-algebraic set $\mathbf{K}$ defined by polynomial inequalities $g_i(\mathbf{x}) \geq 0$, $i = 1, \ldots, m$. By characterization we mean a representation of p in terms of the g_i's such that the nonnegativity of p on $\mathbf{K}$ follows immediately from the latter representation, in the same way that p is obviously nonnegative when it is a sum of squares. In other words, this representation of p can be seen as a *certificate* of nonnegativity of p on $\mathbf{K}$.

In this chapter, we review the key representation theorems for nonnegative (or positive) polynomials. These results form the theoretical basis on which are based the (semidefinite or linear) convex relaxations that we later define for solving the polynomial optimization problem (1).

1.1 Sum of squares representations and semidefinite optimization

In this section we show that if a nonnegative polynomial has a sum of squares representation then one can compute this representation by using semidefinite optimization methods. Given that semidefinite optimization problems are efficiently solved both from a theoretical and a practical point of view, it follows that we can compute a sum of squares decomposition of a nonnegative polynomial, if it exists, efficiently.

Let $\mathbb{R}[\mathbf{x}]$ denote the ring of real polynomials in the variables $\mathbf{x} = (x_1, \dots, x_n)$. A polynomial $p \in \mathbb{R}[\mathbf{x}]$ is a sum of squares (in short s.o.s.) if p can be written as

$$\mathbf{x} \longmapsto p(\mathbf{x}) = \sum_{j \in J} p_j(\mathbf{x})^2, \quad \mathbf{x} \in \mathbb{R}^n,$$

for some finite family of polynomials $(p_j : j \in J) \subset \mathbb{R}[\mathbf{x}]$. Notice that necessarily the degree of p must be even, and the degree of each p_j is bounded by half of that of p.

Denote by $\Sigma[\mathbf{x}] \subset \mathbb{R}[\mathbf{x}]$ the space of s.o.s. polynomials. For any two real symmetric matrices $\mathbf{A}, \mathbf{B}$, recall that $\langle \mathbf{A}, \mathbf{B} \rangle$ stands for trace($\mathbf{AB}$), and the notation $\mathbf{A} \succeq 0$ stands for $\mathbf{A}$ is positive semidefinite. Finally, for a multi-index $\boldsymbol{\alpha} \in \mathbb{N}^n$, let $|\boldsymbol{\alpha}| = \sum_{i=1}^n \alpha_i$. With $\mathbf{x}'$ denoting the transpose of a vector $\mathbf{x}$, consider the vector

$$\begin{aligned} \mathbf{v}_d(\mathbf{x}) &= (\mathbf{x}^{\boldsymbol{\alpha}})_{|\boldsymbol{\alpha}| \le d} \\ &= \left(1, x_1, \dots, x_n, x_1^2, x_1 x_2, \dots, x_{n-1} x_n, x_n^2, \dots, x_1^d, \dots, x_n^d\right)' \end{aligned}$$

of all monomials $\mathbf{x}^{\boldsymbol{\alpha}}$ of degree less than or equal to d, which has dimension $s(d) = \binom{n+d}{d}$. Those monomials form the canonical basis of the vector space $\mathbb{R}[\mathbf{x}]_d$ of polynomials of degree at most d.

Proposition 1.1. *A polynomial $g \in \mathbb{R}[\mathbf{x}]_{2d}$ has a sum of squares decomposition (i.e., is s.o.s.) if and only if there exists a real symmetric and positive semidefinite matrix $\mathbf{Q} \in \mathbb{R}^{s(d) \times s(d)}$ such that $g(\mathbf{x}) = \mathbf{v}_d(\mathbf{x})' \mathbf{Q} \mathbf{v}_d(\mathbf{x})$, for all $\mathbf{x} \in \mathbb{R}^n$.*

Proof. Suppose that there exists a real symmetric $s(d) \times s(d)$ matrix $\mathbf{Q} \succeq \mathbf{0}$ for which $g(\mathbf{x}) = \mathbf{v}_d(\mathbf{x})' \mathbf{Q} \mathbf{v}_d(\mathbf{x})$ for all $\mathbf{x} \in \mathbb{R}^n$. Then $\mathbf{Q} = \mathbf{HH}'$ for some $s(d) \times k$ matrix $\mathbf{H}$, and thus

$$g(\mathbf{x}) = \mathbf{v}_d(\mathbf{x})' \mathbf{HH}' \mathbf{v}_d(\mathbf{x}) = \sum_{i=1}^{k} (\mathbf{H}' \mathbf{v}_d(\mathbf{x}))_i^2, \qquad \forall \mathbf{x} \in \mathbb{R}^n.$$

Since $\mathbf{x} \mapsto (\mathbf{H}' \mathbf{v}_d(\mathbf{x}))_i$ is a polynomial, g is expressed as a sum of squares of the polynomials $\mathbf{x} \mapsto (\mathbf{H}' \mathbf{v}_d(\mathbf{x}))_i$, $i = 1, \dots, k$.

Conversely, suppose that g of degree $2d$ has an s.o.s. decomposition $g(\mathbf{x}) = \sum_{i=1}^k h_i(\mathbf{x})^2$ for some family $\{h_i : i = 1, \dots, k\} \subset \mathbb{R}[\mathbf{x}]$. Then necessarily the degree of each h_i is bounded by d. Let $\mathbf{h}_i$ be the vector of coefficients of the polynomial h_i, i.e., $h_i(\mathbf{x}) = \mathbf{h}_i'\mathbf{v}_d(\mathbf{x})$, $i = 1, \ldots, k$. Thus,

$$g(\mathbf{x}) = \sum_{i=1}^{k} \mathbf{v}_d(\mathbf{x})'\mathbf{h}_i\mathbf{h}_i'\mathbf{v}_d(\mathbf{x}) = \mathbf{v}_d(\mathbf{x})'\mathbf{Q}\mathbf{v}_d(\mathbf{x}), \qquad \forall \mathbf{x} \in \mathbb{R}^n,$$

with $\mathbf{Q} \in \mathbb{R}^{s(d)\times s(d)}$, $\mathbf{Q} = \sum_{i=1}^{k} \mathbf{h}_i\mathbf{h}_i' \succeq \mathbf{0}$, and the proposition follows. □

Given a s.o.s. polynomial $g \in \mathbb{R}[\mathbf{x}]_{2d}$, the identity $g(\mathbf{x}) = \mathbf{v}_d(\mathbf{x})'\mathbf{Q}\mathbf{v}_d(\mathbf{x})$ for all $\mathbf{x}$ provides linear equations that the coefficients of the matrix $\mathbf{Q}$ must satisfy. Hence, if we write

$$\mathbf{v}_d(\mathbf{x})\mathbf{v}_d(\mathbf{x})' = \sum_{\boldsymbol{\alpha}\in\mathbb{N}^n} \mathbf{B}_{\boldsymbol{\alpha}}\, \mathbf{x}^{\boldsymbol{\alpha}}$$

for appropriate $s(d) \times s(d)$ real symmetric matrices $(\mathbf{B}_{\boldsymbol{\alpha}})$, checking whether the polynomial $\mathbf{x} \mapsto g(\mathbf{x}) = \sum_{\boldsymbol{\alpha}} g_{\boldsymbol{\alpha}}\mathbf{x}^{\boldsymbol{\alpha}}$ is s.o.s. reduces to solving the *semidefinite optimization (feasibility) problem:*

$$\begin{aligned} &\text{Find } \mathbf{Q} \in \mathbb{R}^{s(d)\times s(d)} \text{ such that} \\ &\mathbf{Q} = \mathbf{Q}', \quad \mathbf{Q} \succeq 0, \quad \langle \mathbf{Q}, \mathbf{B}_{\boldsymbol{\alpha}} \rangle = g_{\boldsymbol{\alpha}}, \quad \forall \boldsymbol{\alpha} \in \mathbb{N}^n, \end{aligned} \tag{1.1}$$

a tractable convex optimization problem for which efficient software packages are available. Indeed, up to a fixed arbitrary precision $\epsilon > 0$, a semidefinite program can be solved in a computational time that is polynomial in the input size of the problem. Observe that the size $s(d) = \binom{n+d}{n}$ of the semidefinite program (1.1) is bounded by n^d.

On the other hand, *nonnegativity* of a polynomial $g \in \mathbb{R}[\mathbf{x}]$ can be checked also by solving a single semidefinite program. Indeed, if g is nonnegative then it can be written as a sum of squares of rational functions, or equivalently, clearing denominators,

$$hg = f, \tag{1.2}$$

for some s.o.s. polynomials $h, f \in \Sigma[\mathbf{x}]$, $k = 1, \ldots, r$, and there exist bounds on the degree of h and f. Conversely, if there exist $h, f \in \Sigma[\mathbf{x}]$ solving (1.2) then obviously g is nonnegative. Therefore, using (1.1) for h and f in (1.2), finding a certificate $h, f \in \Sigma[\mathbf{x}]$ for nonnegativity of g reduces to solving a single semidefinite program. Unfortunately, the available bounds for the size of this semidefinite program are by far too large for practical implementation. This is what makes the sum of squares property very attractive computationally, in contrast with the weaker nonnegativity property which is much harder to check (if not impossible).

Example 1.2. Consider the polynomial in $\mathbb{R}[\mathbf{x}] = \mathbb{R}[x_1, x_2]$

$$f(\mathbf{x}) = 2x_1^4 + 2x_1^3x_2 - x_1^2x_2^2 + 5x_2^4.$$

Suppose we want to check whether f is a sum of squares. As f is homogeneous, we attempt to write f in the form

$$\begin{aligned} f(x_1,x_2) &= 2x_1^4 + 2x_1^3x_2 - x_1^2x_2^2 + 5x_2^4 \\ &= \begin{pmatrix} x_1^2 \\ x_2^2 \\ x_1x_2 \end{pmatrix}' \begin{bmatrix} q_{11} & q_{12} & q_{13} \\ q_{12} & q_{22} & q_{23} \\ q_{13} & q_{23} & q_{33} \end{bmatrix} \begin{pmatrix} x_1^2 \\ x_2^2 \\ x_1x_2 \end{pmatrix} \\ &= q_{11}\,x_1^4 + q_{22}\,x_2^4 + \big(q_{33} + 2q_{12}\big)\,x_1^2x_2^2 + 2q_{13}\,x_1^3x_2 + 2q_{23}\,x_1x_2^3, \end{aligned}$$

for some $\mathbf{Q} \succeq 0$. In order to have an identity, we obtain

$$\mathbf{Q} \succeq 0; \quad q_{11} = 2, \quad q_{22} = 5, \quad q_{33} + 2q_{12} = -1, \quad 2q_{13} = 2, \quad q_{23} = 0.$$

In this case we easily find the particular solution

$$0 \preceq \mathbf{Q} = \begin{bmatrix} 2 & -3 & 1 \\ -3 & 5 & 0 \\ 1 & 0 & 5 \end{bmatrix} = \mathbf{HH}', \quad \text{for} \quad \mathbf{H} = \frac{1}{\sqrt{2}} \begin{bmatrix} 2 & 0 \\ -3 & 1 \\ 1 & 3 \end{bmatrix},$$

and so

$$f(x_1,x_2) = \frac{1}{2}\big(2x_1^2 - 3x_2^2 + x_1x_2\big)^2 + \frac{1}{2}\big(x_2^2 + 3x_1x_2\big)^2,$$

which is indeed a sum of squares.

Sufficient condition for being s.o.s.

Observe that checking whether a polynomial $g \in \mathbb{R}[\mathbf{x}]_{2d}$ is s.o.s. via (1.1) requires introducing the auxiliary (symmetric matrix) variable $\mathbf{Q} \in \mathbb{R}^{s(d)\times s(d)}$, i.e., we do not have "if and only if" conditions expressed directly in terms of the coefficients $\mathbf{g} = (g_{\boldsymbol{\alpha}})$ of g. In fact such conditions on g exist. They define the set

$$G = \left\{\mathbf{g} = (g_{\boldsymbol{\alpha}}) \in \mathbb{R}^{s(2d)} \; : \; \exists \mathbf{Q} \in \mathbb{R}^{s(d)\times s(d)} \text{ such that (1.1) holds}\right\},$$

which is the orthogonal projection on $\mathbb{R}^{s(2d)}$ of the set of elements $(g, \mathbf{Q})$ that satisfy (1.1), a basic semi-algebraic set of $\mathbb{R}^{s(2d)+s(d)(s(d)+1)/2}$. Indeed the semidefinite constraint $\mathbf{Q} \succeq 0$ can be stated as polynomial inequality constraints on the entries of $\mathbf{Q}$ (use determinants of its principal minors). Hence, by the Projection Theorem of real algebraic geometry, the set G is a semi-algebraic set (but not a basic semi-algebraic set in general); see [16, Theorem 2.2.1].

Hence, in general, G is not a basic semi-algebraic set but a finite union $\bigcup_{i\in I} G_i$ of basic semi-algebraic sets $G_i \subset \mathbb{R}^{s(2d)}$, and it is very hard to obtain the polynomials in the variables $g_{\boldsymbol{\alpha}}$ that define each G_i.

The next result states a sufficient condition for $g \in \mathbb{R}[\mathbf{x}]$ to be s.o.s., directly in terms of the coefficients $(g_{\boldsymbol{\alpha}})$. Let $\mathbb{N}_d^n = \{\boldsymbol{\alpha} \in \mathbb{N}^n \; : \; |\alpha| \leq d\}$ and $\Gamma_d = \{2\boldsymbol{\beta} \; : \; \boldsymbol{\beta} \in \mathbb{N}_d^n\}$.

Theorem 1.3. *Let* $\mathbf{x} \mapsto g(\mathbf{x}) = \sum_{\alpha} g_{\alpha} \mathbf{x}^{\alpha}$ *be a polynomial of degree* $2d$ *with* $d \geq 1$, *and write*

$$g = \sum_{i=1}^{n} g_{id}\, x_i^{2d} + h + g_0,$$

where $h \in \mathbb{R}[\mathbf{x}]$ *contains none of the monomials* $(x_i^{2d})_{i=1}^{n}$. *Then* g *is s.o.s. if*

$$g_0 \geq \sum_{\alpha \notin \Gamma_d} |g_{\alpha}| - \sum_{\alpha \in \Gamma_d} \min[0, g_{\alpha}], \tag{1.3}$$

$$g_{id} \geq \sum_{\alpha \notin \Gamma_d} |g_{\alpha}| \frac{|\alpha|}{2d} - \sum_{\alpha \in \Gamma_d} \min[0, g_{\alpha}]\, \frac{|\alpha|}{2d}, \quad \forall i = 1, \ldots, n. \tag{1.4}$$

On the one hand, the conditions (1.3)–(1.4) are only sufficient, but on the other hand they define a convex polyhedron in the space $\mathbb{R}^{s(2d)}$ of coefficients (g_{α}) of $g \in \mathbb{R}[\mathbf{x}]$. This latter property may be interesting if one has to optimize in the space of s.o.s. polynomials of degree at most d.

1.2 Nonnegative versus s.o.s. polynomials

It is important to compare nonnegative and s.o.s. polynomials because we have just seen that one knows how to check efficiently whether a polynomial is s.o.s. but not whether it is only nonnegative. There are two kinds of results for that comparison, depending on whether or not one keeps the degree fixed.

The first result is rather negative, as it shows that, when the degree is fixed and the number of variables grows, the gap between nonnegative and s.o.s. polynomials increases and is unbounded. Namely, let $\mathcal{P}[\mathbf{x}]_d$ (resp. $\mathcal{H}[\mathbf{x}]_d$) denote the cone of homogeneous and nonnegative (resp. homogeneous and s.o.s.) polynomials of degree $2d$. (Recall that a polynomial $p \in \mathbb{R}[\mathbf{x}]$ of degree d is *homogeneous* if $p(\lambda \mathbf{x}) = \lambda^d p(\mathbf{x})$ for all $\lambda \geq 0$ and all $\mathbf{x} \in \mathbb{R}^n$.)

To compare both sets, we need subsets of finite volume with respect to some appropriate measure μ. So let $\mathcal{H}$ be the hyperplane $\{f \in \mathbb{R}[\mathbf{x}] : \int_{\mathbf{S}^{n-1}} f\, d\mu = 1\}$, where μ is the rotation invariant measure on the unit sphere $\mathbf{S}^{n-1} \subset \mathbb{R}^n$. Finally, let $\widehat{\mathcal{P}}[\mathbf{x}]_d = \mathcal{H} \cap \mathcal{P}[\mathbf{x}]_d$ and $\widehat{\mathcal{H}}[\mathbf{x}]_d = \mathcal{H} \cap \mathcal{H}[\mathbf{x}]_d$.

Theorem 1.4. *There exist constants* $C_1, C_2 > 0$ *depending only on* d *and such that*

$$C_1\, n^{(d/2-1)/2} \leq \frac{\mathrm{vol}(\widehat{\mathcal{P}}[\mathbf{x}]_d)}{\mathrm{vol}(\widehat{\mathcal{H}}[\mathbf{x}]_d)} \leq C_2\, n^{(d/2-1)/2}. \tag{1.5}$$

Therefore, if d is fixed and $n \to \infty$, the gap between $\widehat{\mathcal{P}}[\mathbf{x}]_d$ and $\widehat{\mathcal{H}}[\mathbf{x}]_d$ can become as large as desired.

On the other hand, while a nonnegative polynomial $f \in \mathbb{R}[\mathbf{x}]$ may not be a s.o.s., we next show that we can perturb f to make it a sum of squares. The price to pay is consistent with Theorem 1.4 in that the approximation of f we need to consider does *not* have the same degree as f.

Given $r \in \mathbb{N}$ arbitrary, let $\Theta_r, \theta_r \in \mathbb{R}[\mathbf{x}]$ be the polynomials

$$\Theta_r(\mathbf{x}) = 1 + \sum_{i=1}^n x_i^{2r}; \quad \theta_r(\mathbf{x}) = \sum_{i=1}^n \sum_{k=0}^r \frac{x_i^{2k}}{k!}. \tag{1.6}$$

Given $f \in \mathbb{R}[\mathbf{x}]$, let $\|f\|_1 = \sum_{\alpha \in \mathbb{N}^n} |f_\alpha|$ if $\mathbf{f} = (f_\alpha)$ is the vector of coefficients of f. Next, with $\epsilon > 0$, we define

$$f_{\epsilon,r}^1 = f + \epsilon\, \Theta_r; \quad f_{\epsilon,r}^2 = f + \epsilon\, \theta_r.$$

Theorem 1.5.

(a) *If $f \in \mathbb{R}[\mathbf{x}]$ is nonnegative on $[-1,1]^n$, then for every $\epsilon > 0$ there exists r_ϵ^1 such that $f_{\epsilon,r}^1 \in \Sigma[\mathbf{x}]$ for all $r \geq r_\epsilon^1$ and $\|f - f_{\epsilon,r}^1\|_1 \to 0$ as $\epsilon \downarrow 0$ (and $r \geq r_\epsilon^1$).*

(b) *If $f \in \mathbb{R}[\mathbf{x}]$ is nonnegative, then for every $\epsilon > 0$ there exists r_ϵ^2 such that $f_{\epsilon,r}^2 \in \Sigma[\mathbf{x}]$ for all $r \geq r_\epsilon^2$ and $\|f - f_{\epsilon,r}^2\|_1 \to 0$ as $\epsilon \downarrow 0$ (and $r \geq r_\epsilon^2$).*

Theorem 1.5 is a denseness result with respect to the l_1-norm of coefficients. Indeed, it states that a polynomial f which is nonnegative on $[-1,1]^n$ (resp. on $\mathbb{R}^n$) can be perturbed to a s.o.s. polynomial $f_{\epsilon,r}^1$ (resp. $f_{\epsilon,r}^2$) such that $\|f - f_{\epsilon,r}^1\|_1 \to 0$ (resp. $\|f - f_{\epsilon,r}^2\|_1 \to 0$). It also provides a certificate of nonnegativity of f on $[-1,1]^n$ (resp. on $\mathbb{R}^n$). Indeed, fix $\mathbf{x} \in [-1,1]^n$ (resp. $\mathbf{x} \in \mathbb{R}^n$) and let $\epsilon \to 0$ to obtain that $0 \leq f_{\epsilon,r}^1(\mathbf{x}) \to f(\mathbf{x})$ (resp. $0 \leq f_{\epsilon,r}^2(\mathbf{x}) \to f(\mathbf{x})$).

Concerning Theorem 1.5(b), observe that, in addition to the l_1-norm convergence $\|f - f_{\epsilon r}^2\|_1 \to 0$, the convergence is also *uniform* on compact sets. Notice that a polynomial f nonnegative on the whole $\mathbb{R}^n$ (hence on the box $[-1,1]^n$) can also be approximated by the s.o.s. polynomial $f_{\epsilon,r}^1$ of Theorem 1.5(a), which is simpler than the s.o.s. approximation $f_{\epsilon,r}^2$. However, in contrast to the latter, the s.o.s. approximation $f \approx f_{\epsilon,r}^1$ is *not* uniform on compact sets, and is really more appropriate for polynomials nonnegative on $[-1,1]^n$ only (and indeed the approximation $f \approx f_{\epsilon r}^1$ is uniform on $[-1,1]^n$). In addition, in Theorem 1.5(a) the integer r_ϵ^1 does *not* depend on the *explicit choice* of the polynomial f but only on:

(a) ϵ and the dimension n;

(b) the degree and size of the coefficients of f.

Therefore, if one fixes these four parameters, we find an r such that the statement of Theorem 1.5(a) holds for any f nonnegative on $[-1,1]^n$, whose degree and size of coefficients do not exceed the fixed parameters.

1.3 Representation theorems: Univariate case

In this section, we review the major representation theorems for nonnegative univariate polynomials, for which the results are quite complete. Let $\mathbb{R}[x]$ be the ring of real polynomials in the single variable x, and let $\Sigma[x] \subset \mathbb{R}[x]$ be its subset of polynomials that are sums of squares of elements of $\mathbb{R}[x]$. We first prove that, if $p \in \mathbb{R}[x]$ is nonnegative, then $p \in \Sigma[x]$.

Theorem 1.6. *A polynomial $p \in \mathbb{R}[x]$ of even degree is nonnegative if and only if it can be written as a sum of squares of other polynomials, i.e., $p(x) = \sum_{i=1}^{k} h_i(x)^2$ with $h_i \in \mathbb{R}[x]$, $i = 1, \ldots, k$.*

Proof. Clearly, if $p(x) = \sum_{j=1}^{k} [h_j(x)]^2$, then $p(x) \geq 0$ for all $x \in \mathbb{R}$. Conversely, suppose that a polynomial $p \in \mathbb{R}[x]$ of degree $2d$ (and with highest degree term $p_{2d}x^{2d}$) is nonnegative on the real line $\mathbb{R}$. Then the real roots of p should have even multiplicity; otherwise p would alter its sign in a neighborhood of a root. Let λ_j, $j = 1, \ldots, r$ be its real roots with corresponding multiplicity $2m_j$. Its complex roots can be arranged in conjugate pairs $a_\ell + ib_\ell$, $a_\ell - ib_\ell$, $\ell = 1, \ldots, t$ (with $i^2 = -1$). Then p can be written in the form

$$x \longmapsto p(x) = p_{2d} \prod_{j=1}^{r} (x - \lambda_j)^{2m_j} \prod_{\ell=1}^{t} \left((x - a_\ell)^2 + b_\ell^2\right).$$

Note that the leading coefficient p_{2d} needs to be positive. By expanding the terms in the products, we see that p can be written as a sum of squares of $k = 2^t$ polynomials. (In fact, one may also show that p is a sum of only two squares.) □

We next concentrate on polynomials $p \in \mathbb{R}[x]$ that are nonnegative on an interval $I \subset \mathbb{R}$. Moreover, the three cases $I = (-\infty, b]$, $I = [a, \infty)$ and $I = [a, b]$ (with $a, b \in \mathbb{R}$) all reduce to the basic cases $I = [0, \infty)$ and $I = [-1, +1]$ using the change of variable

$$f(x) = p(b - x), \quad f(x) = p(x - a), \text{ and } f(x) = p\left(\frac{2x - (a + b)}{b - a}\right),$$

respectively.

The representation results that one obtains depend on the particular choice of the polynomials used in the description of the interval. The main result in the one-dimensional case can be summarized in the next theorem.

Theorem 1.7. *Let $p \in \mathbb{R}[x]$ be of degree n.*

(a) *Let $g \in \mathbb{R}[x]$ be the polynomial $x \mapsto g(x) = 1 - x^2$. Then $p \geq 0$ on $[-1, 1]$ if and only if*

$$p = f + gh, \quad f, h \in \Sigma[x],$$

and with both summands of degree less than $2n$.

(b) *Let $x \mapsto g_1(x) = 1 - x$, $x \mapsto g_2(x) = 1 + x$, $x \mapsto g_3(x) = g_1(x)g_2(x)$. Then $p \geq 0$ on $I = [-1, 1]$ if and only if*

$$p = f_0 + g_1 f_1 + g_2 f_2 + g_3 f_3, \quad f_0, f_1, f_2, f_3 \in \Sigma[x].$$

In addition, all the summands have degree at most n, and $f_1, f_2 = 0$ if n is even, whereas $f_0, f_3 = 0$ if n is odd.

The Goursat transform $f \to \hat{f}$ on $\mathbb{R}[x]$ of a polynomial $f \in \mathbb{R}[x]$ is defined by

$$x \longmapsto \hat{f}(\overline{x}) = (1+x)^m f\left(\frac{1-x}{1+x}\right), \quad x \in \mathbb{R}.$$

Using the Goursat transform, the case $[0, \infty)$ reduces to the case $[-1, 1]$ and we have:

Theorem 1.8. *Let $p \in \mathbb{R}[x]$ be nonnegative on $[0, +\infty)$. Then*

$$p = f_0 + x\, f_1$$

for two s.o.s. polynomials $f_0, f_1 \in \Sigma[x]$, and the degree of both summands is bounded by $\deg p$.

It is important to emphasize that Theorem 1.7 (resp. Theorem 1.8) explicitly uses the specific representation of the interval $[-1, 1]$ (resp. $[0, +\infty)$) as the basic semi-algebraic set $\{x : 1 - x \geq 0; 1 + x \geq 0\}$ or $\{x : 1 - x^2 \geq 0\}$ (resp. $\{x : x \geq 0\}$).

In the case where $[-1, 1] = \{x \in \mathbb{R} : h(x) \geq 0\}$, with h not equal to the polynomial $x \mapsto (1+x)(1-x)$, a weaker result is obtained in the next section.

The next result also considers the interval $[-1, 1]$ but provides another representation that does not use s.o.s.

Theorem 1.9. *Let $p \in \mathbb{R}[x]$. Then $p > 0$ on $[-1, 1]$ if and only if*

$$p = \sum_{i+j \leq d} c_{ij} (1-x)^i (1+x)^j, \quad c_{ij} \geq 0, \tag{1.7}$$

for some sufficiently large d.

Notice that Theorem 1.9 leads to a linear optimization feasibility problem to determine the coefficients c_{ij} in the representation (1.7).

1.4 Representation theorems: Multivariate case

In this section we consider the multivariate case. As already mentioned, a nonnegative polynomial $p \in \mathbb{R}[\mathbf{x}]$ does not necessarily have a sum of squares representation. For instance, the two polynomials

$$\mathbf{x} \longmapsto p(\mathbf{x}) = x_1^2 x_2^2 (x_1^2 + x_2^2 - 1) + 1,$$

$$\mathbf{x} \longmapsto p(\mathbf{x}) = x_1^2 x_2^2 (x_1^2 + x_2^2 - 3x_3^2) + 6x_3^6$$

are nonnegative but they do not have a s.o.s. representation. On the other hand, nonnegative quadratic polynomials and nonnegative fourth degree homogeneous polynomials of three variables have a s.o.s. representation.

The next celebrated theorem due to Pólya provides a certificate of positivity for homogeneous polynomials that are positive on the simplex.

Theorem 1.10 (Pólya). *If $p \in \mathbb{R}[\mathbf{x}]$ is homogeneous and $p > 0$ on $\mathbb{R}^n_+ \setminus \{\mathbf{0}\}$, then, for sufficienly large $k \in \mathbb{N}$, all nonzero coefficients of the polynomial $\mathbf{x} \mapsto (x_1 + \cdots + x_n)^k p(\mathbf{x})$ are positive.*

As a consequence of Theorem 1.10, we next obtain the following representation result for nonhomogeneous polynomials that are strictly positive on $\mathbb{R}^n_+$. If $p \in \mathbb{R}[\mathbf{x}]$, we denote by $\tilde{p} \in \mathbb{R}[\mathbf{x}, x_0]$ the homogeneous polynomial associated with p, that is, $\tilde{p}(\mathbf{x}, x_0) = x_0^d p(\mathbf{x}/x_0)$, with d being the degree of p, and denote by $p_d(\mathbf{x}) = \tilde{p}(\mathbf{x}, 0)$ for all $\mathbf{x} \in \mathbb{R}^n$ the homogeneous part of p of degree d.

Theorem 1.11. *Let $\mathbf{x} \mapsto g(\mathbf{x}) = 1 + \sum_{j=1}^n x_j$ and $p \in \mathbb{R}[\mathbf{x}]$ with degree d. If $p > 0$ on $\mathbb{R}^n_+$ and $p_d > 0$ on $\mathbb{R}^n_+ \setminus \{\mathbf{0}\}$, then for sufficienly large $k \in \mathbb{N}$ the polynomial $g^k p$ has positive coefficients.*

Proof. We first prove that $\tilde{p} > 0$ on $\mathbb{R}^{n+1}_+ \setminus \{\mathbf{0}\}$. Let $(\mathbf{x}, x_0) \in \mathbb{R}^{n+1}_+ \setminus \{\mathbf{0}\}$. If $x_0 = 0$, then $\mathbf{x} \in \mathbb{R}^n_+ \setminus \{\mathbf{0}\}$ and $\tilde{p}(\mathbf{x}, 0) = p_d(\mathbf{x}) > 0$. If $x_0 > 0$, then $\mathbf{y} = \mathbf{x}/x_0 \in \mathbb{R}^n_+$, and thus $p(\mathbf{y}) > 0$ implies that $\tilde{p}(\mathbf{x}, x_0) = x_0^d p(\mathbf{y}) > 0$. Therefore, $\tilde{p} > 0$ on $\mathbb{R}^{n+1}_+ \setminus \{\mathbf{0}\}$. Hence by Theorem 1.10, for all sufficiently large $k \in \mathbb{N}$, the polynomial

$$(\mathbf{x}, x_0) \longmapsto q(\mathbf{x}, x_0) = \left(\sum_{j=1}^n x_j + x_0 \right)^k \tilde{p}(\mathbf{x}, x_0),$$

has positive coefficients. But then, as $\tilde{p}(\mathbf{x}, 1) = p(\mathbf{x})$, it follows that the polynomial

$$\mathbf{x} \longmapsto q(\mathbf{x}, 1) = \left(\sum_{j=1}^n x_j + 1 \right)^k \tilde{p}(\mathbf{x}, 1) = g^k p,$$

has positive coefficients, for all sufficiently large $k \in \mathbb{N}$. □

We next characterize when a semi-algebraic set described by polynomial inequalities, equalities and non-equalities is empty. In order to achieve this, we need the following definition.

Definition 1.12. For $F = \{f_1, \ldots, f_m\} \subset \mathbb{R}[\mathbf{x}]$ and a set $J \subseteq \{1, \ldots, m\}$, we denote by $f_J \in \mathbb{R}[\mathbf{x}]$ the polynomial $\mathbf{x} \mapsto f_J(\mathbf{x}) = \prod_{j \in J} f_j(\mathbf{x})$, with the convention that $f_\emptyset = 1$. The set

$$P(f_1, \ldots, f_m) = \left\{ \sum_{J \subseteq \{1,\ldots,m\}} q_J f_J \;:\; q_J \in \Sigma[\mathbf{x}] \right\} \tag{1.8}$$

is called (by algebraic geometers) the *preordering* generated by $f_1, \ldots, f_m$.

We first state a key result from Stengle:

Theorem 1.13 (Stengle). *Let k be a real closed field, and let*

$$F = (f_i)_{i\in I_1},\ G = (g_i)_{i\in I_2},\ H = (h_i)_{i\in I_3} \subset k[\mathbf{x}]$$

be finite families of polynomials. Let

(a) *$P(F)$ be the preordering generated by the family F,*
(b) *$M(G)$ be the set of all finite products of the g_i's, $i \in I_2$ (the empty product being the constant polynomial 1), and*
(c) *$I(H)$ be the ideal generated by H.*

Consider the semi-algebraic set

$$\mathbf{K} = \left\{\mathbf{x} \in k^n \,:\, f_i(\mathbf{x}) \geq 0,\ \forall i \in I_1;\ g_i(\mathbf{x}) \neq 0,\ \forall i \in I_2;\ h_i(\mathbf{x}) = 0,\ \forall i \in I_3\right\}.$$

The set $\mathbf{K}$ is empty if and only if there exist $f \in P(F)$, $g \in M(G)$ and $h \in I(H)$ such that

$$f + g^2 + h = 0. \tag{1.9}$$

The polynomials $f, g, h \in k[\mathbf{x}]$ in (1.9) provide a *Stengle certificate* of $\mathbf{K} = \emptyset$. In (1.9) there is also a (very large) bound on the degree of $g \in M(G)$, the degree of the s.o.s. weights $q_J \in \Sigma[\mathbf{x}]$ in the representation (1.8) of $f \in P(F)$, as well as the degree of the weights $p_j \in \mathbb{R}[\mathbf{x}]$ in the representation of $h = \sum_j p_j h_j$.

Therefore, in view of Proposition 1.1, in principle checking existence of a certificate (f, g, h) requires solving a *single* semidefinite program. Unfortunately, the available degree bound being huge, the size of such a semidefinite program is by far too large for practical implementation.

Stengle and Farkas certificates

The Stengle certificate $f, g, h \in k[\mathbf{x}]$ in (1.9) for $\mathbf{K} = \emptyset$ is a nonlinear generalization of the celebrated Farkas Lemma (or Theorem of the alternative) in linear algebra which provides a certificate of emptyness for a polyhedral set $\mathbf{K} = \{\mathbf{x} : \mathbf{A}\mathbf{x} \leq \mathbf{b}\}$ (for some matrix $\mathbf{A} \in \mathbb{R}^{m\times n}$ and some vector $\mathbf{b} \in \mathbb{R}^m$).

In fact, a Farkas certificate is a particularly simple Stengle certificate for convex polyhedra. Indeed, a Farkas certificate of $\emptyset = \{\mathbf{x} : \mathbf{A}\mathbf{x} \leq \mathbf{b}\}$ is a nonnegative vector $\mathbf{u} \in \mathbb{R}^m_+$ such that $\mathbf{u}'\mathbf{A} = 0$ and $\mathbf{u}'\mathbf{b} < 0$. So let $\mathbf{x} \mapsto f_j(\mathbf{x}) = (\mathbf{b} - \mathbf{A}\mathbf{x})_j$, $j = 1, \ldots, m$, $\mathbf{x} \mapsto g_1(\mathbf{x}) = 1$ (so that $M(G)$ is identical to the constant polynomial 1) and $\mathbf{x} \mapsto h_1(\mathbf{x}) = 0$ (so that $I(H)$ is the 0 polynomial). The polynomial

$$\mathbf{x} \mapsto f(\mathbf{x}) = \frac{-1}{\mathbf{u}'\mathbf{b}} \sum_{j=1}^{m} u_j (\mathbf{b} - \mathbf{A}\mathbf{x})_j$$

is an element of $P(F)$ because $u_j \geq 0$ for every $j = 1, \ldots, m$ and $\mathbf{u}'\mathbf{b} < 0$.

But then

$$1 + f(\mathbf{x}) = 1 - \frac{1}{\mathbf{u}'\mathbf{b}} \sum_{j=1}^{m} u_j(\mathbf{b} - \mathbf{A}\mathbf{x})_j = 1 - \frac{\mathbf{u}'\mathbf{b} - \mathbf{u}'\mathbf{A}\mathbf{x}}{\mathbf{u}'\mathbf{b}} = 0,$$

that is, (1.9) holds.

We next consider basic semi-algebraic sets, that is, semi-algebraic sets defined by inequalities only. As a direct consequence of Theorem 1.13, one obtains:

Theorem 1.14 (Stengle's Positivstellensatz and Nullstellensatz). *Let k be a real closed field, $f \in k[\mathbf{x}]$, and let*

$$\mathbf{K} = \{\mathbf{x} \in k^n \; : \; f_j(\mathbf{x}) \geq 0, \; j = 1, \ldots, m\}.$$

(a) Nichtnegativstellensatz. *$f \geq 0$ on $\mathbf{K}$ if and only if there exist $\ell \in \mathbb{N}$ and $g, h \in P(f_1, \ldots, f_m)$ such that $fg = f^{2\ell} + h$.*

(b) Positivstellensatz. *$f > 0$ on $\mathbf{K}$ if and only if there exist $g, h \in P(f_1, \ldots, f_m)$ such that $fg = 1 + h$.*

(c) Nullstellensatz. *$f = 0$ on $\mathbf{K}$ if and only if there exist $\ell \in \mathbb{N}$ and $g \in P(f_1, \ldots, f_m)$ such that $f^{2\ell} + g = 0$.*

Again, as for Theorem 1.13, there is also a bound on ℓ and the degree of the s.o.s. weights $q_J \in \Sigma[\mathbf{x}]$ in the representation (1.8) of $g, h \in P(f_1, \ldots, f_m)$. This bound depends only on the dimension n and on the degree of the polynomials $(f, f_1, \ldots, f_m)$. Therefore, in principle, checking existence of a certificate (l, g, h) in Theorem 1.14(a)–(c) requires solving a single semidefinite program (but of huge size). In practice, one fixes an *a priori* (much smaller) degree bound and solves the corresponding semidefinite program. If the latter has a feasible solution then one obtains a certificate $g, h \in P(f_1, \ldots, f_m)$. However, such a certificate is *numerical* and so can be obtained only up to some machine precision, because of numerical inaccuracies inherent to semidefinite programming solvers.

1.5 Polynomials positive on a compact basic semi-algebraic set

In this section, we restrict our attention to compact basic semi-algebraic sets $\mathbf{K} \subset \mathbb{R}^n$ and obtain certificates of positivity on $\mathbf{K}$ that have certain algorithmic advantages. In fact, Putinar's Positivstellensatz below is the key result that we will later use extensively to solve polynomial optimization problems.

1.5.1 Representations via sums of squares

The first representation result, known as Schmüdgen's Positivstellensatz, was an important breakthrough as it was the first to provide a simple characterization of polynomials positive on a compact basic semi-algebraic set $\mathbf{K}$, and with *no* additional assumption on $\mathbf{K}$ (or on its description).

Theorem 1.15 (Schmüdgen's Positivstellensatz). *Let $(g_j)_{j=1}^m \subset \mathbb{R}[\mathbf{x}]$ be such that the basic semi-algebraic set*

$$\mathbf{K} = \{\mathbf{x} \in \mathbb{R}^n \,:\, g_j(\mathbf{x}) \geq 0,\ j = 1, \dots, m\} \tag{1.10}$$

is compact. If $f \in \mathbb{R}[\mathbf{x}]$ is strictly positive on $\mathbf{K}$ then $f \in P(g_1, \dots, g_m)$, that is,

$$f = \sum_{J \subseteq \{1,\dots,m\}} f_J\, g_J \tag{1.11}$$

for some s.o.s. $f_J \in \Sigma[\mathbf{x}]$ and with $g_J = \prod_{j \in J} g_j$.

Theorem 1.15 is a very powerful result, but note that the number of terms in (1.11) is exponential in the number of polynomials that define the set $\mathbf{K}$. However, a major improvement is possible under a relatively weak assumption on the polynomials that define the compact set $\mathbf{K}$. Associated with the finite family $(g_j) \subset \mathbb{R}[\mathbf{x}]$, the set

$$Q(g) = \left\{ q_0 + \sum_{j=1}^m q_j g_j \,:\, (q_j)_{j=0}^m \subset \Sigma[\mathbf{x}] \right\} \tag{1.12}$$

is called the *quadratic module* generated by the family (g_j).

Assumption 1.16. Let $(g_j)_{j=1}^m \subset \mathbb{R}[\mathbf{x}]$. There exists $u \in Q(g)$ such that the level set $\{\mathbf{x} \in \mathbb{R}^n \,:\, u(\mathbf{x}) \geq 0\}$ is compact.

Theorem 1.17 (Putinar's Positivstellensatz). *Let $\mathbf{K} \subset \mathbb{R}^n$ be as in* (1.10) *and let Assumption* 1.16 *hold. If $f \in \mathbb{R}[\mathbf{x}]$ is strictly positive on $\mathbf{K}$ then $f \in Q(g)$, that is,*

$$f = f_0 + \sum_{j=1}^m f_j g_j \tag{1.13}$$

for some s.o.s. polynomials $f_j \in \Sigma[\mathbf{x}]$, $j = 0, 1, \dots, m$.

In contrast to Theorem 1.15, the number of terms in the representation (1.13) is linear in the number of polynomials that define $\mathbf{K}$, a crucial improvement from a computational point of view. The condition on $\mathbf{K}$ (Assumption 1.16) is not very restrictive. For instance, it is satisfied in the following cases:

(a) All the g_i's are affine and $\mathbf{K}$ is compact (hence a polytope).

(b) The set $\{\mathbf{x} \in \mathbb{R}^n \,:\, g_j(\mathbf{x}) \geq 0\}$ is compact for some $j \in \{1, \dots, m\}$.

Suppose that we know some $N > 0$ such that $\mathbf{K} \subset \{\mathbf{x} \in \mathbb{R}^n \,:\, \|\mathbf{x}\|^2 \leq N\}$. Let $\mathbf{x} \mapsto g_{m+1}(\mathbf{x}) = N - \|\mathbf{x}\|^2$. Adding the quadratic constraint $g_{m+1}(\mathbf{x}) \geq 0$ in the definition (1.10) of $\mathbf{K}$ does not change $\mathbf{K}$, as this last constraint is redundant. But with this new representation, $\mathbf{K}$ satisfies the required condition in Theorem 1.17.

The following theorem provides further understanding on the condition in Theorem 1.17.

Theorem 1.18. *Let $(g_j)_{j=1}^m \subset \mathbb{R}[\mathbf{x}]$, assume that $\mathbf{K} \subset \mathbb{R}^n$ defined in (1.10) is compact, and let $Q(g)$ be as in (1.12). The following conditions are equivalent:*

(a) *There exist finitely many $p_1, \dots, p_s \in Q(g)$ such that the level set $\{\mathbf{x} \in \mathbb{R}^n : p_j(\mathbf{x}) \geq 0,\ j = 1, \dots, s\}$ which contains $\mathbf{K}$ is compact and $\prod_{j \in J} p_j \in Q(g)$ for all $J \subseteq \{1, \dots, s\}$.*

(b) *Assumption* 1.16 *holds.*

(c) *There exists $N \in \mathbb{N}$ such that the polynomial $\mathbf{x} \mapsto N - \|\mathbf{x}\|^2$ is in $Q(g)$.*

(d) *For all $p \in \mathbb{R}[\mathbf{x}]$, there is some $N \in \mathbb{N}$ such that both polynomials $\mathbf{x} \mapsto N + p(\mathbf{x})$ and $\mathbf{x} \mapsto N - p(\mathbf{x})$ are in $Q(g)$.*

Both Theorems 1.15 and 1.17 have significant computational advantages. Indeed, from Proposition 1.1, given a polynomial $f > 0$ on $\mathbf{K}$, checking whether f has the representation (1.11) or (1.13), and assuming an *a priori* bound on the degree of the unknown s.o.s. polynomials, reduces to solving a semidefinite optimization problem, as we saw in Section 1.3.

Example 1.19. Let $\mathbf{x} \mapsto f(\mathbf{x}) = x_1^3 - x_1^2 + 2x_1x_2 - x_2^2 + x_2^3$ and $\mathbf{K} = \{\mathbf{x} : g_1(\mathbf{x}) = x_1 \geq 0,\ g_2(\mathbf{x}) = x_2 \geq 0,\ g_3(\mathbf{x}) = x_1 + x_2 - 1 \geq 0\}$. To check whether $f \geq 0$ on $\mathbf{K}$, we attempt to write $f = f_0 + \sum_{i=1}^3 f_i g_i$, where $f_i \in \Sigma[\mathbf{x}]$ and each has degree 2, that is,

$$f_i = (1, x_1, x_2)\, \mathbf{Q}_i\, (1, x_1, x_2)', \quad \mathbf{Q}_i \succeq \mathbf{0}, \quad i = 0, 1, 2, 3.$$

Solving the semidefinite feasiblity problem, we find that

$$\mathbf{Q}_0 = \mathbf{0},\ \mathbf{Q}_1 = \begin{bmatrix} 0 & 0 & 0 \\ 0 & 0 & 0 \\ 0 & 0 & 1 \end{bmatrix},\ \mathbf{Q}_2 = \begin{bmatrix} 0 & 0 & 0 \\ 0 & 1 & 0 \\ 0 & 0 & 0 \end{bmatrix},\ \mathbf{Q}_3 = \begin{bmatrix} 0 & 0 & 0 \\ 0 & 1 & -1 \\ 0 & -1 & 1 \end{bmatrix},$$

and so

$$f(\mathbf{x}) = x_2^2 x_1 + x_1^2 x_2 + (x_1 - x_2)^2 (x_1 + x_2 - 1),$$

which proves that $f \geq 0$ on $\mathbf{K}$.

Degree bound

The following result provides a bound on the degree of the weights $(f_j) \subset \Sigma[\mathbf{x}]$ in the representation (1.13). For $f \in \mathbb{R}[\mathbf{x}]$, written $\mathbf{x} \mapsto f(\mathbf{x}) = \sum_\alpha f_\alpha \mathbf{x}^\alpha$, let

$$\|f\| = \max_\alpha \frac{|f_\alpha|}{\binom{|\alpha|}{\alpha}} \quad \text{with} \quad \binom{|\alpha|}{\alpha} = \frac{|\alpha|!}{\alpha_1! \cdots \alpha_n!}.$$

Theorem 1.20 (Degree bound). *Let $\mathbf{K} \subset \mathbb{R}^n$ in* (1.10) *satisfy Assumption* 1.16 *and assume that $\emptyset \neq \mathbf{K} \subset (-1,1)^n$. Then there is some $c > 0$ such that for all $f \in \mathbb{R}[\mathbf{x}]$ of degree d and strictly positive on $\mathbf{K}$ (i.e., such that $f^* = \min\{f(\mathbf{x}) : \mathbf{x} \in \mathbf{K}\} > 0$), the representation* (1.13) *holds with*

$$\deg f_j g_j \leq c \exp\left(\left(d^2 n^d \frac{\|f\|}{f^*}\right)^c\right), \quad \forall j = 1, \ldots, m.$$

1.5.2 A matrix version of Putinar's Positivstellensatz

For $f \in \mathbb{R}[\mathbf{x}]$, written $\mathbf{x} \mapsto f(\mathbf{x}) = \sum_\alpha f_\alpha \mathbf{x}^\alpha$, let

$$\|f\|_0 = \max_\alpha |f_\alpha| \frac{\alpha_1! \cdots \alpha_n!}{|\alpha|!}. \tag{1.14}$$

Definition 1.21.

(a) The *norm* of a real symmetric matrix-polynomial $\mathbf{F} \in \mathbb{R}[\mathbf{x}]^{p\times p}$ is defined by $\|\mathbf{F}\| = \max_{\|\xi\|=1} \|\xi'\mathbf{F}(\mathbf{x})\xi\|_0$.

(b) A real symmetric matrix-polynomial $\mathbf{F} \in \mathbb{R}[\mathbf{x}]^{p\times p}$ is said to be a *sum of squares* (in short s.o.s.) if $\mathbf{F} = \mathbf{L}\mathbf{L}'$ for some $q \in \mathbb{N}$ and some real matrix-polynomial $\mathbf{L} \in \mathbb{R}[\mathbf{x}]^{p\times q}$.

Let $\mathbf{I}$ denote the $p \times p$ identity matrix.

Theorem 1.22. *Let $\mathbf{K} \subset \mathbb{R}^n$ be the basic semi-algebraic set in* (1.10) *and let Assumption* 1.16 *hold. Let $\mathbf{F} \in \mathbb{R}[\mathbf{x}]^{p\times p}$ be a real symmetric matrix-polynomial of degree d. If, for some $\delta > 0$, $\mathbf{F}(\mathbf{x}) \succeq \delta \mathbf{I}$ for all $\mathbf{x} \in \mathbf{K}$, then*

$$\mathbf{F}(\mathbf{x}) = \mathbf{F}_0(\mathbf{x}) + \sum_{j=1}^m \mathbf{F}_j(\mathbf{x})\, g_j(\mathbf{x}) \tag{1.15}$$

for some s.o.s. matrix-polynomials $(\mathbf{F}_j)_{j=0}^m$, and

$$\deg \mathbf{F}_0, \deg \mathbf{F}_1 g_1, \ldots, \deg \mathbf{F}_m g_m \leq c \left(d^2 n^d \frac{\|\mathbf{F}\|}{\delta}\right)^c.$$

Obviously, Theorem 1.22 is a matrix-polynomial analogue of Theorem 1.17. In fact, one might have characterized the property $\mathbf{F}(\mathbf{x}) \succeq \delta\mathbf{I}$ on $\mathbf{K}$ by using Theorem 1.17 as follows. Let $\mathbf{S} = \{\xi \in \mathbb{R}^p : \xi'\xi = 1\}$ and notice that the compact basic semi-algebraic set $\mathbf{K} \times \mathbf{S} \subset \mathbb{R}^n \times \mathbb{R}^p$ satisfies Assumption 1.16 whenever $\mathbf{K}$ does. If $f \in \mathbb{R}[\mathbf{x}, \xi]$ denotes the polynomial $(\mathbf{x}, \xi) \mapsto \xi'\mathbf{F}(\mathbf{x})\xi$, then

$$\mathbf{F} \succeq \delta\mathbf{I} \text{ on } \mathbf{K} \quad \Longleftrightarrow \quad f \geq \delta \text{ on } \mathbf{K} \times \mathbf{S},$$

and so, by Theorem 1.17, $\mathbf{F} \succeq \delta \mathbf{I}$ on $\mathbf{K}$ implies that

$$\xi'\mathbf{F}(\mathbf{x})\xi = \sigma_0(\mathbf{x}, \xi) + \sum_{j=1}^{m} \sigma_j(\mathbf{x}, \xi)\, g_j(\mathbf{x}) + \sigma_{m+1}(\mathbf{x}, \xi)\, (\xi'\xi - 1) \tag{1.16}$$

for some s.o.s. polynomials $(\sigma_0)_0^m \subset \Sigma[\mathbf{x}, \xi]$ and a polynomial $\sigma_{m+1} \in \mathbb{R}[\mathbf{x}, \xi]$.

However, in general, the degree of weights in the representation (1.15) and (1.16) is not the same. It is not clear which one should be preferred.

1.5.3 An alternative representation

We next present an alternative representation not based on s.o.s. polynomials. We make the following assumption:

Assumption 1.23. The set $\mathbf{K}$ in (1.10) is compact and the polynomials $(g_j)_{j=0}^m$ (with $g_0 = 1$) generate the algebra $\mathbb{R}[\mathbf{x}]$.

If $(1, g_j)_{j=1}^m$ do not generate $\mathbb{R}[\mathbf{x}]$, we add redundant inequalities as follows. Let $\underline{x}_k \leq \min\{x_k : \mathbf{x} \in \mathbf{K}\}$ for all $k = 1, \ldots, n$. Then, with $g_{m+k}(\mathbf{x}) = x_k - \underline{x}_k$, we introduce the additional (redundant) constraints $g_{m+k}(\mathbf{x}) \geq 0$, $k = 1, \ldots, n$, in the definition (1.10) of $\mathbf{K}$, and reset $m = m + n$. With this new equivalent definition, Assumption 1.23 holds.

For every $j = 1, \ldots, m$, let $\overline{g}_j = \max_{\mathbf{x} \in \mathbf{K}} g_j(\mathbf{x})$ (well defined because $\mathbf{K}$ is compact), and let $(\widehat{g}_j)_{j=1}^m$ be the polynomials g_j normalized with respect to $\mathbf{K}$, that is, for $j = 1, \ldots, m$,

$$\widehat{g}_j = \begin{cases} g_j/\overline{g}_j & \text{if } \overline{g}_j > 0, \\ g_j & \text{if } \overline{g}_j = 0. \end{cases} \tag{1.17}$$

We next let $G = (0, 1, \widehat{g}_1, \ldots, \widehat{g}_m, 1 - \widehat{g}_1, \ldots, 1 - \widehat{g}_m) \subset \mathbb{R}[\mathbf{x}]$, and let $\Delta_G \subset \mathbb{R}[\mathbf{x}]$ be the set of all products of the form $q_1 \cdots q_k$, for polynomials $(q_j)_{j=1}^k \subset G$ and an integer $k \geq 1$. Denote by C_G the *cone* generated by Δ_G, i.e., $f \in C_G$ if

$$f = \sum_{\boldsymbol{\alpha}, \boldsymbol{\beta} \in \mathbb{N}^m} c_{\boldsymbol{\alpha}\boldsymbol{\beta}}\, \widehat{g}_1^{\alpha_1} \cdots \widehat{g}_m^{\alpha_m} (1 - \widehat{g}_1)^{\beta_1} \cdots (1 - \widehat{g}_m)^{\beta_m},$$

for finitely many nonnegative coefficients $(c_{\boldsymbol{\alpha}\boldsymbol{\beta}}) \subset \mathbb{R}_+$ (and with $\widehat{g}_k^0 = 1$), or, using the vector notation

$$\widehat{\mathbf{g}} = \begin{pmatrix} \widehat{g}_1 \\ \vdots \\ \widehat{g}_m \end{pmatrix}, \quad \mathbf{1} - \widehat{\mathbf{g}} = \begin{pmatrix} 1 - \widehat{g}_1 \\ \vdots \\ 1 - \widehat{g}_m \end{pmatrix},$$

$f \in C_G$ if f has the compact form

$$f = \sum_{\boldsymbol{\alpha}, \boldsymbol{\beta} \in \mathbb{N}^m} c_{\boldsymbol{\alpha}\boldsymbol{\beta}}\, \widehat{\mathbf{g}}^{\boldsymbol{\alpha}} (\mathbf{1} - \widehat{\mathbf{g}})^{\boldsymbol{\beta}}. \tag{1.18}$$

Equivalently, $f \in C_G$ if f is a polynomial of $\mathbb{R}[\widehat{g}_1, \dots, \widehat{g}_m, 1 - \widehat{g}_1, \dots, 1 - \widehat{g}_m]$ with nonnegative coefficients.

Theorem 1.24. *Let $(g_i)_{i=1}^m \subset \mathbb{R}[\mathbf{x}]$, $\mathbf{K} \subset \mathbb{R}^n$ be as in (1.10). Under Assumption 1.23, if $f \in \mathbb{R}[\mathbf{x}]$ is strictly positive on $\mathbf{K}$, then $f \in C_G$. Equivalently, (1.18) holds for finitely many nonnegative coefficients $(c_{\boldsymbol{\alpha\beta}})_{\boldsymbol{\alpha},\boldsymbol{\beta} \in \mathbb{N}^m} \in \mathbb{R}_+$.*

In contrast to Theorems 1.15 and 1.17, the representation (1.18) involves some nonnegative scalar coefficients $(c_{\boldsymbol{\alpha\beta}})$ rather than s.o.s. polynomials in $\Sigma[\mathbf{x}]$. Determining if $f > 0$ on $\mathbf{K}$ using Theorem 1.24 leads to a linear optimization feasibility problem for which extremely efficient software packages are available. On the other hand, it involves products of arbitrary powers of the g_i's and $(1 - g_j)$'s, a highly undesirable feature. In particular, the presence of large binomial coefficients is a source of ill-conditioning and numerical instability.

The case of polytopes

If $\mathbf{K}$ is a convex polytope, then Theorem 1.24 simplifies and we obtain a generalization of Theorem 1.9 for $[-1, 1] \subset \mathbb{R}$ to a polytope in $\mathbb{R}^n$.

Theorem 1.25. *Let $g_j \in \mathbb{R}[\mathbf{x}]$ be affine for every $j = 1, \dots, m$ and assume that $\mathbf{K}$ in (1.10) is compact with a nonempty interior. If $f \in \mathbb{R}[\mathbf{x}]$ is strictly positive on $\mathbf{K}$, then*

$$f = \sum_{\boldsymbol{\alpha} \in \mathbb{N}^m} c_{\boldsymbol{\alpha}} \, g_1^{\alpha_1} \cdots g_m^{\alpha_m} \tag{1.19}$$

for finitely many nonnegative scalars $(c_{\boldsymbol{\alpha}})$.

Notice that Theorem 1.25 is of the same flavor as Theorem 1.24, except that it does not require to introduce the polynomials $1 - \widehat{g}_j$, $j = 1, \dots, m$.

Remarks. There are three features that distinguish the case $n > 1$ from the case $n = 1$ treated in the previous section:

(a) Theorems 1.15, 1.17, and 1.24 all deal with *compact* sets $\mathbf{K}$, whereas Theorem 1.7 can handle the (noncompact) interval $[0, \infty)$.

(b) In Theorems 1.15, 1.17, and 1.24, p is restricted to be *strictly positive* instead of *nonnegative* in Theorem 1.7.

(c) In Theorems 1.15, 1.17, and 1.24, nothing is said on the *degree* of the polynomials involved in (1.11), (1.13), or in (1.18), whereas in Theorem 1.7 the degree is bounded and known in advance. In fact, bounds exist for the representations (1.11) and (1.13); see, e.g., Theorem 1.20. However they are not practical from a computational viewpoint. This is the reason why Theorems 1.15, 1.17, and 1.24 do not lead to a polynomial-time algorithm to check whether a polynomial f is positive on $\mathbf{K}$.

1.6 Polynomials nonnegative on real varieties

In this section, we introduce representations of polynomials on a real variety. The first result considers an arbitrary real variety, whereas the second considers a finite variety associated with a zero-dimensional ideal I of $\mathbb{R}[\mathbf{x}]$ which is radical. For a background on basic definitions and results of algebraic geometry, the interested reader is referred to, e.g., [9].

Let $V \subset \mathbb{R}^n$ be the real variety defined by

$$V = \{ \mathbf{x} \in \mathbb{R}^n \;:\; g_j(\mathbf{x}) = 0,\; j = 1, \ldots, m \}, \tag{1.20}$$

for some family of real polynomials $(g_j) \subset \mathbb{R}[\mathbf{x}]$. Given $f \in \mathbb{R}[\mathbf{x}]$ and $\epsilon > 0$, let $f_{\epsilon r} \in \mathbb{R}[\mathbf{x}]$ be the polynomial

$$f_{\epsilon r} = f + \epsilon \sum_{k=0}^{r} \sum_{i=1}^{n} \frac{x_i^{2k}}{k!}, \quad \epsilon \geq 0, \quad r \in \mathbb{N}. \tag{1.21}$$

Theorem 1.26. *Let $V \subset \mathbb{R}^n$ be as in (1.20), and let $f \in \mathbb{R}[\mathbf{x}]$ be nonnegative on V. Then, for every $\epsilon > 0$, there exist $r_\epsilon \in \mathbb{N}$ and nonnegative scalars $(\lambda_j)_{j=1}^m$ such that, for all $r \geq r_\epsilon$,*

$$f_{\epsilon r} + \sum_{j=1}^{m} \lambda_j \, g_j^2 \quad \text{is s.o.s.} \tag{1.22}$$

In addition, $\|f - f_{\epsilon r}\|_1 \to 0$ as $\epsilon \downarrow 0$ (and $r \geq r_\epsilon$).

Theorem 1.26 is a denseness result, the analogue for varieties of Theorem 1.5(b) for $V = \mathbb{R}^n$, and provides a certificate of nonnegativity of f on V. Notice that, in contrast with Theorems 1.15 and 1.17 (letting an equality constraint being two reverse inequality constraints), Theorem 1.26 makes *no* assumption on the variety V; in addition one has *scalar* multipliers (λ_j) in (1.22) instead of s.o.s. multipliers in (1.11) or (1.13). On the other hand, the former theorems state that if V is compact and f is nonnegative on V, then $f + \epsilon = f_{\epsilon/n,0}$ has the sum of squares representation (1.11) (or (1.13)), and $\|f - f_{\epsilon/n0}\|_\infty \to 0$ as $\epsilon \downarrow 0$, instead of the weaker conclusion $\|f - f_{\epsilon r}\|_1 \to 0$ in Theorem 1.26.

Polynomials nonnegative on a finite variety

We recall that a zero-dimensional ideal $I \subset \mathbb{R}[\mathbf{x}]$ is an ideal[1] such that the associated variety

$$V_{\mathbb{C}}(I) = \{ \mathbf{x} \in \mathbb{C}^n \;:\; g(\mathbf{x}) = 0,\; \forall g \in I \}$$

is *finite*. In such a case, the quotient ring $\mathbb{R}[\mathbf{x}]/I$ is a finite-dimensional $\mathbb{R}$-vector space whose dimension is larger than $|V_{\mathbb{C}}(I)|$ and equal to $|V_{\mathbb{C}}(I)|$ if and only if I is radical.

[1] For definitions and properties of ideal, radical ideal, etc., the reader is referred to, e.g., [1] and [9].

This is an important special case when one deals with discrete sets as in discrete optimization; for instance when the set $\mathbf{K} \subset \mathbb{R}^n$ consists of the grid points (x_{ij}) that are solutions of the polynomial equations

$$\mathbf{K} = \left\{ \mathbf{x} \in \mathbb{R}^n \;:\; \prod_{j=1}^{2r_i} (x_i - x_{ij}) = 0;\; i = 1, \dots, n \right\}. \tag{1.23}$$

Binary optimization deals with the case when $r_i = 1$ for all $i = 1, \dots, n$ and $x_{ij} \in \{0, 1\}$.

Theorem 1.27. *Let $I = \langle g_1, \dots, g_m \rangle$ be a zero-dimensional ideal of $\mathbb{R}[\mathbf{x}]$ with associated (finite) variety $V_{\mathbb{C}}(I) \subset \mathbb{C}^n$. Assume that I is radical ($I = \sqrt{I}$), and let $V_{\mathbb{R}}(I) = V_{\mathbb{C}}(I) \cap \mathbb{R}^n$. Let $(h_i)_{i=1}^r \subset \mathbb{R}[\mathbf{x}]$ and*

$$\mathbf{K} = \left\{ \mathbf{x} \in V_{\mathbb{R}}(I) \;:\; h_i(\mathbf{x}) \geq 0,\; i = 1, \dots, r \right\}. \tag{1.24}$$

Then $f \in \mathbb{R}[\mathbf{x}]$ is nonnegative on $\mathbf{K}$ if and only if

$$f = f_0 + \sum_{i=1}^{r} f_i h_i + \sum_{j=1}^{m} v_i g_i, \tag{1.25}$$

where $(f_i)_{i=0}^r \subset \Sigma[\mathbf{x}]$ and $(v_i)_{i=1}^m \subset \mathbb{R}[\mathbf{x}]$.

In the case where $\mathbf{K}$ is the set (1.23), the degree of the polynomials (f_i) is bounded by $(\sum_{i=1}^n r_i) - n$.

For finite varieties, observe that Theorem 1.27 provides a representation result stronger than that of Theorem 1.17. In particular, f is not required to be strictly positive and sometimes a degree bound is also available.

1.7 Representations with sparsity properties

In this section, we introduce sparse representations for polynomials f nonnegative on a basic semi-algebraic set $\mathbf{K}$, when there is *weak coupling* between some subsets of variables in the polynomials g_j that define the set $\mathbf{K}$, and f. By weak coupling (to be detailed later) we mean that (a) each polynomial in the definition of $\mathbf{K}$ contains a few variables only, and (b) the polynomial f is a sum of polynomials, each containing also a few variables only. This sparse representation is computationally important, as it translates to smaller semidefinite programs for computing the s.o.s. polynomials that define the representation. In fact, given the current state of semidefinite optimization, it is absolutely critical to exploit sparsity in order to solve problems involving a large number of variables.

With $\mathbb{R}[\mathbf{x}] = \mathbb{R}[x_1, \dots, x_n]$, let the index set $I_0 = \{1, \dots, n\}$ be the union $\bigcup_{k=1}^p I_k$ of p subsets of indices I_k, $k = 1, \dots, p$ (with possible overlaps). Let us write $n_k = |I_k|$. Let $\mathbb{R}[\mathbf{x}(I_k)]$ denote the ring of polynomials in the n_k variables $\mathbf{x}(I_k) = \{x_i : i \in I_k\}$, and so $\mathbb{R}[\mathbf{x}(I_0)] = \mathbb{R}[\mathbf{x}]$.

Assumption 1.28. Let $\mathbf{K} \subset \mathbb{R}^n$ be as in (1.10). A scalar $M > 0$ is known such that $\|\mathbf{x}\|_\infty < M$ for all $\mathbf{x} \in \mathbf{K}$.

Note that, under Assumption 1.28, we have $\sum_{i\in I_k} x_i^2 \leq n_k M^2$, $k = 1, \ldots, p$, and therefore, in the Definition (1.10) of $\mathbf{K}$, we add the p redundant quadratic constraints

$$0 \leq g_{m+k}(\mathbf{x}) = n_k M^2 - \sum_{i\in I_k} x_i^2, \quad k = 1, \ldots, p, \tag{1.26}$$

and set $m' = m + p$, so that $\mathbf{K}$ is now defined by

$$\mathbf{K} = \{\mathbf{x} \in \mathbb{R}^n \;:\; g_j(\mathbf{x}) \geq 0, \; j = 1, \ldots, m'\}. \tag{1.27}$$

Note that $g_{m+k} \in \mathbb{R}[\mathbf{x}(I_k)]$ for all $k = 1, \ldots, p$.

Assumption 1.29. Let $\mathbf{K} \subset \mathbb{R}^n$ be as in (1.27). The index set $J = \{1, \ldots, m'\}$ is partitioned into p disjoint sets J_k, $k = 1, \ldots, p$, and the collections $\{I_k\}$ and $\{J_k\}$ satisfy:

(a) For every $j \in J_k$, $g_j \in \mathbb{R}[\mathbf{x}(I_k)]$, that is, for every $j \in J_k$, the constraint $g_j(\mathbf{x}) \geq 0$ only involves the variables $\mathbf{x}(I_k) = \{x_i \;:\; i \in I_k\}$.

(b) The objective function $f \in \mathbb{R}[\mathbf{x}]$ can be written as

$$f = \sum_{k=1}^{p} f_k, \quad \text{with } f_k \in \mathbb{R}[\mathbf{x}(I_k)], \quad k = 1, \ldots, p. \tag{1.28}$$

The main result about sparsity is as follows.

Theorem 1.30. *Let* $\mathbf{K} \subset \mathbb{R}^n$ *be as in* (1.27) (*i.e.,* $\mathbf{K}$ *as in* (1.10) *with the additional redundant quadratic constraints* (1.26)). *Let Assumptions* 1.28 *and* 1.29 *hold and, in addition, assume that, for every* $k = 1, \ldots, p-1$,

$$\left(I_{k+1} \cap \left(\cup_{j=1}^{k} I_j\right)\right) \subseteq I_s \quad \text{for some } s \leq k. \tag{1.29}$$

If $f \in \mathbb{R}[\mathbf{x}]$ *is strictly positive on* $\mathbf{K}$, *then*

$$f = \sum_{k=1}^{p} \left(q_k + \sum_{j\in J_k} q_{jk}\, g_j \right), \tag{1.30}$$

for some sums of squares polynomials $(q_k, q_{jk}) \subset \mathbb{R}[\mathbf{x}(I_k)]$, $k = 1, \ldots, p$.

The key property (1.29) that allows the sparse representation (1.30) is called the *running intersection property*. Under this property, the absence of coupling of variables in the original data is preserved in the representation (1.30). So Theorem 1.30 provides a representation that is more specific than that of Theorem 1.17. Let us illustrate Theorem 1.30 with an example.

Example 1.31. Let $\mathbf{x} = (x_1, \ldots, x_5)$ and

$$\mathbf{x} \longmapsto g_1(\mathbf{x}) = 1 - x_1^2 - x_2^2$$

$$\mathbf{x} \longmapsto g_2(\mathbf{x}) = 1 - x_2^4 x_3^4 - x_3^2$$

$$\mathbf{x} \longmapsto g_3(\mathbf{x}) = 1 - x_3^2 - x_4^2 - x_5^2$$

$$\mathbf{x} \longmapsto f(\mathbf{x}) = 1 + x_1^2 x_2^2 + x_2^2 x_3^2 + x_3^2 - x_3^2 x_4^2 - x_3^2 x_5^2$$

$$\mathbf{K} = \big\{\mathbf{x} \,:\, g_1(x_1, x_2),\, g_2(x_2, x_3),\, g_3(x_3, x_4, x_5) \geq 0\big\}.$$

Then $I_1 = \{1, 2\}$, meaning that the polynomial g_1 only involves variables x_1, x_2, $I_2 = \{2, 3\}$ and $I_3 = \{3, 4, 5\}$. Moreover, $f = f_1 + f_2 + f_3$, with $f_1 = 1 + x_1^2 x_2^2$, $f_2 = x_2^2 x_3^2$ and $f_3 = x_3^2 - x_3^2 x_4^2 - x_3^2 x_5^2$. Let us check property (1.29). For $k = 1$, $I_2 \cap I_1 = \{2\} \subset I_1$. For $k = 2$, $I_3 \cap (I_1 \cup I_2) = \{3\} \subset I_2$ and thus property (1.29) holds. Hence, Theorem 1.30 allows the sparse representation

$$f(\mathbf{x}) = 1 + x_1^4 x_2^2 + x_1^2 x_2^4 + x_1^2 x_2^2 \, g_1(\mathbf{x}) + x_2^6 x_3^6 + x_2^2 x_3^4 + x_2^2 x_3^2 \, g_2(\mathbf{x}) + x_3^4 + x_3^2 \, g_3(\mathbf{x}).$$

1.8 Notes and sources

1.2 For a nice exposition on degree bounds for Hilbert's 17th problem, see Schmid (1998) [109]. Theorem 1.4 is from Blekherman (2006) [15] whereas Theorem 1.5(a) is from Lasserre and Netzer (2007) [78] and provides an explicit construction of an approximating sequence of sums of squares for the denseness result of Berg (1987) [12]. Theorem 1.5(b) is from Lasserre (2006) [72] and Theorem 1.3 is from Lasserre (2007) [73].

1.3 Most of the material for the one-dimensional case is taken from Powers and Reznick (2000) [101]. Theorem 1.7(a) is due to Fekete (1935) [33] whereas Theorem 1.7(b) is attributed to F. Lukács. Theorem 1.8 is from Pólya and Szegö (1976) [100]. Theorem 1.9 is due to Hausdorff (1915) [40] and Bernstein (1921) [13].

1.4 The important Theorem 1.10 is due to Pólya (1974) [99]. The Positivstellensatz (Theorem 1.14) is credited to Stengle (1974) [122] but was proved earlier by Krivine (1964) [56]. For a complete and recent exposition of representation of positive polynomials, the reader is referred to Prestel and Delzell (2001) [102]; see also Kuhlmann et al. (2005) [60] and the more recent Helton and Putinar (2007) [44], Scheiderer (2008) [107] and Marshall (2008) [85]. For a nice discussion on historical aspects on Hilbert's 17th problem, the reader is referred to Reznick (2000) [104].

1.5 Theorem 1.15 is due to Schmüdgen (1991) [110]. The machinery uses the spectral theory of self-adjoint operators in Hilbert spaces. The important Theorem 1.17 is due to Putinar (1993) [103] and Jacobi and Prestel (2001)

[48], whereas Theorem 1.18 is due to Schmüdgen (1991) [110]. Concerning degree bounds for s.o.s. terms that appear in those representation results, Theorem 1.20 is from Nie and Schweighofer (2007) [93]; see also Marshall (2009) [86] and Schweighofer (2005) [114]. For the noncompact case, negative results are provided in Scheiderer (2008) [107]. However, nice representation results have been obtained in some specific cases; see, e.g., Marshall (2009) [87]. Theorem 1.25 for the case of polytopes is due to Cassier (1984) [21] and Handelman (1988) [38], and Theorem 1.24 for compact semi-algebraic sets follows from a result due to Krivine (1964) [56, 57] and restated later in Becker and Schwartz (1983) [10] and Vasilescu (2003) [127]. Theorem 1.22, the matrix-polynomial version of Putinar's Positivstellensatz, was first proved in Scherer and Hol (2004) [108] and Kojima and Maramatsu (2007) [52] independently (with no degree bound). The version with degree bound is from Helton and Nie (2009) [43].

1.6 Theorem 1.26 is from Lasserre (2005) [68] whereas Theorem 1.27 is from Parrilo (2002) [96], who extended previous results in Lasserre (2002) [66] for the grid case.

1.7 Theorem 1.30 was first proved in Lasserre (2006) [70] under the assumption that the feasible set $\mathbf{K}$ has a nonempty interior. This assumption was later removed in Kojima and Maramatsu (2009) [53]. For extensions of Theorem 1.30 to some noncompact cases, see the recent work of Kuhlmann and Putinar (2007, 2009) [58, 59].

Chapter 2

Moments

Most results of this chapter are *dual* analogues of those described in Chapter 1. Indeed, the problem of representing polynomials that are positive on a set $\mathbf{K} \subset \mathbb{R}^n$ has a dual facet which is the problem of characterizing sequences of reals that are moment sequences of some finite Borel measure supported on $\mathbf{K}$. Moreover, as we shall see, this beautiful duality is nicely captured by standard duality in convex optimization, applied to some appropriate convex cones of $\mathbb{R}[\mathbf{x}]$. We review basic results in the moment problem and also particularize to some specific important cases like in Chapter 1.

Let $\overline{\mathbf{z}} \in \mathbb{C}^n$ denote the complex conjugate vector of $\mathbf{z} \in \mathbb{C}^n$. Let

$$\left\{\mathbf{z}^{\boldsymbol{\alpha}}\, \overline{\mathbf{z}}^{\boldsymbol{\beta}}\right\}_{\mathbf{z}\in\mathbb{C}^n,\, \boldsymbol{\alpha},\boldsymbol{\beta}\in\mathbb{N}^n}$$

be the basis of monomials for the ring $\mathbb{C}[\mathbf{z},\overline{\mathbf{z}}] = \mathbb{C}[z_1,\dots,z_n,\overline{z}_1,\dots,\overline{z}_n]$ of polynomials of the $2n$ variables $\{z_j,\overline{z}_j\}$, with coefficients in $\mathbb{C}$. Recall that for every $\mathbf{z} \in \mathbb{C}^n$, $\boldsymbol{\alpha} \in \mathbb{N}^n$, the notation $\mathbf{z}^{\boldsymbol{\alpha}}$ stands for the monomial $z_1^{\alpha_1}\cdots z_n^{\alpha_n}$ of $\mathbb{C}[\mathbf{z}]$.

The *support* of a Borel measure μ on $\mathbb{R}^n$ is a closed set, the complement in $\mathbb{R}^n$ of the largest open set $O \subset \mathbb{R}^n$ such that $\mu(O) = 0$ (and $\mathbb{C}^n$ may be identified with $\mathbb{R}^{2n}$).

Definition 2.1.

(a) *The full moment problem.* Let $(g_i)_{i=1}^m \subset \mathbb{C}[\mathbf{z},\overline{\mathbf{z}}]$ be such that each $g_i(\mathbf{z},\overline{\mathbf{z}})$ is real-valued, and let $\mathbf{K} \subset \mathbb{C}^n$ be the set defined by

$$\mathbf{K} = \left\{\mathbf{z} \in \mathbb{C}^n \;:\; g_i(\mathbf{z},\overline{\mathbf{z}}) \geq 0,\; i = 1,\dots,m\right\}.$$

Given an infinite sequence $\mathbf{y} = (y_{\boldsymbol{\alpha}\boldsymbol{\beta}})$ of complex numbers, where $\boldsymbol{\alpha},\boldsymbol{\beta} \in \mathbb{N}^n$, is $\mathbf{y}$ a $\mathbf{K}$-moment sequence, i.e., does there exist a measure μ supported on $\mathbf{K}$ such that

$$y_{\boldsymbol{\alpha}\boldsymbol{\beta}} = \int_{\mathbf{K}} \mathbf{z}^{\boldsymbol{\alpha}}\, \overline{\mathbf{z}}^{\boldsymbol{\beta}}\, d\mu, \quad \forall \boldsymbol{\alpha},\boldsymbol{\beta} \in \mathbb{N}^n\,? \tag{2.1}$$

(b) *The truncated moment problem.* Given (g_i), $\mathbf{K} \subset \mathbb{C}^n$ as above, a finite subset $\Delta \subset \mathbb{N}^n \times \mathbb{N}^n$, and a finite sequence $\mathbf{y} = (y_{\alpha\beta})_{(\alpha,\beta)\in\Delta} \subset \mathbb{C}$ of complex numbers, is $\mathbf{y}$ a $\mathbf{K}$-moment sequence, i.e., does there exist a measure μ supported on $\mathbf{K}$, such that

$$y_{\alpha\beta} = \int_{\mathbf{K}} \mathbf{z}^{\alpha}\, \overline{\mathbf{z}}^{\beta}\, d\mu, \quad \forall(\alpha,\beta) \in \Delta? \tag{2.2}$$

Definition 2.2. In both full and truncated cases, a measure μ as in (2.1) or (2.2) is said to be a *representing* measure of the sequence $\mathbf{y}$.

If the representing measure μ is unique, then μ is said to be *determinate* (i.e., determined by its moments), and *indeterminate* otherwise.

Example 2.3. For instance, the probability measure μ on the real line $\mathbb{R}$ with density with respect to the Lebesgue measure given by

$$x \longmapsto f(x) = \begin{cases} \left(x\sqrt{2\pi}\right)^{-1} \exp\left(-\ln(x)^2/2\right) & \text{if } x > 0, \\ 0 \ \text{otherwise}, \end{cases}$$

called the log-normal distribution, is *not* determinate. Indeed, for each a with $-1 \leq a \leq 1$, the probability measure with density

$$x \longmapsto f_a(x) = f(x)\left[1 + a\sin\left(2\pi \ln x\right)\right]$$

has exactly the same moments as μ.

The above moment problem encompasses all the classical one-dimensional moment problems of the 20th century:

(a) The Hamburger problem refers to $\mathbf{K} = \mathbb{R}$ and $(y_\alpha)_{\alpha\in\mathbb{N}} \subset \mathbb{R}$.

(b) The Stieltjes problem refers to $\mathbf{K} = \mathbb{R}_+$ and $(y_\alpha)_{\alpha\in\mathbb{N}} \subset \mathbb{R}_+$.

(c) The Hausdorff problem refers to $\mathbf{K} = [a, b]$ and $(y_\alpha)_{\alpha\in\mathbb{N}} \subset \mathbb{R}$.

(d) The Toeplitz problem refers to $\mathbf{K}$ being the unit circle in $\mathbb{C}$ and $(y_\alpha)_{\alpha\in\mathbb{Z}} \subset \mathbb{C}$.

In this book, we only consider *real* moment problems, that is, moment problems characterized with

- a set $\mathbf{K} \subset \mathbb{R}^n$, and
- a sequence $\mathbf{y} = (y_{\boldsymbol{\alpha}}) \subset \mathbb{R}$, $\boldsymbol{\alpha} \in \mathbb{N}^n$.

The multi-dimensional moment problem is significantly more difficult than the one-dimensional case, for which the results are fairly complete. This is because, in view of Theorem 2.4 below, obtaining conditions for a sequence to be moments of a representing measure with support on a given subset $\Omega \subseteq \mathbb{R}^n$, is related to characterizing polynomials that are nonnegative on Ω. When the latter characterization is available, it will translate into conditions on the sequence. But as we have seen in Chapter 1, and in contrast to the univariate case, polynomials that are nonnegative on a given set $\Omega \subseteq \mathbb{R}^n$ have no simple characterization, except for compact basic semi-algebraic sets as detailed in Section 1.5. Thus, for instance, the full multi-dimensional $\mathbf{K}$-moment problem is still unsolved for general sets $\mathbf{K} \subset \mathbb{C}^n$, including $\mathbf{K} = \mathbb{C}$.

Before we proceed further, we first state the important Riesz–Haviland theorem. Let $\mathbf{y} = (y_{\boldsymbol{\alpha}}) \subset \mathbb{R}$ be an infinite sequence, and let $L_{\mathbf{y}} : \mathbb{R}[\mathbf{x}] \to \mathbb{R}$ be the linear functional

$$f(\mathbf{x}) = \sum_{\boldsymbol{\alpha} \in \mathbb{N}^n} f_{\boldsymbol{\alpha}} \mathbf{x}^{\boldsymbol{\alpha}} \longmapsto L_{\mathbf{y}}(f) = \sum_{\boldsymbol{\alpha} \in \mathbb{N}^n} f_{\boldsymbol{\alpha}} y_{\boldsymbol{\alpha}}. \tag{2.3}$$

Theorem 2.4 (Riesz–Haviland). *Let $\mathbf{y} = (y_{\boldsymbol{\alpha}})_{\boldsymbol{\alpha} \in \mathbb{N}^n} \subset \mathbb{R}$ and let $\mathbf{K} \subset \mathbb{R}^n$ be closed. There exists a finite Borel measure μ on $\mathbf{K}$ such that*

$$\int_{\mathbf{K}} \mathbf{x}^{\boldsymbol{\alpha}} \, d\mu = y_{\boldsymbol{\alpha}}, \quad \forall \boldsymbol{\alpha} \in \mathbb{N}^n, \tag{2.4}$$

if and only if $L_{\mathbf{y}}(f) \geq 0$ for all polynomials $f \in \mathbb{R}[\mathbf{x}]$ nonnegative on $\mathbf{K}$.

Note that Theorem 2.4 is not very practical, since we do not have any explicit characterization of polynomials that are nonnegative on a general closed set $\mathbf{K} \subset \mathbb{R}^n$. However, we have seen in Chapter 1 some nice representations for the subclass of compact basic semi-algebraic sets $\mathbf{K} \subset \mathbb{R}^n$. Theorem 2.4 will serve as our primary proof tool in the next sections.

2.1 One-dimensional moment problems

Given an infinite sequence $\mathbf{y} = (y_j) \subset \mathbb{R}$, we introduce the Hankel matrices $\mathbf{H}_n(\mathbf{y})$, $\mathbf{B}_n(\mathbf{y})$ and $\mathbf{C}_n(\mathbf{y}) \in \mathbb{R}^{(n+1)\times(n+1)}$, defined by

$$\mathbf{H}_n(\mathbf{y})(i,j) = y_{i+j-2}; \quad \mathbf{B}_n(\mathbf{y})(i,j) = y_{i+j-1}; \quad \mathbf{C}_n(\mathbf{y})(i,j) = y_{i+j},$$

for all $i, j \in \mathbb{N}$, with $1 \leq i, j \leq n+1$. The Hankel matrix $\mathbf{H}_n(\mathbf{y})$ is the one-dimensional (or univariate) version of what we later call a *moment* matrix in Subsection 2.2.1.

2.1.1 The full moment problem

Recall that, for any two real-valued square symmetric matrices $\mathbf{A}, \mathbf{B}$, the notation $\mathbf{A} \succeq \mathbf{B}$ (resp. $\mathbf{A} \succ \mathbf{B}$) stands for $\mathbf{A} - \mathbf{B}$ being positive semidefinite (resp. $\mathbf{A} - \mathbf{B}$ being positive definite).

For the full Hamburger, Stieltjes, and Hausdorff moment problems, we have:

Theorem 2.5. *Let* $\mathbf{y} = (y_j)_{j\in\mathbb{N}} \subset \mathbb{R}$. *Then:*

(a) (Hamburger) $\mathbf{y}$ *has a representing Borel measure* μ *on* $\mathbb{R}$ *if and only if the quadratic form*

$$\mathbf{x} \longmapsto s_n(\mathbf{x}) = \sum_{i,j=0}^{n} y_{i+j} x_i x_j \tag{2.5}$$

is positive semidefinite for all $n \in \mathbb{N}$. *Equivalently,* $\mathbf{H}_n(\mathbf{y}) \succeq \mathbf{0}$ *for all* $n \in \mathbb{N}$.

(b) (Stieltjes) $\mathbf{y}$ *has a representing Borel measure* μ *on* $\mathbb{R}_+$ *if and only if the quadratic forms* (2.5) *and*

$$\mathbf{x} \longmapsto u_n(\mathbf{x}) = \sum_{i,j=0}^{n} y_{i+j+1} x_i x_j \tag{2.6}$$

are positive semidefinite for $n \in \mathbb{N}$. *Equivalently,* $\mathbf{H}_n(\mathbf{y}) \succeq \mathbf{0}$ *and* $\mathbf{B}_n(\mathbf{y}) \succeq \mathbf{0}$ *for all* $n \in \mathbb{N}$.

(c) (Hausdorff) $\mathbf{y}$ *has a representing Borel measure* μ *on* $[a, b]$ *if and only if the quadratic forms* (2.5) *and*

$$\mathbf{x} \longmapsto v_n(\mathbf{x}) = \sum_{i,j=0}^{n} \left(-y_{i+j+2} + (b+a)\, y_{i+j+1} - ab\, y_{i+j}\right) x_i x_j \tag{2.7}$$

are positive semidefinite for $n \in \mathbb{N}$. *Equivalently,* $\mathbf{H}_n(\mathbf{y}) \succeq \mathbf{0}$ *and* $-\mathbf{C}_n(\mathbf{y}) + (a+b)\,\mathbf{B}_n(\mathbf{y}) - ab\,\mathbf{H}_n(\mathbf{y}) \succeq \mathbf{0}$ *for all* $n \in \mathbb{N}$.

Proof. (a) If $y_n = \int_{\mathbb{R}} z^n \, d\mu$, then

$$s_n(\mathbf{x}) = \sum_{i,j=0}^{n} x_i x_j \int_{\mathbb{R}} z^{i+j} d\mu = \int_{\mathbb{R}} \left(\sum_{i=0}^{n} x_i z^i \right)^2 d\mu \geq 0.$$

Conversely, we assume that (2.5) holds, or, equivalently, $\mathbf{H}_n(\mathbf{y}) \succeq \mathbf{0}$, for all $n \in \mathbb{N}$. Therefore, for every $\mathbf{q} \in \mathbb{R}^{n+1}$ we have $\mathbf{q}'\mathbf{H}_n(\mathbf{y})\mathbf{q} \geq 0$. Let $p \in \mathbb{R}[x]$ be nonnegative on $\mathbb{R}$, so that it is s.o.s. and can be written as $p = \sum_{j=1}^{r} q_j^2$ for some $r \in \mathbb{N}$ and some polynomials $(q_j)_{j=1}^{r} \subset \mathbb{R}[x]$. Then

$$\sum_{k=0}^{2n} p_k y_k = L_{\mathbf{y}}(p) = L_{\mathbf{y}}\left(\sum_{j=1}^{r} q_j^2 \right) = \sum_{j=1}^{r} \mathbf{q}_j' \mathbf{H}_n(\mathbf{y}) \mathbf{q}_j \geq 0,$$

where $\mathbf{q}_j$ is the vector of coefficients of the polynomial $q_j \in \mathbb{R}[x]$. As $p \geq 0$ was arbitrary, by Theorem 2.4, (2.4) holds with $\mathbf{K} = \mathbb{R}$.

(b) This is similar to part (a).

(c) One direction is immediate. For the converse, we assume that both (2.5) and (2.7) hold. Then, for every $\mathbf{q} \in \mathbb{R}^{n+1}$ we have

$$-\mathbf{q}'\left[\mathbf{C}_n(\mathbf{y}) + (a+b)\,\mathbf{B}_n(\mathbf{y}) - ab\,\mathbf{H}_n(\mathbf{y})\right]\mathbf{q} \geq 0.$$

Let $p \in \mathbb{R}[x]$ be nonnegative on $[a,b]$ of even degree $2n$ and thus, by Theorem 1.7(b), it can be written $x \mapsto p(x) = u(x) + (b-x)(x-a)q(x)$ with both polynomials u, q being s.o.s. with $\deg q \leq 2n-2$, $\deg u \leq 2n$. If $\deg p = 2n-1$ then, again by Theorem 1.7(b), p can be written $x \mapsto p(x) = v(x)(x-a) + w(x)(b-x)$ for some s.o.s. polynomials v, w of degree less than $2n-2$. But then

$$p(x) = ((x-a)p(x) + (b-x)p(x))/(b-a) = u(x) + (b-x)(x-a)q(x)$$

for some s.o.s. polynomials u, q with degree less than $2n$.

Thus, in both even and odd cases, writing $u = \sum_j u_j^2$ and $q = \sum_k q_k^2$ for some polynomials (u_j, q_k) of degree less than n, and with associated vectors of coefficients $(\mathbf{u}_j, \mathbf{q}_k) \subset \mathbb{R}^{n+1}$, one obtains

$$L_{\mathbf{y}}(p) = \sum_j \mathbf{u}_j'\mathbf{H}_n(\mathbf{y})\mathbf{u}_j + \sum_k (-\mathbf{q}_k'\left[\mathbf{C}_n(\mathbf{y}) + (a+b)\,\mathbf{B}_n(\mathbf{y}) - ab\,\mathbf{H}_n(\mathbf{y})\right]\mathbf{q}_k) \geq 0.$$

Therefore $L_{\mathbf{y}}(p) \geq 0$ for every polynomial p nonnegative on $[a,b]$. By Theorem 2.4, Eq. (2.4) holds with $\mathbf{K} = [a,b]$. □

Observe that Theorem 2.5 provides a criterion directly in terms of the sequence (y_n).

2.1.2 The truncated moment problem

We now state the analogue of Theorem 2.5 for the truncated moment problem for a sequence $\mathbf{y} = (y_k)_{k=0}^{2n}$ (even case), and $\mathbf{y} = (y_k)_{k=0}^{2n+1}$ (odd case). We first introduce some notation.

For an infinite sequence $\mathbf{y} = (y_j)_{j\in\mathbb{N}}$, write the Hankel moment matrix $\mathbf{H}_n(\mathbf{y})$ in the form

$$\mathbf{H}_n(\mathbf{y}) = [\mathbf{v}_0, \mathbf{v}_1, \ldots, \mathbf{v}_n] \tag{2.8}$$

where $(\mathbf{v}_j) \subset \mathbb{R}^{n+1}$ denote the column vectors of $\mathbf{H}_n(\mathbf{y})$. The Hankel rank of $\mathbf{y} = (y_j)_{j=0}^{2n}$, denoted by $\operatorname{rank}(\mathbf{y})$, is the smallest integer $1 \leq i \leq n$ such that $\mathbf{v}_i \in \operatorname{span}\{\mathbf{v}_0, \ldots, \mathbf{v}_{i-1}\}$. If $\mathbf{H}_n(\mathbf{y})$ is nonsingular, then $\operatorname{rank}(\mathbf{y}) = n+1$. Given an $m \times n$ matrix $\mathbf{A}$, $\operatorname{Range}(\mathbf{A})$ denotes the image space of $\mathbf{A}$, i.e., $\operatorname{Range}(\mathbf{A}) = \{\mathbf{Au},\, \mathbf{u} \in \mathbb{R}^n\}$.

Theorem 2.6 (The even case). *Let $\mathbf{y} = (y_j)_{0\leq j\leq 2n} \subset \mathbb{R}$. Then:*

(a) *$\mathbf{y}$ has a representing Borel measure μ on $\mathbb{R}$ if and only if $\mathbf{H}_n(\mathbf{y}) \succeq \mathbf{0}$ and* $\operatorname{rank}(\mathbf{H}_n(\mathbf{y})) = \operatorname{rank}(\mathbf{y})$.

(b) $\mathbf{y}$ *has a representing Borel measure* μ *on* $\mathbb{R}_+$ *if and only if* $\mathbf{H}_n(\mathbf{y}) \succeq 0$, $\mathbf{B}_{n-1}(\mathbf{y}) \succeq \mathbf{0}$, *and the vector* $(y_{n+1}, \ldots, y_{2n})$ *is in* $\text{Range}(\mathbf{B}_{n-1}(\mathbf{y}))$.

(c) $\mathbf{y}$ *has a representing Borel measure* μ *on* $[a, b]$ *if and only if* $\mathbf{H}_n(\mathbf{y}) \succeq \mathbf{0}$ *and* $(a + b)\, \mathbf{B}_{n-1}(\mathbf{y}) \succeq ab\, \mathbf{H}_{n-1}(\mathbf{y}) + \mathbf{C}_{n-1}(\mathbf{y})$.

Theorem 2.7 (The odd case). *Let* $\mathbf{y} = (y_j)_{0 \leq j \leq 2n+1} \subset \mathbb{R}$. *Then:*

(a) $\mathbf{y}$ *has a representing Borel measure* μ *on* $\mathbb{R}$ *if and only if* $\mathbf{H}_n(\mathbf{y}) \succeq \mathbf{0}$ *and* $\mathbf{v}_{n+1} \in \text{Range}(\mathbf{H}_n(\mathbf{y}))$.

(b) $\mathbf{y}$ *has a representing Borel measure* μ *on* $\mathbb{R}_+$ *if and only if* $\mathbf{H}_n(\mathbf{y}) \succeq 0$, $\mathbf{B}_n(\mathbf{y}) \succeq \mathbf{0}$, *and the vector* $(y_{n+1}, \ldots, y_{2n+1})$ *is in* $\text{Range}(\mathbf{H}_n(\mathbf{y}))$.

(c) $\mathbf{y}$ *has a representing Borel measure* μ *on* $[a, b]$ *if and only if* $b\, \mathbf{H}_n(\mathbf{y}) \succeq \mathbf{B}_n(\mathbf{y})$ *and* $\mathbf{B}_n(\mathbf{y}) \succeq a\, \mathbf{H}_n(\mathbf{y})$.

Example 2.8. In the univariate case $n = 1$, let $\mathbf{y} \in \mathbb{R}^5$ be the truncated sequence $\mathbf{y} = (1, 1, 1, 1, 2)$, hence with associated Hankel moment matrix

$$\mathbf{H}_2(\mathbf{y}) = \begin{bmatrix} 1 & 1 & 1 \\ 1 & 1 & 1 \\ 1 & 1 & 2 \end{bmatrix}.$$

One may easily check that $\mathbf{H}_2(\mathbf{y}) \succeq 0$, but it turns out that $\mathbf{y}$ has no representing Borel measure μ on the real line $\mathbb{R}$. However, observe that, for sufficiently small $\epsilon > 0$, the perturbed sequence $\mathbf{y}_\epsilon = (1, 1, 1 + \epsilon, 1, 1, 2)$ satisfies $\mathbf{H}_2(\mathbf{y}_\epsilon) \succ 0$ and so, by Theorem 2.6(a), $\mathbf{y}_\epsilon$ has a finite Borel representing measure μ_ϵ. But then, necessarily, there is no compact interval $[a, b]$ such that μ_ϵ is supported on $[a, b]$ for every $\epsilon > 0$.

In truncated moment problems, i.e., given a finite sequence $\mathbf{y} = (y_k)_{k=0}^n$, the basic issue is to find conditions under which we may extend the sequence $\mathbf{y}$ so as to be able to build up positive semidefinite moment matrices of higher orders. These higher-order moment matrices are called *positive extensions* or *flat extensions* when their rank does not increase with size. The rank and range conditions in Theorems 2.6–2.7 are such conditions.

2.2 The multi-dimensional moment problem

Most of the applications considered in later chapters of this book refer to real (and not complex) moment problems. Correspondingly, we introduce the basic concepts of moment and localizing matrices in the real case $\mathbb{R}^n$. However, these concepts also have their natural counterparts in $\mathbb{C}^n$, with the usual scalar product $\langle \mathbf{u}, \mathbf{v} \rangle = \sum_j \overline{u}_j v_j$.

As already mentioned, the multi-dimensional case is significantly more difficult because of the lack of a nice characterization of polynomials that are nonnegative on a given subset $\Omega \subseteq \mathbb{R}^n$. Fortunately, we have seen in Section 1.5 that such a characterization exists for the important case of compact basic semi-algebraic sets.

For an integer $r \in \mathbb{N}$, let $\mathbb{N}^n_r = \{\boldsymbol{\alpha} \in \mathbb{N}^n \,:\, |\boldsymbol{\alpha}| \leq r\}$ with $|\boldsymbol{\alpha}| = \sum_{i=1}^n \alpha_i \leq r$. Recall that

$$\mathbf{v}_r(\mathbf{x}) = \left(1, x_1, x_2, \ldots x_n, x_1^2, x_1x_2, \ldots, x_1x_n, \ldots, x_1^r, \ldots, x_n^r\right)' \tag{2.9}$$

denotes the canonical basis of the real vector space $\mathbb{R}[\mathbf{x}]_r$ of real-valued polynomials of degree at most r (and let $s(r) = \binom{n+r}{n}$ denote its dimension). Then, a polynomial $p \in \mathbb{R}[\mathbf{x}]_r$ is written as

$$\mathbf{x} \longmapsto p(\mathbf{x}) = \sum_{\boldsymbol{\alpha} \in \mathbb{N}^n} p_{\boldsymbol{\alpha}} \mathbf{x}^{\boldsymbol{\alpha}} = \langle \mathbf{p}, \mathbf{v}_r(\mathbf{x}) \rangle,$$

where $\mathbf{p} = \{p_{\boldsymbol{\alpha}}\} \in \mathbb{R}^{s(r)}$ denotes its vector of coefficients in the basis (2.9). And so we may identify $p \in \mathbb{R}[\mathbf{x}]$ with its vector of coefficients $\mathbf{p} \in \mathbb{R}^{s(r)}$.

2.2.1 Moment and localizing matrix

We next define the important notions of moment and localizing matrices.

Moment matrix

Given a $s(2r)$-sequence $\mathbf{y} = (y_{\boldsymbol{\alpha}})$, let $\mathbf{M}_r(\mathbf{y})$ be the *moment* matrix of dimension $s(r)$, with rows and columns labeled by (2.9), and constructed as follows:

$$\mathbf{M}_r(\mathbf{y})(\boldsymbol{\alpha}, \boldsymbol{\beta}) = L_{\mathbf{y}}\left(\mathbf{x}^{\boldsymbol{\alpha}}\mathbf{x}^{\boldsymbol{\beta}}\right) = y_{\boldsymbol{\alpha}+\boldsymbol{\beta}}, \quad \forall \boldsymbol{\alpha}, \boldsymbol{\beta} \in \mathbb{N}^n_r, \tag{2.10}$$

with $L_{\mathbf{y}}$ defined in (2.3). Equivalently, $\mathbf{M}_r(\mathbf{y}) = L_{\mathbf{y}}(\mathbf{v}_r(\mathbf{x})\mathbf{v}_r(\mathbf{x})')$, where the latter notation means that we apply $L_{\mathbf{y}}$ to each entry of the matrix $\mathbf{v}_r(\mathbf{x})\mathbf{v}_r(\mathbf{x})'$.

Let us consider an example with $n = r = 2$. In this case, $\mathbf{M}_2(\mathbf{y})$ becomes

$$\mathbf{M}_2(\mathbf{y}) = \left[\begin{array}{c|cc|ccc} y_{00} & y_{10} & y_{01} & y_{20} & y_{11} & y_{02} \\ \hline y_{10} & y_{20} & y_{11} & y_{30} & y_{21} & y_{12} \\ y_{01} & y_{11} & y_{02} & y_{21} & y_{12} & y_{03} \\ \hline y_{20} & y_{30} & y_{21} & y_{40} & y_{31} & y_{22} \\ y_{11} & y_{21} & y_{12} & y_{31} & y_{22} & y_{13} \\ y_{02} & y_{12} & y_{03} & y_{22} & y_{13} & y_{04} \end{array}\right].$$

In general, $\mathbf{M}_r(\mathbf{y})$ defines a bilinear form $\langle\,\cdot\,,\,\cdot\,\rangle_{\mathbf{y}}$ on $\mathbb{R}[\mathbf{x}]_r$ as follows:

$$\langle p, q \rangle_{\mathbf{y}} = L_{\mathbf{y}}(pq) = \langle \mathbf{p}, \mathbf{M}_r(\mathbf{y})\mathbf{q} \rangle = \mathbf{p}'\mathbf{M}_r(\mathbf{y})\mathbf{q}, \quad \forall \mathbf{p}, \mathbf{q} \in \mathbb{R}^{s(r)},$$

where again $p, q \in \mathbb{R}[\mathbf{x}]_r$, and $\mathbf{p}, \mathbf{q} \in \mathbb{R}^{s(r)}$ denote their vector of coefficients.

Recall from Definition 2.2 that, if $\mathbf{y}$ is a sequence of moments for some measure μ, then μ is called a representing measure for $\mathbf{y}$, and if unique, then μ is said to be determinate, and indeterminate otherwise, whereas $\mathbf{y}$ is called a determinate (resp. indeterminate) moment sequence. In addition, for every $q \in \mathbb{R}[\mathbf{x}]$,

$$\langle \mathbf{q}, \mathbf{M}_r(\mathbf{y})\mathbf{q} \rangle = L_{\mathbf{y}}(q^2) = \int q^2 \, d\mu \geq 0, \tag{2.11}$$

so that $\mathbf{M}_r(\mathbf{y}) \succeq \mathbf{0}$. It is also immediate to check that if the polynomial q^2 is expanded as $q(\mathbf{x})^2 = \sum_{\boldsymbol{\alpha} \in \mathbb{N}^n} h_{\boldsymbol{\alpha}} \mathbf{x}^{\boldsymbol{\alpha}}$, then

$$\langle \mathbf{q}, \mathbf{M}_r(\mathbf{y})\mathbf{q} \rangle = L_{\mathbf{y}}(q^2) = \sum_{\boldsymbol{\alpha} \in \mathbb{N}^n} h_{\boldsymbol{\alpha}} y_{\boldsymbol{\alpha}}.$$

- Every measure with *compact* support (say $\mathbf{K} \subset \mathbb{R}^n$) is determinate, because by the Stone–Weierstrass theorem, the space of polynomials is dense (for the sup-norm) in the space of continuous functions on $\mathbf{K}$.
- Not every sequence $\mathbf{y}$ that satisfies $\mathbf{M}_i(\mathbf{y}) \succeq \mathbf{0}$ has a representing measure μ on $\mathbb{R}^n$. This is in contrast with the one-dimensional case where, by Theorem 2.5(a), a full sequence $\mathbf{y}$ such that $\mathbf{H}_n(\mathbf{y}) \succeq \mathbf{0}$ for all $n = 0, 1, \ldots$, has a representing measure. (Recall that, in the one-dimensional case, the moment matrix is just the Hankel matrix $\mathbf{H}_n(\mathbf{y})$ in (2.8).) However, we have the following useful result:

Proposition 2.9. *Let* $\mathbf{y}$ *be a sequence indexed in the basis* $\mathbf{v}_\infty(\mathbf{x})$*, which satisfies* $\mathbf{M}_i(\mathbf{y}) \succeq \mathbf{0}$*, for all* $i = 0, 1, \ldots$.

(a) *If the sequence* $\mathbf{y}$ *satisfies*

$$\sum_{k=1}^{\infty} \left[L_{\mathbf{y}}(x_i^{2k}) \right]^{-1/2k} = +\infty, \quad i = 1, \ldots, n, \tag{2.12}$$

then $\mathbf{y}$ *has a determinate representing measure on* $\mathbb{R}^n$.

(b) *If there exist* $c, a > 0$ *such that*

$$|\, y_{\boldsymbol{\alpha}} \,| \leq c\, a^{|\boldsymbol{\alpha}|}, \quad \forall \boldsymbol{\alpha} \in \mathbb{N}^n, \tag{2.13}$$

then $\mathbf{y}$ *has a determinate representing measure with support contained in the box* $[-a, a]^n$.

In the one-dimensional case (2.12) is called Carleman's condition. The moment matrix also has the following properties:

Proposition 2.10. *Let* $d \geq 1$*, and let* $\mathbf{y} = (y_{\boldsymbol{\alpha}}) \subset \mathbb{R}$ *be such that* $M_d(\mathbf{y}) \succeq 0$*. Then*

$$|\, y_{\boldsymbol{\alpha}} \,| \leq \max \left[y_0, \, \max_{i=1,\ldots,n} L_{\mathbf{y}}\left(x_i^{2d}\right) \right], \quad \forall \boldsymbol{\alpha} \in \mathbb{N}^n_{2d}.$$

In addition, rescaling $\mathbf{y}$ *so that* $y_0 = 1$*, and letting* $\tau_d = \max_{i=1,\ldots,n} L_{\mathbf{y}}(x_i^{2d})$,

$$|\,y_{\boldsymbol{\alpha}}\,|^{\frac{1}{|\boldsymbol{\alpha}|}} \leq \tau_d^{\frac{1}{2d}}, \quad \forall \boldsymbol{\alpha} \in \mathbb{N}^n_{2d}.$$

Localizing matrix

Given a polynomial $u \in \mathbb{R}[\mathbf{x}]$ with coefficient vector $\mathbf{u} = \{u_\gamma\}$, we define the *localizing* matrix with respect to $\mathbf{y}$ and u to be the matrix $\mathbf{M}_r(u\,\mathbf{y})$ with rows and columns indexed by (2.9), and obtained from $\mathbf{M}_r(\mathbf{y})$ by:

$$\mathbf{M}_r(u\,\mathbf{y})(\boldsymbol{\alpha}, \boldsymbol{\beta}) = L_{\mathbf{y}}\big(u(\mathbf{x})\mathbf{x}^{\boldsymbol{\alpha}}\mathbf{x}^{\boldsymbol{\beta}}\big) = \sum_{\gamma \in \mathbb{N}^n} u_\gamma y_{\gamma+\boldsymbol{\alpha}+\boldsymbol{\beta}}, \quad \forall \boldsymbol{\alpha}, \boldsymbol{\beta} \in \mathbb{N}^n_r. \tag{2.14}$$

Equivalently, $\mathbf{M}_r(u\,\mathbf{y}) = L_{\mathbf{y}}(u\,\mathbf{v}_r(\mathbf{x})\mathbf{v}_r(\mathbf{x})')$, where the previous notation means that $L_{\mathbf{y}}$ is applied entrywise. For instance, when $n = 2$, with

$$\mathbf{M}_1(\mathbf{y}) = \begin{bmatrix} y_{00} & y_{10} & y_{01} \\ y_{10} & y_{20} & y_{11} \\ y_{01} & y_{11} & y_{02} \end{bmatrix} \quad \text{and} \quad \mathbf{x} \longmapsto u(\mathbf{x}) = a - x_1^2 - x_2^2,$$

we obtain

$$\mathbf{M}_1(u\,\mathbf{y}) = \begin{bmatrix} ay_{00} - y_{20} - y_{02} & ay_{10} - y_{30} - y_{12} & ay_{01} - y_{21} - y_{03} \\ ay_{10} - y_{30} - y_{12} & ay_{20} - y_{40} - y_{22} & ay_{11} - y_{31} - y_{13} \\ ay_{01} - y_{21} - y_{03} & ay_{11} - y_{31} - y_{13} & ay_{02} - y_{22} - y_{04} \end{bmatrix}.$$

Similarly to (2.11), we have

$$\langle \mathbf{p}, \mathbf{M}_r(u\,\mathbf{y})\mathbf{q} \rangle = L_{\mathbf{y}}(u\,pq)$$

for all polynomials $p, q \in \mathbb{R}[\mathbf{x}]_r$ with coefficient vectors $\mathbf{p}, \mathbf{q} \in \mathbb{R}^{s(r)}$. In particular, if $\mathbf{y}$ has a representing measure μ, then

$$\langle \mathbf{q}, \mathbf{M}_r(u\,\mathbf{y})\mathbf{q} \rangle = L_{\mathbf{y}}(u\,q^2) = \int u\,q^2\,d\mu \tag{2.15}$$

for each polynomial $q \in \mathbb{R}[\mathbf{x}]$ with coefficient vector $\mathbf{q} \in \mathbb{R}^{s(r)}$. Hence, $\mathbf{M}_r(u\,\mathbf{y}) \succeq \mathbf{0}$ whenever μ has its support contained in the set $\{\mathbf{x} \in \mathbb{R}^n \,:\, u(\mathbf{x}) \geq 0\}$.

It is also immediate to check that if the polynomial uq^2 is expanded as $u(\mathbf{x})\,q(\mathbf{x})^2 = \sum_{\boldsymbol{\alpha} \in \mathbb{N}^n} h_{\boldsymbol{\alpha}} \mathbf{x}^{\boldsymbol{\alpha}}$ then

$$\langle \mathbf{q}, \mathbf{M}_r(u\,\mathbf{y})\mathbf{q} \rangle = \sum_{\boldsymbol{\alpha} \in \mathbb{N}^n} h_{\boldsymbol{\alpha}} y_{\boldsymbol{\alpha}} = L_{\mathbf{y}}(u\,q^2). \tag{2.16}$$

2.2.2 Positive and flat extensions of moment matrices

We next discuss the notion of positive extension for moment matrices.

Definition 2.11. Given a finite sequence $\mathbf{y} = (y_{\boldsymbol{\alpha}})_{|\boldsymbol{\alpha}|\leq 2r}$ with $\mathbf{M}_r(\mathbf{y}) \succeq \mathbf{0}$, the *moment extension problem* is defined as follows: extend the sequence $\mathbf{y}$ with new scalars $y_{\boldsymbol{\beta}}$, $2r < |\boldsymbol{\beta}| \leq 2(r+1)$, so as to obtain a new finite sequence $(y_{\boldsymbol{\alpha}})_{|\boldsymbol{\alpha}|\leq 2(r+1)}$ such that $\mathbf{M}_{r+1}(\mathbf{y}) \succeq \mathbf{0}$.

If such an extension $\mathbf{M}_{r+1}(\mathbf{y})$ is possible, it is called a *positive extension* of $\mathbf{M}_r(\mathbf{y})$. If, in addition, rank $\mathbf{M}_{r+1}(\mathbf{y})$ = rank $\mathbf{M}_r(\mathbf{y})$, then $\mathbf{M}_{r+1}(\mathbf{y})$ is called a *flat extension* of $\mathbf{M}_r(\mathbf{y})$.

For truncated moment problems, flat extensions play an important role. We first introduce the notion of an atomic measure. An s-atomic measure is a measure with s *atoms*, that is, a linear positive combination of s Dirac measures.

Theorem 2.12 (Flat extension). *Let* $\mathbf{y} = (y_{\boldsymbol{\alpha}})_{|\boldsymbol{\alpha}|\leq 2r}$. *Then the sequence* $\mathbf{y}$ *admits a* rank $\mathbf{M}_r(\mathbf{y})$*-atomic representing measure* μ *on* $\mathbb{R}^n$ *if and only if* $\mathbf{M}_r(\mathbf{y}) \succeq \mathbf{0}$ *and* $\mathbf{M}_r(\mathbf{y})$ *admits a flat extension* $\mathbf{M}_{r+1}(\mathbf{y}) \succeq 0$.

Theorem 2.12 is useful as it provides a simple numerical means to check whether a finite sequence has a representing measure.

Example 2.13. Let μ be the measure on $\mathbb{R}$ defined to be $\mu = \delta_0 + \delta_1$, that is, μ is the sum of two Dirac measures at the points $\{0\}$ and $\{1\}$. Then

$$\mathbf{M}_1(\mathbf{y}) = \begin{bmatrix} 2 & 1 \\ 1 & 1 \end{bmatrix}, \quad \mathbf{M}_2(\mathbf{y}) = \begin{bmatrix} 2 & 1 & 1 \\ 1 & 1 & 1 \\ 1 & 1 & 1 \end{bmatrix},$$

and, obviously, rank $\mathbf{M}_2(\mathbf{y})$ = rank $\mathbf{M}_1(\mathbf{y}) = 2$.

2.3 The K-moment problem

The (real) $\mathbf{K}$-moment problem identifies those sequences $\mathbf{y}$ that are moment sequences of a measure with support contained in a set $\mathbf{K} \subset \mathbb{R}^n$.

Given m polynomials $g_i \in \mathbb{R}[\mathbf{x}]$, $i = 1, \dots, m$, let $\mathbf{K} \subset \mathbb{R}^n$ be the basic semi-algebraic set

$$\mathbf{K} = \{\mathbf{x} \in \mathbb{R}^n \ : \ g_i(\mathbf{x}) \geq 0, \ i = 1, \dots, m\}. \tag{2.17}$$

For notational convenience, we also define $g_0 \in \mathbb{R}[\mathbf{x}]$ to be the constant polynomial with value 1 (i.e., $g_0 = 1$).

Recall that, given a family $(g_j)_{j=1}^m \subset \mathbb{R}[\mathbf{x}]$, we denote by g_J, $J \subseteq \{1, \dots, m\}$, the polynomial $\mathbf{x} \mapsto g_J(\mathbf{x}) = \prod_{j\in J} g_j(\mathbf{x})$. In particular, when $J = \emptyset$, $g_\emptyset = 1$.

Let $\mathbf{y} = (y_{\boldsymbol{\alpha}})_{\boldsymbol{\alpha}\in\mathbb{N}^n}$ be a given infinite sequence. For every $r \in \mathbb{N}$ and every $J \subseteq \{1, \dots, m\}$, let $\mathbf{M}_r(g_J\,\mathbf{y})$ be the localizing matrix of order r with respect to the polynomial $g_J = \prod_{j \in J} g_j$ (so that, with $J = \emptyset$, $\mathbf{M}_r(g_\emptyset\,\mathbf{y}) = \mathbf{M}_r(\mathbf{y})$ is the moment matrix (of order r) associated with $\mathbf{y}$).

As we have already seen, there is a duality between the theory of moments and the representation of positive polynomials. The following important theorem, which is the dual facet of Theorem 1.15 and Theorem 1.17, makes this statement more precise.

Theorem 2.14. *Let $\mathbf{y} = (y_{\boldsymbol{\alpha}}) \subset \mathbb{R}$, $\boldsymbol{\alpha} \in \mathbb{N}^n$, be a given infinite sequence in $\mathbb{R}$, $L_{\mathbf{y}} \colon \mathbb{R}[\mathbf{x}] \to \mathbb{R}$ be the Riesz-functional introduced in (2.3), and let $\mathbf{K}$ be as in (2.17), assumed to be compact. Then:*

(a) *The sequence $\mathbf{y}$ has a finite Borel representing measure with support contained in $\mathbf{K}$ if and only if*

$$L_{\mathbf{y}}(f^2\,g_J) \geq 0, \quad \forall J \subseteq \{1,\ldots,m\}, \quad \forall f \in \mathbb{R}[\mathbf{x}], \tag{2.18}$$

or, equivalently, if and only if

$$\mathbf{M}_r(g_J\,\mathbf{y}) \succeq \mathbf{0}, \quad \forall J \subseteq \{1,\ldots,m\}, \quad \forall r \in \mathbb{N}. \tag{2.19}$$

(b) *Assume that there exists $u \in \mathbb{R}[\mathbf{x}]$ of the form*

$$u = u_0 + \sum_{j=1}^{m} u_i g_i, \quad u_i \in \Sigma[\mathbf{x}], \quad i = 0, 1, \ldots, m,$$

and such that the level set $\{\mathbf{x} \in \mathbb{R}^n \;:\; u(\mathbf{x}) \geq 0\}$ is compact. Then $\mathbf{y}$ has a finite Borel representing measure with support contained in $\mathbf{K}$ if and only if

$$L_{\mathbf{y}}(f^2\,g_j) \geq 0, \quad \forall j = 0, 1, \ldots, m, \quad \forall f \in \mathbb{R}[\mathbf{x}], \tag{2.20}$$

or, equivalently, if and only if

$$\mathbf{M}_r(g_j\,\mathbf{y}) \succeq \mathbf{0}, \quad \forall j = 0, 1, \ldots, m, \quad \forall r \in \mathbb{N}. \tag{2.21}$$

Proof. (a) For every $J \subseteq \{1,\ldots,m\}$ and $f \in \mathbb{R}[\mathbf{x}]_r$, the polynomial $f^2 g_J$ is nonnegative on $\mathbf{K}$. Therefore, if $\mathbf{y}$ is the sequence of moments of a measure μ supported on $\mathbf{K}$, then $\int f^2 g_J\, d\mu \geq 0$. Equivalently, $L_{\mathbf{y}}(f^2\,g_J) \geq 0$, or, in view of (2.16), $\mathbf{M}_r(g_J\,\mathbf{y}) \succeq \mathbf{0}$. Hence (2.18)–(2.19) hold.

Conversely, assume that (2.18) or equivalently (2.19) holds. As $\mathbf{K}$ is compact, it follows from Theorem 2.4 that $\mathbf{y}$ is the moment sequence of a measure with support contained in $\mathbf{K}$ if and only if $\sum_{\boldsymbol{\alpha} \in \mathbb{N}^n} f_{\boldsymbol{\alpha}} y_{\boldsymbol{\alpha}} \geq 0$ for all polynomials $f \geq 0$ on $\mathbf{K}$. Let $f > 0$ on $\mathbf{K}$, so that, by Theorem 1.15,

$$f = \sum_{J \subseteq \{1,\ldots,m\}} p_J\, g_J \tag{2.22}$$

for some polynomials $\{p_J\} \subset \mathbb{R}[\mathbf{x}]$, all sums of squares. Hence, since $p_J \in \Sigma[\mathbf{x}]$, from (2.18) and from the linearity of $L_{\mathbf{y}}$ we have $L_{\mathbf{y}}(f) \geq 0$. Hence, for all polynomials $f > 0$ on $\mathbf{K}$, we have $L_{\mathbf{y}}(f) = \sum_{\boldsymbol{\alpha}} f_{\boldsymbol{\alpha}} y_{\boldsymbol{\alpha}} \geq 0$. Next, let $f \in \mathbb{R}[\mathbf{x}]$ be nonnegative on $\mathbf{K}$. Then for arbitrary $\epsilon > 0$, $f + \epsilon > 0$ on $\mathbf{K}$, and thus, $L_{\mathbf{y}}(f+\epsilon) = L_{\mathbf{y}}(f) + \epsilon y_0 \geq 0$. As $\epsilon > 0$ was arbitrary, $L_{\mathbf{y}}(f) \geq 0$ follows. Therefore, $L_{\mathbf{y}}(f) \geq 0$ for all $f \in \mathbb{R}[\mathbf{x}]$ nonnegative on $\mathbf{K}$, which, by Theorem 2.4, implies that $\mathbf{y}$ is the moment sequence of some measure with support contained in $\mathbf{K}$.

(b) The proof is similar to part (a) and is left as an exercise. □

Note that the conditions (2.19) and (2.21) of Theorem 2.14 are stated in terms of positive semidefiniteness of the localizing matrices associated with the polynomials g_J and g_j involved in the definition (2.17) of the compact set $\mathbf{K}$. Alternatively, we also have:

Theorem 2.15. *Let $\mathbf{y} = (y_{\boldsymbol{\alpha}}) \subset \mathbb{R}$, $\boldsymbol{\alpha} \in \mathbb{N}^n$, be an infinite sequence, $L_{\mathbf{y}} \colon \mathbb{R}[\mathbf{x}] \to \mathbb{R}$ be the Riesz-functional introduced in* (2.3), *and let $\mathbf{K}$ be as in* (2.17), *assumed to be compact. Let $C_G \subset \mathbb{R}[\mathbf{x}]$ be the convex cone defined in* (1.18), *and let Assumption* 1.23 *hold. Then $\mathbf{y}$ has a representing measure μ with support contained in $\mathbf{K}$ if and only if*

$$L_{\mathbf{y}}(f) \geq 0, \quad \forall f \in C_G, \tag{2.23}$$

or, equivalently,

$$L_{\mathbf{y}}\left(\widehat{g}^{\boldsymbol{\alpha}}\,(1-\widehat{g})^{\boldsymbol{\beta}}\right) \geq 0, \quad \forall \boldsymbol{\alpha}, \boldsymbol{\beta} \in \mathbb{N}^m \tag{2.24}$$

with $\widehat{g}$ as in (1.17).

Proof. If $\mathbf{y}$ is the moment sequence of some measure with support contained in $\mathbf{K}$, then (2.24) follows directly from Theorem 2.4, because $\widehat{g}_j \geq 0$ and $1 - \widehat{g}_j \geq 0$ on $\mathbf{K}$, for all $j = 1, \ldots, m$.

Conversely, let (2.24) (and so (2.23)) hold, and let $f \in \mathbb{R}[\mathbf{x}]$, with $f > 0$ on $\mathbf{K}$. By Theorem 1.24, $f \in C_G$, and so f can be written as in (1.18), from which $L_{\mathbf{y}}(f) \geq 0$ follows. Finally, let $f \geq 0$ on $\mathbf{K}$, so that $f + \epsilon > 0$ on $\mathbf{K}$ for every $\epsilon > 0$. Therefore, $0 \leq L_{\mathbf{y}}(f+\epsilon) = L_{\mathbf{y}}(f) + \epsilon y_0$ because $f + \epsilon \in C_G$. As $\epsilon > 0$ was arbitrary, we obtain $L_{\mathbf{y}}(f) \geq 0$. Therefore, $L_{\mathbf{y}}(f) \geq 0$ for all $f \in \mathbb{R}[\mathbf{x}]$, nonnegative on $\mathbf{K}$, which by Theorem 2.4, implies that $\mathbf{y}$ is the moment sequence of some measure with support contained in $\mathbf{K}$. □

Exactly as Theorem 2.14 was the dual facet of Theorems 1.15 and 1.17, Theorem 2.15 is the dual facet of Theorem 1.24.

Note that Eqs. (2.24) reduce to countably many *linear* conditions on the sequence $\mathbf{y}$. Indeed, for fixed $\boldsymbol{\alpha}, \boldsymbol{\beta} \in \mathbb{N}^m$, we write

$$\widehat{g}^{\boldsymbol{\alpha}}\,(1-\widehat{g})^{\boldsymbol{\beta}} = \sum_{\gamma \in \mathbb{N}^m} q_\gamma(\boldsymbol{\alpha}, \boldsymbol{\beta})\, \mathbf{x}^\gamma,$$

for finitely many coefficients $(q_\gamma(\boldsymbol{\alpha},\boldsymbol{\beta}))$. Then, (2.24) becomes

$$\sum_{\gamma\in\mathbb{N}^m} q_\gamma(\boldsymbol{\alpha},\boldsymbol{\beta})\, y_\gamma \geq 0, \quad \forall \boldsymbol{\alpha},\boldsymbol{\beta}\in\mathbb{N}^m. \tag{2.25}$$

Eq. (2.25) is to be contrasted with the positive semidefiniteness conditions (2.20) of Theorem 2.14.

In the case where all the g_j's in (2.17) are affine (so that $\mathbf{K}$ is a convex polytope), we also have a specialized version of Theorem 2.15.

Theorem 2.16. *Let* $\mathbf{y} = (y_{\boldsymbol{\alpha}}) \subset \mathbb{R}$, $\boldsymbol{\alpha} \in \mathbb{N}^n$, *be a given infinite sequence, and let* $L_{\mathbf{y}}\colon \mathbb{R}[\mathbf{x}] \to \mathbb{R}$ *be the Riesz-functional introduced in* (2.3). *Assume that* $\mathbf{K}$ *is compact with nonempty interior, and all the* g_j*'s in* (2.17) *are affine, so that* $\mathbf{K}$ *is a convex polytope. Then* $\mathbf{y}$ *has a finite Borel representing measure with support contained in* $\mathbf{K}$ *if and only if*

$$L_{\mathbf{y}}(g^{\boldsymbol{\alpha}}) \geq 0, \quad \forall \boldsymbol{\alpha}\in\mathbb{N}^m. \tag{2.26}$$

A sufficient condition for the truncated K-moment problem

Finally, we present a very important sufficient condition for the truncated $\mathbf{K}$-moment problem. That is, we provide a condition on a finite sequence $\mathbf{y} = (y_{\boldsymbol{\alpha}})$ to admit a finite Borel representing measure supported on $\mathbf{K}$. Moreover, this condition can be checked numerically by standard techniques from linear algebra.

Theorem 2.17. *Let* $\mathbf{K} \subset \mathbb{R}^n$ *be the basic semi-algebraic set*

$$\mathbf{K} = \{\mathbf{x}\in\mathbb{R}^n \;:\; g_j(\mathbf{x}) \geq 0,\ j = 1,\dots,m\},$$

for some polynomials $g_j \in \mathbb{R}[\mathbf{x}]$ *of degree* $2v_j$ *or* $2v_j - 1$, *for all* $j = 1,\dots,m$. *Let* $\mathbf{y} = (y_{\boldsymbol{\alpha}})$ *be a finite sequence with* $|\alpha| \leq 2r$, *and let* $v = \max_j v_j$. *Then* $\mathbf{y}$ *has a* rank $\mathbf{M}_{r-v}(\mathbf{y})$*-atomic representing measure* μ *with support contained in* $\mathbf{K}$ *if and only if:*

(a) $\mathbf{M}_r(\mathbf{y}) \succeq \mathbf{0}$, $\mathbf{M}_{r-v}(g_j\mathbf{y}) \succeq \mathbf{0}$, $j = 1,\dots,m$, *and*
(b) $\operatorname{rank} \mathbf{M}_r(\mathbf{y}) = \operatorname{rank} \mathbf{M}_{r-v}(\mathbf{y})$.

In addition, μ *has* $\operatorname{rank} \mathbf{M}_r(\mathbf{y}) - \operatorname{rank} \mathbf{M}_{r-v}(g_i\mathbf{y})$ *atoms* $\mathbf{x}\in\mathbb{R}^n$ *that satisfy* $g_i(\mathbf{x}) = 0$, *for all* $i = 1,\dots,m$.

Note that, in Theorem 2.17, the set $\mathbf{K}$ is *not* required to be compact. The rank condition can be checked by standard techniques from numerical linear algebra. However, it is also important to remember that computing the rank is sensitive to numerical imprecisions.

2.4 Moment conditions for bounded density

Let $\mathbf{K} \subset \mathbb{R}^n$ be a Borel set with associated Borel σ-field $\mathcal{B}$, and let μ be a Borel measure on $\mathbf{K}$; so $(\mathbf{K}, \mathcal{B}, \mu)$ is a measure space. With f being a Borel measurable function on $(\mathbf{K}, \mathcal{B}, \mu)$, the notation $\|f\|_\infty$ stands for the *essential supremum* of f on $\mathbf{K}$, that is,

$$\|f\|_\infty = \inf \big\{c \in \overline{\mathbb{R}} \,:\, \mu\{\omega : f(\omega) > c\} = 0\big\}.$$

The (Lebesgue) space $L_\infty(\mathbf{K}, \mu)$ is the space of functions f such that $\|f\|_\infty < \infty$; it is a Banach space. And so $f \in L_\infty(\mathbf{K}, \mu)$ if f is essentially bounded, that is, bounded outside of a set of measure 0. (See, e.g., [7, p. 89].)

In this section, we consider the $\mathbf{K}$-moment problem with bounded density. That is, given a finite Borel measure μ on $\mathbf{K} \subseteq \mathbb{R}^n$ with moment sequence $\mathbf{y} = (y_{\boldsymbol{\alpha}})$, $\boldsymbol{\alpha} \in \mathbb{N}^n$, under what conditions on $\mathbf{y}$ do we have

$$y_{\boldsymbol{\alpha}} = \int_{\mathbf{K}} \mathbf{x}^{\boldsymbol{\alpha}} \, h \, d\mu, \quad \forall \boldsymbol{\alpha} \in \mathbb{N}^n, \tag{2.27}$$

for some bounded density $0 \leq h \in L_\infty(\mathbf{K}, \mu)$? (The measure μ is called the *reference* measure.)

The measure $d\nu = h\, d\mu$ is said to be *uniformly absolutely continuous* with respect to μ (denoted $\nu \ll \mu$) and h is called the Radon–Nikodym derivative of ν with respect to μ. This is a refinement of the general $\mathbf{K}$-moment problem, where one only asks for existence of some finite Borel representing measure on $\mathbf{K}$ (not necessarily with a density with respect to some reference measure μ).

Recall that, for two finite measures μ, ν on a σ-algebra $\mathcal{B}$, one has the natural partial order $\nu \leq \mu$ if and only if $\nu(B) \leq \mu(B)$ for every $B \in \mathcal{B}$; and observe that $\nu \leq \mu$ obviously implies $\nu \ll \mu$ but the converse does not hold.

2.4.1 The compact case

We first consider the case where the support of the reference measure μ is a compact basic semi-algebraic set $\mathbf{K} \subset \mathbb{R}^n$.

Theorem 2.18. *Let $\mathbf{K} \subset \mathbb{R}^n$ be compact and defined as in (2.17). Let $\mathbf{z} = (z_\alpha)$ be the moment sequence of a finite Borel measure μ on $\mathbf{K}$. Then:*

(a) *A sequence $\mathbf{y} = (y_\alpha)$ has a finite Borel representing measure on $\mathbf{K}$, uniformly absolutely continuous with respect to μ, if and only if there is some scalar κ such that*

$$0 \leq L_{\mathbf{y}}(f^2 g_J) \leq \kappa\, L_{\mathbf{z}}(f^2 g_J), \quad \forall f \in \mathbb{R}[\mathbf{x}], \quad \forall J \subseteq \{1, \ldots, m\}. \tag{2.28}$$

(b) *In addition, if the polynomial $N - \|\mathbf{x}\|^2$ belongs to the quadratic module $Q(g)$ then one may replace (2.28) with the weaker condition*

$$0 \leq L_{\mathbf{y}}(f^2 g_j) \leq \kappa\, L_{\mathbf{z}}(f^2 g_j), \quad \forall f \in \mathbb{R}[\mathbf{x}], \quad \forall j = 0, \ldots, m \tag{2.29}$$

(with the convention $g_0 = 1$).

(c) *Suppose that the g_j's are normalized so that*

$$0 \leq g_j \leq 1 \text{ on } \mathbf{K}, \quad \forall j = 1, \ldots, m,$$

and that the family $(0, 1, \{g_j\})$ generates the algebra $\mathbb{R}[\mathbf{x}]$. Then a sequence $\mathbf{y} = (y_\alpha)$ has a finite Borel representing measure on $\mathbf{K}$, uniformly absolutely continuous with respect to μ, if and only if there is some scalar κ such that

$$0 \leq L_{\mathbf{y}}(g^{\boldsymbol{\alpha}}(1-g)^{\boldsymbol{\beta}}) \leq \kappa\, L_{\mathbf{z}}(g^{\boldsymbol{\alpha}}(1-g)^{\boldsymbol{\beta}}), \quad \forall \boldsymbol{\alpha}, \boldsymbol{\beta} \in \mathbb{N}^m. \tag{2.30}$$

Proof. We only prove (a) as (b) and (c) can be proved with very similar arguments. The *only if* part: Let $d\nu = h\, d\mu$ for some $0 \leq h \in L_\infty(\mathbf{K}, \mu)$, and let $\kappa = \|h\|_\infty$. Observe that $g_J \geq 0$ on $\mathbf{K}$ for all $J \subseteq \{1, \ldots, m\}$. Therefore, for every $J \subseteq \{1, \ldots, m\}$ and all $f \in \mathbb{R}[\mathbf{x}]$,

$$L_{\mathbf{y}}(f^2 g_J) = \int_{\mathbf{K}} f^2 g_J\, d\nu = \int_{\mathbf{K}} f^2 g_J h\, d\mu \leq \kappa \int_{\mathbf{K}} f^2 g_J\, d\mu = \kappa\, L_{\mathbf{z}}(f^2 g_J),$$

and so (2.28) is satisfied.

The *if* part: Let $\mathbf{y}$ and $\mathbf{z}$ be such that (2.28) holds true. Then, by Theorem 2.14, $\mathbf{y}$ has a finite Borel representing measure ν on $\mathbf{K}$. In addition, let $\boldsymbol{\gamma} = (\gamma_{\boldsymbol{\alpha}})$ with $\gamma_{\boldsymbol{\alpha}} = \kappa z_{\boldsymbol{\alpha}} - y_{\boldsymbol{\alpha}}$ for all $\boldsymbol{\alpha} \in \mathbb{N}^n$. From (2.28), one has

$$L_{\boldsymbol{\gamma}}(f^2 g_J) \geq 0, \quad \forall f \in \mathbb{R}[\mathbf{x}], \quad \forall J \subseteq \{1, \ldots, m\},$$

and so, by Theorem 2.14 again, $\boldsymbol{\gamma}$ has a finite Borel representing measure ψ on $\mathbf{K}$. Moreover, from the definition of the sequence $\boldsymbol{\gamma}$,

$$\int f\, d(\psi + \nu) = \int f\, \kappa\, d\mu, \quad \forall f \in \mathbb{R}[\mathbf{x}],$$

and therefore, as measures on compact sets are moment determinate, $\psi + \nu = \kappa\mu$. Hence $\kappa\mu \geq \nu$, which shows that $\nu \ll \mu$ and so one may write $d\nu = h\, d\mu$ for some $0 \leq h \in L_1(\mathbf{K}, \mu)$. From $\nu \leq \kappa\mu$ one obtains

$$\int_A (h - \kappa)\, d\mu \leq 0, \quad \forall A \in \mathcal{B}(\mathbf{K})$$

(where $\mathcal{B}(\mathbf{K})$ is the Borel σ-algebra associated with $\mathbf{K}$). So $0 \leq h \leq \kappa$, μ-almost everywhere on $\mathbf{K}$. Equivalently, $\|h\|_\infty \leq \kappa$, the desired result. ☐

Notice that, using moment and localizing matrices,

$$(2.28) \Longleftrightarrow 0 \preceq \mathbf{M}_i(g_J\, \mathbf{y}) \preceq \kappa\, \mathbf{M}_i(g_J\, \mathbf{z}), \quad i = 1, \ldots; \quad J \subseteq \{1, \ldots, m\};$$

$$(2.29) \Longleftrightarrow 0 \preceq \mathbf{M}_i(g_j\, \mathbf{y}) \preceq \kappa\, \mathbf{M}_i(g_j\, \mathbf{z}), \quad i = 1, \ldots; \quad j = 0, 1, \ldots, m.$$

2.4.2 The non-compact case

Let the reference measure μ be a finite Borel measure on $\mathbb{R}^n$, not supported on a compact set. As one wishes to find moment conditions, it is natural to consider the case where all the moments $\mathbf{z} = (z_{\boldsymbol{\alpha}})$ of μ are finite, and a simple sufficient condition is that μ satisfies the generalized Carleman condition (2.12) of Nussbaum.

Theorem 2.19. *Let $\mathbf{z} = (z_{\boldsymbol{\alpha}})$ be the moment sequence of a finite Borel measure μ on $\mathbb{R}^n$ which satisfies the generalized Carleman condition, i.e.,*

$$\sum_{k=1}^{\infty} L_{\mathbf{z}}\big(x_i^{2k}\big)^{-1/2k} = +\infty, \quad \forall i = 1, \ldots, n. \tag{2.31}$$

A sequence $\mathbf{y} = (y_{\boldsymbol{\alpha}})$ has a finite Borel representing measure ν on $\mathbb{R}^n$, uniformly absolutely continuous with respect to μ, if there exists a scalar $0 < \kappa$ such that, for all $i = 1, \ldots, n$,

$$0 \leq L_{\mathbf{y}}\big(f^2\big) \leq \kappa\, L_{\mathbf{z}}\big(f^2\big), \quad \forall f \in \mathbb{R}[\mathbf{x}]. \tag{2.32}$$

Proof. For every $i = 1, \ldots, n$, (2.32) with $\mathbf{x} \mapsto f(\mathbf{x}) = x_i^k$ yields

$$L_{\mathbf{y}}\big(x_i^{2k}\big)^{-1/2k} \geq \kappa^{-1/2k}\, L_{\mathbf{z}}\big(x_i^{2k}\big)^{-1/2k}, \quad \forall k = 1, \ldots,$$

and so, using (2.31), one obtains

$$\sum_{k=1}^{\infty} L_{\mathbf{y}}\big(x_i^{2k}\big)^{-1/2k} \geq \sum_{k=1}^{\infty} \kappa^{-1/2k}\, L_{\mathbf{z}}\big(x_i^{2k}\big)^{-1/2k} = +\infty,$$

for every $i = 1, \ldots, n$, i.e., the generalized Carleman condition (2.12) holds for the sequence $\mathbf{y}$. Combining this with the first inequality in (2.32) yields that $\mathbf{y}$ has a unique finite Borel representing measure ν on $\mathbb{R}^n$. It remains to prove that $\nu \ll \mu$ and its density h is in $L_\infty(\mathbb{R}^n, \mu)$.

Let $\boldsymbol{\gamma} = (\gamma_{\boldsymbol{\alpha}})$ with $\gamma_{\boldsymbol{\alpha}} = \kappa z_{\boldsymbol{\alpha}} - y_{\boldsymbol{\alpha}}$ for all $\boldsymbol{\alpha} \in \mathbb{N}^n$. Then the second inequality in (2.32) yields

$$L_{\boldsymbol{\gamma}}\big(f^2\big) \geq 0, \quad \forall f \in \mathbb{R}[\mathbf{x}]. \tag{2.33}$$

Next, observe that, from (2.32), for every $i = 1, \ldots, n$ and every $k = 0, 1, \ldots,$

$$L_{\boldsymbol{\gamma}}\big(x_i^{2k}\big) \leq \kappa\, L_{\mathbf{z}}\big(x_i^{2k}\big),$$

which implies that

$$L_{\boldsymbol{\gamma}}\big(x_i^{2k}\big)^{-1/2k} \geq \kappa^{-1/2k}\, L_{\mathbf{z}}\big(x_i^{2k}\big)^{-1/2k}, \tag{2.34}$$

and so, for every $i = 1, \ldots, n$,

$$\sum_{k=1}^{\infty} L_{\boldsymbol{\gamma}}\big(x_i^{2k}\big)^{-1/2k} \geq \sum_{k=1}^{\infty} \kappa^{-1/2k}\, L_{\mathbf{z}}\big(x_i^{2k}\big)^{-1/2k} = +\infty,$$

i.e., γ satisfies the generalized Carleman condition. In view of (2.33), γ has a (unique) finite Borel representing measure ψ on $\mathbb{R}^n$. Next, from the definition of γ, one has

$$\int f\, d(\psi + \nu) = \kappa \int f\, d\mu, \quad \forall f \in \mathbb{R}[\mathbf{x}].$$

But as μ (and so $\kappa\mu$) satisfies the Carleman condition (2.31), $\kappa\mu$ is moment determinate and therefore $\kappa\mu = \psi + \nu$.

Hence $\nu \ll \mu$ follows from $\nu \leq \kappa\mu$. Finally, writing $d\nu = h\, d\mu$ for some nonnegative $h \in L_1(\mathbb{R}^n, \mu)$, and using $\nu \leq \kappa\mu$, one obtains

$$\int_A (h - \kappa)\, d\mu \leq 0, \quad \forall A \in \mathcal{B}(\mathbb{R}^n),$$

and so $0 \leq h \leq \kappa$, μ-almost everywhere on $\mathbb{R}^n$. Equivalently, $\|h\|_\infty \leq \kappa$, the desired result. □

Observe that (2.32) is extremely simple as it is equivalent to stating that

$$\kappa\, \mathbf{M}_r(\mathbf{z}) \succeq \mathbf{M}_r(\mathbf{y}) \succeq 0, \quad \forall r = 0, 1, \ldots$$

Checking whether (2.31) holds is not easy in general. However, the condition

$$\int \exp |x_i|\, d\mu < \infty, \quad \forall i = 1, \ldots, n,$$

which is simpler to verify, ensures that (2.31) holds.

2.5 Notes and sources

Moment problems have a long and rich history. For historical remarks and details on various approaches for the moment problem, the interested reader is referred to Landau (1987) [61]. See also Akhiezer (1965) [2], Curto and Fialkow (2000) [27], and Simon (1998) [121]. Example 2.3 is from Feller (1966) [34, p. 227] whereas Example 2.8 is from Laurent (2008) [83]. Theorem 2.4 was first proved by M. Riesz for closed sets $\mathbf{K} \subset \mathbb{R}$, and subsequently generalized to closed sets $\mathbf{K} \subset \mathbb{R}^n$ by Haviland (1935, 1936) [41, 42].

2.1 Most of this section is from Curto and Fialkow (1991) [24]. Theorem 2.6(c) and Theorem 2.7 were proved by Krein and Nudel'man [55], who also gave the sufficient conditions $\mathbf{H}_n(\gamma), \mathbf{B}_{n-1}(\gamma) \succ \mathbf{0}$ for Theorem 2.6(b) and the sufficient condition $\mathbf{H}_n(\gamma), \mathbf{B}_n(\gamma) \succ 0$ for Theorem 2.7(b).

2.2 The localizing matrix was introduced in Curto and Fialkow (2000) [27] and Berg (1987) [12]. The multivariate condition in Proposition 2.9, that generalizes an earlier result of Carleman (1926) [20] in one dimension, is stated in Berg (1987) [12], and was proved by Nussbaum (1966) [95]. Proposition 2.10

is taken from Lasserre (2007) [73]. The infinite and truncated moment matrices (and in particular their kernel) have a lot of very interesting properties. For more details, the interested reader is referred to Laurent (2008) [83].

2.3 Concerning the solution of the **K**-moment problem, Theorem 2.14(a) was proved by Schmüdgen (1991) [110] with a nice interplay between real algebraic geometry and functional analysis. Indeed, the proof uses Stengle's Positivstellensatz (Theorem 1.14) and the spectral theory of self-adjoint operators in Hilbert spaces. Its refinement (b) is due to Putinar (1993) [103], and Jacobi and Prestel (2001) [48]. Incidentally, in Schmüdgen (1991) [110] the Positivstellensatz Theorem 1.15 appears as a corollary of Theorem 2.14(a). Theorem 1.25 is due to Cassier (1984) [21] and Handelman (1988) [38], and appears prior to the more general Theorem 2.15 due to Vasilescu (2003) [127]. Theorems 2.12 and 2.17 are due to Curto and Fialkow (1991, 1996, 1998) [24, 25, 26], where the results are stated for the complex plane $\mathbb{C}$, but generalize to $\mathbb{C}^n$ and $\mathbb{R}^n$. An alternative proof of some of these results can be found in Laurent (2005) [80]; for instance, Theorem 2.17 follows from Laurent (2005) [80, Theorem 5.23].

2.4 This section is from Lasserre (2006) [71]. The moment problem with bounded density was initially studied by Markov on the interval $[0,1]$ with μ the Lebesgue measure, a refinement of the Hausdorff moment problem where one only asks for existence of some finite Borel representing measure ν on $[0,1]$. For an interesting discussion with historical details, the interested reader is referred to Diaconis and Freedman (2006) [31], where, in particular, the authors have proposed a simplified proof as well as conditions for existence of density in $L_p([0,1],\mu)$ with a similar flavor.

Chapter 3

Polynomial Optimization

In this chapter, we consider the following polynomial optimization problem:

$$f^* = \min \{ f(\mathbf{x}) \;:\; \mathbf{x} \in \mathbf{K} \}, \tag{3.1}$$

where $f \in \mathbb{R}[\mathbf{x}]$ is a real-valued polynomial and $\mathbf{K} \subset \mathbb{R}^n$ is the basic semi-algebraic set defined by

$$\mathbf{K} = \{ \mathbf{x} \in \mathbb{R}^n \;:\; g_i(\mathbf{x}) \geq 0,\; i = 1, \ldots, m \}, \tag{3.2}$$

with $g_i \in \mathbb{R}[\mathbf{x}]$, $i = 1, \ldots, m$.

Whenever $\mathbf{K} \neq \mathbb{R}^n$, the set $\mathbf{K}$ is assumed to be compact but we do not assume that $\mathbf{K}$ is convex or even connected. This is a rather rich modeling framework that includes linear, quadratic, 0/1, and mixed 0/1 optimization problems as special cases. In particular, constraints of type $x_i \in \{0, 1\}$ can be written as $x_i^2 - x_i \geq 0$ and $x_i - x_i^2 \geq 0$.

When $n = 1$, we have seen in Chapter 1 that a univariate polynomial nonnegative on $\mathbf{K} = \mathbb{R}$ is a sum of squares and that a univariate polynomial nonnegative on an interval $\mathbf{K} = (-\infty, b]$, $\mathbf{K} = [a, b]$ or $\mathbf{K} = [a, \infty)$ can be written in a specific form involving sums of squares whose degree is known. We will see that this naturally leads to reformulating problem (3.1) as a single semidefinite optimization problem, for which efficient algorithms and software packages are available. It is quite remarkable that a nonconvex problem can be reformulated as a convex one and underscores the importance of the representation theorems from Chapter 1.

On the other hand, the multivariate case radically differs from the univariate case, because not every polynomial nonnegative on $\mathbf{K} = \mathbb{R}^n$ can be written as a sum of squares of polynomials. Moreover, problem (3.1) (with $\mathbf{K} = \mathbb{R}^n$) involving a polynomial f of degree greater than or equal to four on n variables is $\mathcal{NP}$-hard. However, we will see that problem (3.1) with $\mathbf{K} = \mathbb{R}^n$ can be approximated as closely as desired (and often can be obtained exactly) by solving a finite sequence of semidefinite optimization problems of the same type as in the one-dimensional case. A similar conclusion also holds when $\mathbf{K}$ in (3.2) is compact.

3.1 The primal and dual points of view

Recall from the Introduction that problem (3.1) can be re-written either as the infinite-dimensional linear program

$$f^* = \inf_{\mu \in \mathcal{M}(\mathbf{K})_+} \left\{ \int_{\mathbf{K}} f \, d\mu \; : \; \mu(\mathbf{K}) = 1 \right\} \tag{3.3}$$

or as its dual

$$f^* = \sup_{\lambda} \left\{ \lambda \; : \; f(\mathbf{x}) - \lambda \geq 0, \; \forall \mathbf{x} \in \mathbf{K} \right\}. \tag{3.4}$$

Moreover, observe that, if we write $\mathbf{x} \mapsto f(\mathbf{x}) = \sum_{\boldsymbol{\alpha}} f_{\boldsymbol{\alpha}} \mathbf{x}^{\boldsymbol{\alpha}}$, then (3.3) also reads

$$f^* = \inf_{\mathbf{y}} \left\{ \sum_{\boldsymbol{\alpha} \in \mathbb{N}^n} f_{\boldsymbol{\alpha}} \, y_{\boldsymbol{\alpha}} \; : y_{\mathbf{0}} = 1, \right. \tag{3.5}$$
$$\left. y_{\boldsymbol{\alpha}} = \int_{\mathbf{K}} \mathbf{x}^{\boldsymbol{\alpha}} \, d\mu, \; \boldsymbol{\alpha} \in \mathbb{N}^n, \text{ for some } \mu \in \mathcal{M}(\mathbf{K})_+ \right\},$$

so that the unknown measure μ only appears through its moments $y_{\boldsymbol{\alpha}} = \int_{\mathbf{K}} \mathbf{x}^{\boldsymbol{\alpha}} d\mu$, $\boldsymbol{\alpha} \in \mathbb{N}^n$.

Hence, in fact, both linear programs express two dual points of view. The primal (3.5) considers problem (3.1) from the point of view of *moments* as one searches for the sequence $\mathbf{y}$ of moments of a Borel probability measure μ supported on $\mathbf{K}$. Ideally, if there is a unique global minimizer $\mathbf{x}^* \in \mathbf{K}$, then one searches for the vector $\mathbf{y} = \mathbf{v}(\mathbf{x}^*)$ (with $\mathbf{v}(\mathbf{x})$ as in Section 1.1) of moments of the Dirac measure $\delta_{\mathbf{x}^*}$ at $\mathbf{x}^*$. The dual (3.4) considers problem (3.1) from the point of view of *positive polynomials* as one characterizes f^* as being the largest scalar such that $f - f^*$ is nonnegative on $\mathbf{K}$.

As we next see, the representation results of Chapter 1 can be exploited to solve (or approximate) (3.4), and, similarly, their dual counterparts of Chapter 2 can be exploited to solve (or approximate) (3.5).

In doing so, and for $\mathbf{K} = \mathbb{R}^n$, we obtain a *single* semidefinite relaxation, which is exact if and only if the nonnegative polynomial $\mathbf{x} \mapsto f(\mathbf{x}) - f^*$ is a sum of squares. For $\mathbf{K}$ as in (3.2) and compact, we obtain a hierarchy of semidefinite relaxations, depending on the degree d permitted in the representation (1.13). The larger the degree, the better the optimal value λ. As the degree increases, asymptotic convergence is guaranteed by a simple application of Theorem 1.17. We may also consider the alternative representation of $f(\mathbf{x}) - \lambda$ as an element of the cone C_G in (1.18) to obtain a hierarchy of *linear* relaxations instead of semidefinite relaxations. In this case, asymptotic convergence is now guaranteed by invoking Theorem 1.24.

3.2 Unconstrained polynomial optimization

If $\mathbf{K} = \mathbb{R}^n$, the only interesting case is when $\deg f$ is even, otherwise necessarily $f^* = -\infty$. So let $2r$ be the degree of $f \in \mathbb{R}[\mathbf{x}]$, let $i \geq r$, and consider the semidefinite program

$$\rho_i = \inf_{\mathbf{y}} \left\{ L_{\mathbf{y}}(f) \; : \; \mathbf{M}_i(\mathbf{y}) \succeq \mathbf{0}, \; y_0 = 1 \right\}, \tag{3.6}$$

where $\mathbf{M}_i(\mathbf{y})$ is the moment matrix of order i associated with the sequence $\mathbf{y}$, already defined in Subsection 2.2.1.

Writing $\mathbf{M}_i(\mathbf{y}) = \sum_{|\boldsymbol{\alpha}| \leq 2i} y_{\boldsymbol{\alpha}} \mathbf{B}_{\boldsymbol{\alpha}}$ for appropriate symmetric matrices $\{\mathbf{B}_{\boldsymbol{\alpha}}\}$, the dual of (3.6) is the semidefinite program

$$\rho_i^* = \sup_{\lambda, \mathbf{X}} \left\{ \lambda \; : \; \mathbf{X} \succeq \mathbf{0}; \; \langle \mathbf{X}, \mathbf{B}_{\boldsymbol{\alpha}} \rangle = f_{\boldsymbol{\alpha}} - \lambda \delta_{\boldsymbol{\alpha}=0}, \; \forall \, |\boldsymbol{\alpha}| \leq 2i \right\}, \tag{3.7}$$

where $\delta_{\boldsymbol{\alpha}=0}$ is the Kronecker symbol at $\boldsymbol{\alpha} = 0$. Recall the definition of the vector $\mathbf{v}_d(\mathbf{x})$ in Section 1.1. Multiplying each side of the constraint by $\mathbf{x}^{\boldsymbol{\alpha}}$ and summing up yields

$$\begin{aligned} \sum_{\boldsymbol{\alpha} \in \mathbb{N}^n} f_{\boldsymbol{\alpha}} \mathbf{x}^{\boldsymbol{\alpha}} - \lambda &= \Big\langle \mathbf{X}, \sum_{\boldsymbol{\alpha} \in \mathbb{N}^n} \mathbf{B}_{\boldsymbol{\alpha}} \mathbf{x}^{\boldsymbol{\alpha}} \Big\rangle = \langle \mathbf{X}, \mathbf{v}_i(\mathbf{x}) \mathbf{v}_i(\mathbf{x})' \rangle \\ &= \sum_{k=1}^{s} \langle \mathbf{q}_k \mathbf{q}_k', \mathbf{v}_i(\mathbf{x}) \mathbf{v}_i(\mathbf{x})' \rangle = \sum_{k=1}^{s} \left(\mathbf{q}_k' \mathbf{v}_i(\mathbf{x}) \right)^2 = \sum_{k=1}^{s} q_k(\mathbf{x})^2, \end{aligned}$$

where we have used the spectral decomposition $\mathbf{X} = \sum_k \mathbf{q}_k \mathbf{q}_k'$ of the positive semidefinite matrix $\mathbf{X}$, and interpreted $\mathbf{q}_k$ as the vector of coefficients of the polynomial $q_k \in \mathbb{R}[\mathbf{x}]_i$. Hence, (3.7) has the equivalent compact formulation:

$$\rho_i^* = \sup_{\lambda} \left\{ \lambda \; : \; f - \lambda \in \Sigma(\mathbf{x}) \right\}. \tag{3.8}$$

Therefore, $\rho_i^* = \rho_r^*$ for every $i \geq r$ because, obviously, if f has degree $2r$ then $f - \lambda$ cannot be a sum of squares of polynomials with degree larger than r.

Proposition 3.1. *There is no duality gap, that is, $\rho_r = \rho_r^*$. And, if $\rho_r > -\infty$, then* (3.7) *has an optimal solution.*

Proof. The result follows from standard duality in conic optimization if we can prove that there is a strictly feasible solution $\mathbf{y}$ of problem (3.6), i.e., such that $\mathbf{M}_r(\mathbf{y}) \succ \mathbf{0}$ (Slater condition). So let μ be a probability measure on $\mathbb{R}^n$ with a strictly positive density f with respect to the Lebesgue measure and with all its moments finite; that is, μ is such that

$$y_{\boldsymbol{\alpha}} = \int_{\mathbb{R}^n} \mathbf{x}^{\boldsymbol{\alpha}} \, d\mu = \int_{\mathbb{R}^n} \mathbf{x}^{\boldsymbol{\alpha}} f(\mathbf{x}) \, d\mathbf{x} < \infty, \quad \forall \boldsymbol{\alpha} \in \mathbb{N}^n.$$

Then $\mathbf{M}_r(\mathbf{y})$, with $\mathbf{y}$ as above, is such that $\mathbf{M}_r(\mathbf{y}) \succ \mathbf{0}$. To see this, recall that for every polynomial $q \in \mathbb{R}[\mathbf{x}]$ of degree at most r, and vector of coefficients $\mathbf{q} \in \mathbb{R}^{s(r)}$,

$$\begin{aligned} \langle q, q \rangle_{\mathbf{y}} = \langle \mathbf{q}, \mathbf{M}_r(\mathbf{y})\mathbf{q} \rangle &= \int_{\mathbb{R}^n} q^2 \, d\mu && \text{[from (2.11)]} \\ &= \int_{\mathbb{R}^n} q(\mathbf{x})^2 f(\mathbf{x}) \, d\mathbf{x} \\ &> 0, \text{ whenever } q \neq 0 && (\text{as } f > 0). \end{aligned}$$

Therefore, $\mathbf{y}$ is strictly feasible for problem (3.6), i.e., $\mathbf{M}_r(\mathbf{y}) \succ \mathbf{0}$, the desired result. □

We next prove the central result of this section.

Theorem 3.2. *Let $f \in \mathbb{R}[\mathbf{x}]$ be a $2r$-degree polynomial with global minimum f^*.*

(a) *If the nonnegative polynomial $f - f^*$ is s.o.s., then problem (3.1) is equivalent to the semidefinite optimization problem (3.6), i.e., $f^* = \rho_r^* = \rho_r$ and if $\mathbf{x}^* \in \mathbb{R}^n$ is a global minimizer of (3.1), then the moment vector*

$$\mathbf{y}^* = \left(x_1^*, \dots, x_n^*, (x_1^*)^2, x_1^* x_2^*, \dots, (x_1^*)^{2r}, \dots, (x_n^*)^{2r}\right) \tag{3.9}$$

is a minimizer of problem (3.6).

(b) *If problem (3.7) has a feasible solution and $f^* = \rho_r^*$, then $f - f^*$ is s.o.s.*

Proof. (a) Let $f - f^*$ be s.o.s., that is,

$$f(\mathbf{x}) - f^* = \sum_{i=1}^k q_i(\mathbf{x})^2, \quad \mathbf{x} \in \mathbb{R}^n, \tag{3.10}$$

for some polynomials $\{q_i\}_{i=1}^k \subset \mathbb{R}[\mathbf{x}]_r$ with coefficient vectors $\mathbf{q}_i \in \mathbb{R}^{s(r)}$, where $i = 1, 2, \dots, k$ and $s(r) = \binom{n+r}{n}$. Equivalently, with $\mathbf{v}_r(\mathbf{x})$ as in (2.9),

$$f(\mathbf{x}) - f^* = \langle \mathbf{X}, \mathbf{M}_r(\mathbf{y}) \rangle, \quad \mathbf{x} \in \mathbb{R}^n, \tag{3.11}$$

with $\mathbf{X} = \sum_{i=1}^k \mathbf{q}_i \mathbf{q}_i' \succeq 0$ and $\mathbf{y} = \mathbf{v}_{2r}(\mathbf{x})$. From Eq. (3.11) it follows that

$$\langle \mathbf{X}, \mathbf{B}_0 \rangle = f_0 - f^*; \quad \langle \mathbf{X}, \mathbf{B}_\alpha \rangle = f_\alpha, \quad \text{for all } 0 \neq \alpha, \ |\alpha| \leq 2r,$$

so that (as $\mathbf{X} \succeq \mathbf{0}$) $\mathbf{X}$ is feasible for problem (3.7) with value $\lambda = f^*$. Since $\mathbf{y}^*$ in (3.9) is feasible for problem (3.6) with value f^* and $\rho_r^* = \rho_r$, it follows that $\mathbf{y}^*$ and $\mathbf{X}$ are optimal solutions to problems (3.6) and (3.7), respectively.

(b) Suppose that problem (3.7) has a feasible solution and $f^* = \rho_r^*$. Then, by Proposition 3.1, problem (3.7) has an optimal solution $(\mathbf{X}^*, f^*)$, with $\mathbf{X}^* \succeq \mathbf{0}$, and there is no duality gap, that is, $\rho_r = \rho_r^*$. As $\mathbf{X}^* \succeq \mathbf{0}$, we use its spectral decomposition to write $\mathbf{X}^* = \sum_{i=1}^k \mathbf{q}_i\mathbf{q}_i'$ for some vectors $(\mathbf{q}_i) \subset \mathbb{R}^{s(r)}$. Using the feasibility of $(\mathbf{X}^*, f^*)$ in (3.7), we obtain

$$\left\langle \mathbf{X}^*, \sum_{\alpha} \mathbf{B}_\alpha \mathbf{x}^\alpha \right\rangle = f(\mathbf{x}) - f^*,$$

which, using $\mathbf{X}^* = \sum_{i=1}^k \mathbf{q}_i\mathbf{q}_i'$ and

$$\sum_{\alpha} \mathbf{B}_\alpha \mathbf{x}^\alpha = \mathbf{M}_r(\mathbf{v}_{2r}(\mathbf{x})) = \mathbf{v}_r(\mathbf{x})\mathbf{v}_r(\mathbf{x})',$$

yields the desired sum of squares

$$f(\mathbf{x}) - f^* = \sum_{i=1}^k \langle \mathbf{q}_i\mathbf{q}_i', \mathbf{v}_r(\mathbf{x})\mathbf{v}_r(\mathbf{x})' \rangle = \sum_{i=1}^k \langle \mathbf{q}_i, \mathbf{v}_r(\mathbf{x}) \rangle^2 = \sum_{i=1}^k q_i(\mathbf{x})^2,$$

with the polynomials $\mathbf{x} \mapsto q_i(\mathbf{x}) = \langle \mathbf{q}_i, \mathbf{v}_r(\mathbf{x}) \rangle$, for all $i = 1, \ldots, k$. □

From the proof of Theorem 3.2, it is obvious that if $f^* = \rho_r^*$ then any global minimizer $\mathbf{x}^*$ of f is a zero of each polynomial q_i, where $\mathbf{X}^* = \sum_{i=1}^k \mathbf{q}_i\mathbf{q}_i'$ at an optimal solution $\mathbf{X}^*$ of problem (3.7). When $f - f^*$ is s.o.s., solving problem (3.7) provides the polynomials q_i of such a decomposition. As a corollary, we obtain the following.

Corollary 3.3. *Let $f \in \mathbb{R}[\mathbf{x}]$ be of degree $2r$. Assume that problem (3.7) has a feasible solution. Then*

$$f(\mathbf{x}) - f^* = \sum_{i=1}^k q_i(\mathbf{x})^2 - [f^* - \rho_r^*], \quad \mathbf{x} \in \mathbb{R}^n, \tag{3.12}$$

for some real-valued polynomials $q_i \in \mathbb{R}[\mathbf{x}]$, $i = 1, 2, \ldots, k$, of degree at most r.

The proof is the same as that of Theorem 3.2(b), except that now we may not have $f^* = \rho_r^*$, but instead $\rho_r^* \leq f^*$. Hence, ρ_r^* always provides a lower bound on f^*. Corollary 3.3 states that, up to some constant, one may always write $f - f^*$ as a s.o.s. whenever problem (3.7) has a feasible solution. The previous development leads to the following algorithm for either solving problem (3.1) or providing a lower bound on its optimal value f^*.

Algorithm 3.4. (Unconstrained polynomial optimization)

Input: A polynomial $\mathbf{x} \mapsto f(\mathbf{x}) = \sum_{\alpha \in \mathbb{N}^n} f_\alpha \mathbf{x}^\alpha$ of degree $2r$.

Output: The value $f^* = \min_{\mathbf{x} \in \mathbb{R}^n} f(\mathbf{x})$ or a lower bound ρ_r on f^*.

Algorithm:

1. Solve the semidefinite optimization problem (3.6) with optimal value ρ_r and optimal solution $\mathbf{y}^*$ (if $\mathbf{y}^*$ exists).

2. If $\operatorname{rank} \mathbf{M}_{r-1}(\mathbf{y}^*) = \operatorname{rank} \mathbf{M}_r(\mathbf{y}^*)$, then $f^* = \rho_r$ and there are at least $\operatorname{rank} \mathbf{M}_r(\mathbf{y}^*)$ global minimizers, which are extracted using Algorithm 3.13.

3. Otherwise, ρ_r only provides a lower bound $\rho_r \le f^*$.

We next show that Algorithm 3.4 correctly determines whether ρ_r is the exact solution value f^* or a lower bound.

Theorem 3.5. *Let $f \in \mathbb{R}[\mathbf{x}]$ with degree $2r$, and suppose that the optimal value ρ_r of problem (3.6) is attained at some optimal solution $\mathbf{y}^*$. If $\operatorname{rank} \mathbf{M}_{r-1}(\mathbf{y}^*) = \operatorname{rank} \mathbf{M}_r(\mathbf{y}^*)$, then $f^* = \rho_r$ and there are at least $\operatorname{rank} \mathbf{M}_r(\mathbf{y}^*)$ global minimizers.*

Proof. One has already shown that $\rho_r \le f^*$. If $\operatorname{rank} \mathbf{M}_{r-1}(\mathbf{y}^*) = \operatorname{rank} \mathbf{M}_r(\mathbf{y}^*)$, we let $s = \operatorname{rank} \mathbf{M}_r(\mathbf{y}^*)$. Then $\mathbf{M}_r(\mathbf{y}^*)$ is a flat extension of $\mathbf{M}_{r-1}(\mathbf{y}^*)$, and so, by Theorem 2.12, $\mathbf{y}^*$ is the vector of moments up to order $2r$, of some s-atomic probability measure μ^* on $\mathbb{R}^n$. Therefore, $\rho_r = L_{\mathbf{y}}(f) = \int f\, d\mu^*$, which proves that μ^* is an optimal solution of (3.3), and $\rho_r = f^*$ because we always have $\rho_r \le f^*$. We next show that each of the s atoms of μ^* is a global minimizer of f. Indeed, being s-atomic, there is a family $(\mathbf{x}(k))_{k=1}^s \subset \mathbb{R}^n$ and a family $(\beta_k)_{k=1}^s \subset \mathbb{R}$ such that

$$\mu^* = \sum_{k=1}^s \beta_k\, \delta_{\mathbf{x}(k)}, \quad \beta_k > 0, \quad \forall k = 1, \ldots, s; \quad \sum_{k=1}^s \beta_k = 1,$$

Hence,

$$f^* = \rho_r = \int_{\mathbb{R}^n} f\, d\mu^* = \sum_{k=1}^s \beta_k\, f(\mathbf{x}(k)),$$

which, in view of $f(\mathbf{x}(k)) \ge f^*$, for all $k = 1, \ldots, s$, implies the desired result $f(\mathbf{x}(k)) = f^*$, for all $k = 1, \ldots, s$. □

As an illustration, suppose that $\mathbf{y}^*$ satisfies $\operatorname{rank} \mathbf{M}_r(\mathbf{y}^*) = 1$. Therefore, $\mathbf{M}_r(\mathbf{y}^*) = \mathbf{v}_r(\mathbf{x}^*)\mathbf{v}_r(\mathbf{x}^*)'$ for some $\mathbf{x}^* \in \mathbb{R}^n$, that is, $\mathbf{y}^*$ is the vector of moments up to order $2r$ of the Dirac measure at $\mathbf{x}^*$, and one *reads* an optimal solution $\mathbf{x}^*$ from the subvector of first "moments" $y^*_{\boldsymbol{\alpha}}$ with $|\boldsymbol{\alpha}| = 1$. Note also that, when $n = 1$, $f - f^*$ is always a s.o.s., so that we expect that $\rho_r = f^*$, where $2r$ is the degree of f.

In other words, the global minimization of a univariate polynomial is a convex optimization problem, the semidefinite program (3.6).

Therefore, in view of Section 1.4, we also expect that $\rho_r = f^*$ for quadratic polynomials, and bivariate polynomials of degree 4. Let us illustrate these properties with an example.

Example 3.6. We consider the polynomial f on two variables

$$\mathbf{x} \longmapsto f(\mathbf{x}) = \left(x_1^2 + 1\right)^2 + \left(x_2^2 + 1\right)^2 + \left(x_1 + x_2 + 1\right)^2.$$

Note that in this case $r = 2$ and $f - f^*$ is a bivariate polynomial of degree 4, and therefore it is a sum of squares. We thus expect that $\rho_2 = f^*$. Solving problem

(3.6) for $i = 2$ yields a minimum value of $\rho_2 = -0.4926$. In this case, it turns out that $\mathbf{M}_2(\mathbf{y}^*)$ has rank one, and from the optimal solution $\mathbf{y}^*$, we check that

$$\mathbf{y} = \left(1, x_1^*, x_2^*, (x_1^*)^2, x_1^* x_2^*, (x_2^*)^2, \ldots, (x_1^*)^4, \ldots, (x_2^*)^4\right),$$

with $x_1^* = x_2^* \approx -0.2428$. We observe that the solution $\mathbf{x}^*$ is a good approximation of a global minimizer of problem (3.1), since the gradient vector verifies

$$\left.\frac{\partial f}{\partial x_1}\right|_{\mathbf{x}=\mathbf{x}^*} = \left.\frac{\partial f}{\partial x_2}\right|_{\mathbf{x}=\mathbf{x}^*} = 4 \cdot 10^{-9}.$$

In this example, it follows that the semidefinite relaxation is exact. The reason why we have $\nabla f(\mathbf{x}^*) \approx \mathbf{0}$, but not exactly $\mathbf{0}$, is likely due to unavoidable numerical inaccuracies when using an SDP solver.

The next example discusses the effect of perturbations.

Example 3.7. Consider the bivariate polynomial $\mathbf{x} \mapsto f(\mathbf{x}) = x_1^2 x_2^2 (x_1^2 + x_2^2 - 1)$. We have already seen that $f + 1$ is positive, but not a sum of squares. Note that the global optimum is $f^* = -1/27$ and there are four global minimizers $\mathbf{x}^* = (\pm\sqrt{3}/3, \pm\sqrt{3}/3)$. If we consider the problem $f^* = \min f(x_1, x_2)$ and apply Algorithm 3.4 for $r = 3$, we obtain that $\rho_3 = -\infty$, that is, the bound is uninformative.

However, consider the perturbed problem of minimizing the polynomial $f_\epsilon = f(\mathbf{x}) + \epsilon(x_1^{10} + x_2^{10})$ with $\epsilon = 0.001$. Applying Algorithm 3.4 by means of the software GloptiPoly, we find that $\rho_5 \approx f^*$ and four optimal solutions extracted $(x_1, x_2) \approx \mathbf{x}^*$. In other words, while the approach fails when applied to the original polynomial f, it succeeds when applied to the perturbed polynomial f_ϵ.

3.3 Constrained polynomial optimization: SDP-relaxations

In this section, we address problem (3.1) where $f \in \mathbb{R}[\mathbf{x}]$ and the set $\mathbf{K}$ defined in (3.2) is assumed to be a compact basic semi-algebraic subset of $\mathbb{R}^n$. We suppose that Assumption 1.16 holds, which, as we discussed in Section 1.5, is verified in many cases, for example if there is one polynomial g_k such that $\{\mathbf{x} \in \mathbb{R}^n : g_k(\mathbf{x}) \geq 0\}$ is compact; see the discussion following Theorem 1.17 and Theorem 1.18.

3.3.1 A hierarchy of semidefinite relaxations

Recall that, by Theorem 1.17, every polynomial $f \in \mathbb{R}[\mathbf{x}]$ strictly positive on $\mathbf{K}$ can be written in the form

$$f = \sigma_0 + \sum_{j=1}^{m} \sigma_j \, g_j \tag{3.13}$$

for some s.o.s. polynomials $\sigma_j \in \Sigma[\mathbf{x}]$, for all $j = 0, 1, \ldots, m$.

For each $j = 1, \dots, m$, and depending on its parity, let $2v_j$ or $2v_j - 1$ denote the degree of the polynomial g_j in the definition (3.2) of the set $\mathbf{K}$. Similarly, let $2v_0$ or $2v_0 - 1$ be the degree of f. For $i \geq \max_{j=0,\dots,m} v_j$, consider the following semidefinite program:

$$\begin{aligned} \rho_i = \inf_{\mathbf{y}} \; & L_{\mathbf{y}}(f) \\ \text{s.t.} \; & \mathbf{M}_i(\mathbf{y}) \succeq \mathbf{0}, \\ & \mathbf{M}_{i-v_j}(g_j\mathbf{y}) \succeq \mathbf{0}, \quad j = 1, \dots, m, \\ & y_0 = 1. \end{aligned} \tag{3.14}$$

Obviously, (3.14) is a relaxation of (3.1). Indeed, let $\mathbf{x}^* \in \mathbf{K}$ be a global minimizer (guaranteed to exist as $\mathbf{K}$ is compact) and let $\mathbf{y} = \mathbf{v}_{2i}(\mathbf{x}^*)$. That is, $\mathbf{y}$ is the vector of moments (up to order $2i$) of the Dirac measure $\mu = \delta_{\mathbf{x}^*}$ at $\mathbf{x}^* \in \mathbf{K}$. Therefore $\mathbf{y}$ is a feasible solution of (3.14) with value $L_{\mathbf{y}}(f) = f(\mathbf{x}^*) = f^*$, which proves that $\rho_i \leq f^*$ for every i. The dual of (3.14) is the semidefinite program:

$$\begin{aligned} \rho_i^* = \; & \sup_{\lambda, \{\sigma_j\}} \; \lambda \\ \text{s.t.} \; & f - \lambda = \sigma_0 + \sum_{j=1}^m \sigma_j \, g_j, \\ & \deg \sigma_0 \leq 2i; \quad \deg \sigma_j \leq 2i - 2v_j, \; j = 1, \dots, m. \end{aligned} \tag{3.15}$$

Note that, if in the definition of $\mathbf{K}$ there is an equality constraint $g_j(\mathbf{x}) = 0$ (i.e., two opposite inequality constraints $g_j(\mathbf{x}) \geq 0$ and $-g_j(\mathbf{x}) \geq 0$), then one has the equality constraint $\mathbf{M}_{i-v_j}(g_j \, \mathbf{y}) = 0$ in (3.14) and, accordingly, in (3.15) the weight $\sigma_j \in \mathbb{R}[\mathbf{x}]$ is not required to be a s.o.s. polynomial.

The overall algorithm is as follows:

Algorithm 3.8. (Constrained polynomial optimization)

Input: A polynomial $\mathbf{x} \mapsto f(\mathbf{x})$ of degree $2v_0$ or $2v_0 - 1$; a set $\mathbf{K} = \{\mathbf{x} \in \mathbb{R}^n : g_j(\mathbf{x}) \geq 0, \; j = 1, \dots, m\}$, where the polynomials g_j are of degree $2v_j$ or $2v_j - 1$, $j = 1, \dots, m$; a number k of the highest relaxation.

Output: The value $f^* = \min_{\mathbf{x} \in \mathbf{K}} f(\mathbf{x})$ and a list of global minimizers or a lower bound ρ_k on f^*.

Algorithm:

1. Solve the semidefinite optimization problem (3.14) with optimal value ρ_i, and optimal solution $\mathbf{y}^*$ (if it exists).
2. If there is no optimal solution $\mathbf{y}^*$ then ρ_k only yields a lower bound $\rho_k \leq f^*$. If $i < k$, increase i by one and go to Step 1; otherwise stop and output ρ_k.
3. If $\operatorname{rank} \mathbf{M}_{i-v}(\mathbf{y}^*) = \operatorname{rank} \mathbf{M}_i(\mathbf{y}^*)$ (with $v = \max_j v_j$), then $\rho_i = f^*$ and there are at least $\operatorname{rank} \mathbf{M}_i(\mathbf{y}^*)$ global minimizers, which can be extracted.

4. If $\operatorname{rank} \mathbf{M}_{i-v}(\mathbf{y}^*) \neq \operatorname{rank} \mathbf{M}_i(\mathbf{y}^*)$ and $i < k$, then increase i by one and go to Step 1; otherwise, stop and output ρ_k, which only provides a lower bound $\rho_k \leq f^*$.

And we have the following convergence result:

Theorem 3.9. *Let Assumption* 1.16 *hold and consider the semidefinite relaxation* (3.14) *with optimal value* ρ_i.

(a) $\rho_i \uparrow f^*$ *as* $i \to \infty$.

(b) *Assume that* (3.1) *has a unique global minimizer* $\mathbf{x}^* \in \mathbf{K}$ *and let* $\mathbf{y}^i$ *be a nearly optimal solution of* (3.14) *with value* $L_{\mathbf{y}}(f) \leq \rho_i + 1/i$. *Then, as* $i \to \infty$, $L_{\mathbf{y}^i}(x_j) \to x_j^*$ *for every* $j = 1, \ldots, n$.

Sketch of the proof. As $\mathbf{K}$ is compact, $f^* > -\infty$, and with $\epsilon > 0$ arbitrary, the polynomial $f - (f^* - \epsilon)$ is strictly positive on $\mathbf{K}$. Therefore, by Theorem 1.17,

$$f - (f^* - \epsilon) = \sigma_{0\epsilon} + \sum_{j=1}^{m} \sigma_{j\epsilon}\, g_j$$

for some s.o.s. polynomials $(\sigma_{j\epsilon}) \subset \Sigma[\mathbf{x}]$. So $(f^* - \epsilon, (\sigma_{j\epsilon}))$ is a feasible solution of (3.15) as soon as r is large enough. Hence $\rho_r^* \geq f^* - \epsilon$. On the other hand, by weak duality one has $\rho_i^* \leq \rho_i$ for every i and we have seen that $\rho_i \leq f^*$, which yields $f^* - \epsilon \leq \rho_i^* \leq \rho_i \leq f^*$. Hence, as $\epsilon > 0$ was arbitrary, (a) follows.

For (b) one first proves that a subsequence $\mathbf{y}^{i_k}$ converges pointwise to a sequence $\mathbf{y}$ which is shown to be the moment sequence of a measure μ supported on $\mathbf{K}$ and with value $L_{\mathbf{y}}(f) = \lim_{k\to\infty} L_{\mathbf{y}^{i_k}}(f) \leq f^*$. Hence μ is the Dirac measure $\delta_{\mathbf{x}^*}$ at the unique global minimizer $\mathbf{x}^* \in \mathbf{K}$. Therefore the whole sequence $\mathbf{y}^i$ converges pointwise to the vector $\mathbf{v}(\mathbf{x}^*)$ of moments of μ. In particular, for every $j = 1, \ldots, n$, $L_{\mathbf{y}^i}(x_j) \to x_j^*$ as $i \to \infty$. □

3.3.2 Obtaining global minimizers

As for the unconstrained case, after solving the semidefinite relaxation (3.14) for some value of $i \in \mathbb{N}$, we are left with two issues:

(a) How do we know that $\rho_i < f^*$, or $\rho_i = f^*$?

(b) If $\rho_i = f^*$, can we get at least one global minimizer $\mathbf{x}^* \in \mathbf{K}$?

Again, an easy case is when (3.14) has an optimal solution $\mathbf{y}^*$ which satisfies $\operatorname{rank} \mathbf{M}_i(\mathbf{y}^*) = 1$, and so, necessarily, $\mathbf{M}_i(\mathbf{y}^*) = \mathbf{v}_i(\mathbf{x}^*)\mathbf{v}_i(\mathbf{x}^*)'$ for some $\mathbf{x}^* \in \mathbb{R}^n$. In addition, the constraints $\mathbf{M}_{i-v_j}(g_j\mathbf{y}^*) \succeq 0$ imply that $\mathbf{x}^* \in \mathbf{K}$. That is, $\mathbf{y}^*$ is the vector of moments up to order $2i$ of the Dirac measure at $\mathbf{x}^* \in \mathbf{K}$, and one reads the optimal solution $\mathbf{x}^*$ from the subvector of first "moments" $y_{\boldsymbol{\alpha}}^*$ with $|\boldsymbol{\alpha}| = 1$.

For the case of multiple global minimizers, we have the following sufficient condition which is implemented at step 3 of Algorithm 3.8:

Theorem 3.10. *Let $f \in \mathbb{R}[\mathbf{x}]$, and suppose that the optimal value ρ_i of problem* (3.14) *is attained at some optimal solution $\mathbf{y}^*$. If $\operatorname{rank}\mathbf{M}_{i-v}(\mathbf{y}^*) = \operatorname{rank}\mathbf{M}_i(\mathbf{y}^*)$ with $v = \max_{j=1,\ldots,m} v_j$, then $\rho_i = f^*$ and there are at least $s = \operatorname{rank}\mathbf{M}_i(\mathbf{y}^*)$ global minimizers, and they can be extracted by Algorithm* 3.13.

Proof. We know that $\rho_i \leq f^*$ for all i. If $s = \operatorname{rank}\mathbf{M}_{i-v}(\mathbf{y}^*) = \operatorname{rank}\mathbf{M}_i(\mathbf{y}^*)$, then, by Theorem 2.17, $\mathbf{y}^*$ is the moment vector of some s-atomic probability measure μ^* on $\mathbf{K}$. We then argue that each of the s atoms of μ^* is a global minimizer of f on $\mathbf{K}$. Indeed, being s-atomic, there is a family $(\mathbf{x}(k))_{k=1}^s \subset \mathbf{K}$ and a family $(\lambda_k)_{k=1}^s \subset \mathbb{R}$ such that

$$\mu^* = \sum_{k=1}^s \lambda_k \, \delta_{\mathbf{x}(k)}, \quad \lambda_k > 0, \quad \forall k = 1, \ldots, s, \quad \sum_{k=1}^s \lambda_k = 1.$$

Hence,

$$f^* \geq \rho_i = L_{\mathbf{y}^*}(f) = \int_{\mathbf{K}} f \, d\mu^* = \sum_{k=1}^s \lambda_k \, f(\mathbf{x}(k)) \geq f^*,$$

which clearly implies $f^* = \rho_i$, and $f(\mathbf{x}(k)) = f^*$ for all $k = 1, \ldots, s$, the desired result. □

The rank-test of Theorem 3.10 implies that, if it is satisfied, we can conclude that there are $\operatorname{rank}\mathbf{M}_i(\mathbf{y}^*)$ global minimizers encoded in the optimal solution $\mathbf{y}^*$ of (3.14). In order to extract these solutions from $\mathbf{y}^*$, we apply Algorithm 3.13. This extraction algorithm is implemented in the GloptiPoly software.

Example 3.11. Let $n = 2$ and consider the optimization problem

$$\begin{aligned} f^* = \min \; & x_1^2 x_2^2 (x_1^2 + x_2^2 - 1) \\ \text{s.t.} \; & x_1^2 + x_2^2 \leq 4, \end{aligned}$$

which is the same as in Example 3.7, but with the additional ball constraint $\|\mathbf{x}\|^2 \leq 4$. Therefore the optimal value is $f^* = -1/27$ with global minimizers $\mathbf{x}^* = (\pm\sqrt{3}/3, \pm\sqrt{3}/3)$. Applying Algorithm 3.8 for $k = 4$ and using GloptyPoly, the optimal value is obtained with $\rho_4 \approx -1/27$ and the four optimal solutions $(\pm 0.5774, \pm 0.5774)' \approx \mathbf{x}^*$ are extracted.

When we add the additional nonconvex constraint $x_1 x_2 \geq 1$, we find that the optimal value is obtained with $\rho_3 \approx 1$ and the two optimal solutions $(x_1^*, x_2^*)' = (-1, -1)'$ and $(x_1^*, x_2^*)' = (1, 1)'$ are extracted. In both cases, the rank-test is satisfied for an optimal solution $\mathbf{y}^*$ and a global optimality certificate is provided due to Theorem 3.10.

Example 3.12. Let $n = 2$ and $f \in \mathbb{R}[\mathbf{x}]$ be the concave[1] polynomial

$$f(\mathbf{x}) = -(x_1 - 1)^2 - (x_1 - x_2)^2 - (x_2 - 3)^2$$

[1]A real function $f \colon \mathbb{R}^n \to \mathbb{R}$ is *concave* if $-f$ is convex, that is, $f(\lambda\mathbf{x} + (1-\lambda)\mathbf{y}) \geq \lambda f(\mathbf{x}) + (1-\lambda) f(\mathbf{y})$ for all $\mathbf{x}, \mathbf{y} \in \mathbb{R}^n$ and all $\lambda \in [0, 1]$.

and let $\mathbf{K} \subset \mathbb{R}^2$ be the set

$$\mathbf{K} = \left\{\mathbf{x} \in \mathbb{R}^2 \,:\, 1-(x_1-1)^2 \geq 0,\ 1-(x_1-x_2)^2 \geq 0,\ 1-(x_2-3)^2 \geq 0\right\}.$$

The optimal value is $f^* = -2$. Solving problem (3.14) for $i = 1$ yields $\rho_1 = -3$ instead of the desired value -2. On the other hand, solving problem (3.14) for $i = 2$ yields $\rho_2 \approx -2$ and the three optimal solutions $(x_1^*, x_2^*) = (1,2), (2,2), (2,3)$ are extracted. Hence, with polynomials of degree 4 instead of 2, we obtain (a good approximation of) the correct value. Note that there exist scalars $\lambda_j = 1 \geq 0$ such that

$$f(\mathbf{x}) + 3 = 0 + \sum_{j=1}^{3} \lambda_j g_j(\mathbf{x}),$$

but $f(\mathbf{x}) - f^*$ $(= f(\mathbf{x}) + 2)$ cannot be written in this way (otherwise ρ_1 would be the optimal value -2).

For quadratically constrained nonconvex quadratic problems, the semidefinite program (3.14) with $i = 1$ is a well-known relaxation. But ρ_1 which sometimes provides directly the exact global minimum value, is only a lower bound in general.

An algorithm for extraction of global minimizers

We now describe Algorithm 3.13, a procedure that *extracts* global minimizers of problem (3.1) when the semidefinite relaxation (3.14) is exact at some step s of the hierarchy, i.e., with $\rho_s = f^*$, and the rank condition

$$\operatorname{rank} \mathbf{M}_{s-v}(\mathbf{y}^*) = \operatorname{rank} \mathbf{M}_s(\mathbf{y}^*) \tag{3.16}$$

in Theorem 3.10 holds at an optimal solution $\mathbf{y}^*$ of (3.14). The main steps of the extraction algorithm can be sketched as follows.

Cholesky factorization

As condition (3.16) holds, $\mathbf{y}^*$ is the vector of a $\operatorname{rank} \mathbf{M}_s(\mathbf{y}^*)$-atomic Borel probability measure μ supported on $\mathbf{K}$. That is, there are r $(= \operatorname{rank} \mathbf{M}_s(\mathbf{y}^*))$ points $(\mathbf{x}(k))_{k=1}^r \subset \mathbf{K}$ such that

$$\mu = \sum_{j=1}^{r} \kappa_j^2\, \delta_{\mathbf{x}(j)}, \quad \kappa_j \neq 0,\ \forall j; \quad \sum_{j=1}^{r} \kappa_j^2 = y_0 = 1, \tag{3.17}$$

with $\delta_{\mathbf{x}}$ being the Dirac measure at the point $\mathbf{x}$. Hence, by construction of $\mathbf{M}_s(\mathbf{y}^*)$,

$$\mathbf{M}_s(\mathbf{y}^*) = \sum_{j=1}^{r} \kappa_j^2\, \mathbf{v}_s(\mathbf{x}^*(j)) \left(\mathbf{v}_s(\mathbf{x}^*(j))\right)' = \mathbf{V}^*\mathbf{D}(\mathbf{V}^*)' \tag{3.18}$$

where $\mathbf{V}^*$ is written columnwise as

$$\mathbf{V}^* = \left[\ \mathbf{v}_s(\mathbf{x}^*(1)) \quad \mathbf{v}_s(\mathbf{x}^*(2)) \quad \cdots \quad \mathbf{v}_s(\mathbf{x}^*(r))\ \right]$$

with $\mathbf{v}_s(\mathbf{x})$ as in (2.9), and $\mathbf{D}$ is a $r \times r$ diagonal matrix with entries $D(i,i) = d_i^2$, $i = 1, \ldots, r$.

In fact, the weights $(\kappa_j)_{j=1}^r$ do not play any role in the sequel. As long as $\kappa_j \neq 0$ for all j, the rank of the moment matrix $\mathbf{M}_s(\mathbf{y}^*)$ associated with the Borel measure μ defined in (3.17) does *not* depend on the weights κ_j. The extraction procedure with another matrix $\mathbf{M}_s(\widetilde{\mathbf{y}})$ written as in (3.18) but with different weights $\widetilde{\kappa}_j$ would yield the same global minimizers $(\mathbf{x}^*(j))_{j=1}^r$. Of course, the new associated vector $\widetilde{\mathbf{y}}$ would also be an optimal solution of the semidefinite relaxation with value $\rho_s = f^*$.

Extract a Cholesky factor $\mathbf{V}$ of the positive semidefinite moment matrix $\mathbf{M}_s(\mathbf{y}^*)$, i.e., a matrix $\mathbf{V}$ with r columns satisfying

$$\mathbf{M}_s(\mathbf{y}^*) = \mathbf{V}\mathbf{V}'. \tag{3.19}$$

Such a Cholesky factor[2] can be obtained via singular value decomposition, or any cheaper alternative; see, e.g., [37].

The matrices $\mathbf{V}$ and $\mathbf{V}^*$ span the same linear subspace, so the solution extraction algorithm consists of transforming $\mathbf{V}$ into $\mathbf{V}^*$ by suitable column operations. This is described in the sequel.

Column echelon form. Reduce $\mathbf{V}$ to column echelon form

$$\mathbf{U} = \begin{bmatrix} 1 & & & & \\ \star & & & & \\ 0 & 1 & & & \\ 0 & 0 & 1 & & \\ \star & \star & \star & & \\ & \vdots & & \ddots & \\ 0 & 0 & 0 & \cdots & 1 \\ \star & \star & \star & \cdots & \star \\ & \vdots & & & \vdots \\ \star & \star & \star & \cdots & \star \end{bmatrix}$$

by Gaussian elimination with column pivoting. By construction of the moment matrix, each row in $\mathbf{U}$ is indexed by a monomial $\mathbf{x}^{\boldsymbol{\alpha}}$ in the basis $\mathbf{v}_s(\mathbf{x})$. Pivot elements in $\mathbf{U}$ (i.e., the first nonzero elements in each column) correspond to monomials $\mathbf{x}^{\beta_j}$, $j = 1, 2, \ldots, r$, of the basis generating the r solutions. In other words, if

$$\mathbf{w}(\mathbf{x}) = \left[\mathbf{x}^{\beta_1}\ \mathbf{x}^{\beta_2}\ \cdots\ \mathbf{x}^{\beta_r}\right]' \tag{3.20}$$

denotes this generating basis, then

$$\mathbf{v}_s(\mathbf{x}) = \mathbf{U}\mathbf{w}(\mathbf{x}) \tag{3.21}$$

---

[2] *Cholesky factor* is a slight abuse of terminology, since rigorously it should apply for an invertible matrix, in which case $\mathbf{V}$ can be made triangular by a change of basis.

for all solutions $\mathbf{x} = \mathbf{x}^*(j)$, $j = 1, 2, \ldots, r$. In summary, extracting the solutions amounts to solving the system of polynomial equations (3.21).

Solving the system of polynomial equations (3.21)

Once a generating monomial basis $\mathbf{w}(\mathbf{x})$ is available, it turns out that extracting solutions of the system of polynomial equations (3.21) reduces to solving the linear algebra problem described below.

1. Multiplication matrices

For each degree one monomial x_i, $i = 1, 2, \ldots, n$, extract from $\mathbf{U}$ the so-called $r \times r$ *multiplication* (by x_i) matrix $\mathbf{N}_i$ containing the coefficients of monomials $x_i \mathbf{x}^{\beta_j}$, $j = 1, 2, \ldots, r$, in the generating basis (3.20), i.e., such that

$$\mathbf{N}_i \mathbf{w}(\mathbf{x}) = x_i \mathbf{w}(\mathbf{x}), \quad i = 1, 2, \ldots, n. \tag{3.22}$$

The entries of global minimizers $\mathbf{x}^*(j)$, $j = 1, 2, \ldots, r$ are all eigenvalues of multiplication matrices $\mathbf{N}_i$, $i = 1, 2, \ldots, n$. That is,

$$\mathbf{N}_i \mathbf{w}(\mathbf{x}^*(j)) = x_i^*(j)\, \mathbf{w}(\mathbf{x}^*(j)), \quad i = 1, \ldots, n, \quad j = 1, \ldots, r.$$

But how to reconstruct the solutions $(\mathbf{x}^*(j))$ from knowledge of the eigenvalues of the $\mathbf{N}_i$s? Indeed, from the n r-tuples of eigenvalues one could build up r^n possible vectors of $\mathbb{R}^n$ whereas we are looking for only r of them.

2. Common eigenspaces

Observe that, for every $j = 1, \ldots, r$, the vector $\mathbf{w}(\mathbf{x}^*(j))$ is an eigenvector common to *all* matrices $\mathbf{N}_i$, $i = 1, \ldots, n$. Therefore, in order to compute $(\mathbf{x}^*(j))$, one builds up a random combination

$$\mathbf{N} = \sum_{i=1}^{n} \lambda_i \mathbf{N}_i$$

of the multiplication matrices $\mathbf{N}_i$s, where λ_i, $i = 1, 2, \ldots, n$, are nonnegative real numbers summing up to one. Then, with probability 1, the eigenvalues of $\mathbf{N}$ are all distinct and so $\mathbf{N}$ is nonderogatory, i.e., all its eigenspaces are one-dimensional (and spanned by the vectors $\mathbf{w}(\mathbf{x}^*(j))$, $j = 1, \ldots, r$).

Then, compute the ordered Schur decomposition

$$\mathbf{N} = \mathbf{Q}\,\mathbf{T}\,\mathbf{Q}' \tag{3.23}$$

where $\mathbf{Q} = [\mathbf{q}_1\ \mathbf{q}_2\ \cdots\ \mathbf{q}_r]$ is an orthogonal matrix (i.e., $\mathbf{q}_i'\mathbf{q}_i = 1$ and $\mathbf{q}_i'\mathbf{q}_j = 0$ for $i \neq j$) and $\mathbf{T}$ is upper-triangular with eigenvalues of $\mathbf{N}$ sorted in increasing order along the diagonal. Finally, the ith entry $x_i^*(j)$ of $\mathbf{x}^*(j) \in \mathbb{R}^n$ is given by

$$x_i^*(j) = \mathbf{q}_j' \mathbf{N}_i \mathbf{q}_j, \quad i = 1, 2, \ldots, n, \quad j = 1, 2, \ldots, r. \tag{3.24}$$

In summary, here is the extraction algorithm:

Algorithm 3.13. (Extraction algorithm)

Input: The moment matrix $\mathbf{M}_s(\mathbf{y}^*)$ of rank r.

Output: The r points $\mathbf{x}^*(i) \in \mathbf{K}$, $i = 1, \ldots, r$, support of an optimal solution of the moment problem.

Algorithm:

1. Get the *Cholesky factorization* $\mathbf{VV}'$ of $\mathbf{M}_s(\mathbf{y}^*)$.
2. Reduce $\mathbf{V}$ to an *echelon form* $\mathbf{U}$.
3. Extract from $\mathbf{U}$ the multiplication matrices $\mathbf{N}_i$, $i = 1, \ldots, n$.
4. Compute $\mathbf{N} = \sum_{i=1}^n \lambda_i \mathbf{N}_i$ with randomly generated coefficients λ_i, and the Schur decomposition $\mathbf{N} = \mathbf{QTQ}'$. Compute $\mathbf{Q} = [\mathbf{q}_1\ \mathbf{q}_2\ \cdots\ \mathbf{q}_r]$ and $x_i^*(j) = \mathbf{q}_j'\mathbf{N}_i\mathbf{q}_j$, for every $i = 1, 2, \ldots, n$, and every $j = 1, 2, \ldots, r$.

Example 3.14. Consider the bivariate optimization problem with data

$$\mathbf{x} \longmapsto f(\mathbf{x}) = -(x_1 - 1)^2 - (x_1 - x_2)^2 - (x_2 - 3)^2,$$

and $\mathbf{K} = \{\mathbf{x} \in \mathbb{R}^2 : 1 - (x_1 - 1)^2 \geq 0;\ 1 - (x_1 - x_2)^2 \geq 0;\ 1 - (x_2 - 3)^2 \geq 0\}$.

The first ($i = 1$) semidefinite relaxation yields $\rho_1 = -3$ and $\operatorname{rank} \mathbf{M}_1(\mathbf{y}^*) = 3$, whereas the second ($i = 2$) yields $\rho_2 = -2$ and $\operatorname{rank} \mathbf{M}_1(\mathbf{y}^*) = \operatorname{rank} \mathbf{M}_2(\mathbf{y}^*) = 3$. Hence, the rank condition (3.16) is satisfied, which implies that $-2 = \rho_2 = f^*$. The moment matrix $\mathbf{M}_2(\mathbf{y}^*)$ reads

$$\mathbf{M}_2(\mathbf{y}^*) = \begin{bmatrix} 1.0000 & 1.5868 & 2.2477 & 2.7603 & 3.6690 & 5.2387 \\ 1.5868 & 2.7603 & 3.6690 & 5.1073 & 6.5115 & 8.8245 \\ 2.2477 & 3.6690 & 5.2387 & 6.5115 & 8.8245 & 12.7072 \\ 2.7603 & 5.1073 & 6.5115 & 9.8013 & 12.1965 & 15.9960 \\ 3.6690 & 6.5115 & 8.8245 & 12.1965 & 15.9960 & 22.1084 \\ 5.2387 & 8.8245 & 12.7072 & 15.9960 & 22.1084 & 32.1036 \end{bmatrix}$$

and the monomial basis is

$$\mathbf{v}_2(\mathbf{x}) = \begin{bmatrix} 1 & x_1 & x_2 & x_1^2 & x_1x_2 & x_2^2 \end{bmatrix}'.$$

The Cholesky factor (3.19) of $\mathbf{M}_2(\mathbf{y}^*)$ is given by

$$\mathbf{V} = \begin{bmatrix} -0.9384 & -0.0247 & 0.3447 \\ -1.6188 & 0.3036 & 0.2182 \\ -2.2486 & -0.1822 & 0.3864 \\ -2.9796 & 0.9603 & -0.0348 \\ -3.9813 & 0.3417 & -0.1697 \\ -5.6128 & -0.7627 & -0.1365 \end{bmatrix},$$

whose column echelon form reads (after rounding)

$$\mathbf{U} = \begin{bmatrix} 1 & & \\ 0 & 1 & \\ 0 & 0 & 1 \\ -2 & 3 & 0 \\ -4 & 2 & 2 \\ -6 & 0 & 5 \end{bmatrix}.$$

Pivot entries correspond to the following generating basis (3.20):

$$\mathbf{w}(\mathbf{x}) = [1 \quad x_1 \quad x_2]'.$$

From the subsequent rows in matrix $\mathbf{U}$ we deduce from (3.21) that all solutions $\mathbf{x}$ must satisfy the three polynomial equations

$$\begin{aligned} x_1^2 &= -2 + 3x_1 \\ x_1 x_2 &= -4 + 2x_1 + 2x_2 \\ x_2^2 &= -6 + 5x_2. \end{aligned}$$

Multiplication matrices (3.22) (by x_1 and x_2) in the basis $\mathbf{w}(\mathbf{x})$ are readily extracted from rows in $\mathbf{U}$:

$$\mathbf{N}_1 = \begin{bmatrix} 0 & 1 & 0 \\ -2 & 3 & 0 \\ -4 & 2 & 2 \end{bmatrix}, \quad \mathbf{N}_2 = \begin{bmatrix} 0 & 0 & 1 \\ -4 & 2 & 2 \\ -6 & 0 & 5 \end{bmatrix}.$$

A randomly chosen convex combination of $\mathbf{N}_1$ and $\mathbf{N}_2$ yields

$$\mathbf{N} = 0.6909\,\mathbf{N}_1 + 0.3091\,\mathbf{N}_2 = \begin{bmatrix} 0 & 0.6909 & 0.3091 \\ -2.6183 & 2.6909 & 0.6183 \\ -4.6183 & 1.3817 & 2.9274 \end{bmatrix}$$

with orthogonal matrix in Schur decomposition (3.23) given by

$$\mathbf{Q} = \begin{bmatrix} 0.4082 & 0.1826 & -0.8944 \\ 0.4082 & -0.9129 & -0.0000 \\ 0.8165 & 0.3651 & 0.4472 \end{bmatrix}.$$

From equations (3.24), we obtain the three optimal solutions

$$\mathbf{x}^*(1) = \begin{bmatrix} 1 \\ 2 \end{bmatrix}, \quad \mathbf{x}^*(2) = \begin{bmatrix} 2 \\ 2 \end{bmatrix}, \quad \mathbf{x}^*(3) = \begin{bmatrix} 2 \\ 3 \end{bmatrix}.$$

3.3.3 The univariate case

If we consider the univariate case $n = 1$ with $\mathbf{K} = [a, b]$ or $\mathbf{K} = [0, \infty)$, the corresponding sequence of semidefinite relaxations (3.14) simply reduces to a *single* relaxation. In other words, the minimization of a univariate polynomial on an interval of the real line (bounded or not) is a convex optimization problem and reduces to solving a semidefinite program.

Indeed, consider for instance the case where f has degree $2r$ and $\mathbf{K} \subset \mathbb{R}$ is the interval $[-1, 1]$. By Theorem 1.7(b),

$$f(x) - f^* = f_0(x) + f_3(x)(1 - x^2), \quad x \in \mathbb{R}, \tag{3.25}$$

for some s.o.s. polynomials $f_0, f_3 \in \Sigma[x]$ such that the degree of the summands is less than $2r$.

Theorem 3.15. *Let $f \in \mathbb{R}[x]$ be a univariate polynomial of degree $2r$. The semidefinite relaxation* (3.14) *for $i = r$ of the problem*

$$f^* = \min_x \left\{ f(x) \;:\; 1 - x^2 \geq 0 \right\}$$

is exact, i.e., $\rho_r = \rho_r^* = f^*$. In addition, both (3.14) *and* (3.15) *have an optimal solution.*

Proof. From (3.25), (f^*, f_0, f_3) is a feasible solution of (3.15) with $i = r$, and so optimal, because $\rho_r^* = f^*$. Therefore, from $\rho_r^* \leq \rho_r \leq f^*$ we also obtain $\rho_r = f^*$, which in turn implies that $\mathbf{y}^* = \mathbf{v}_{2r}(x^*)$ is an optimal solution, for any global minimizer $x^* \in \mathbf{K}$. □

A similar argument holds if $f \in \mathbb{R}[x]$ has odd degree $2r - 1$, in which case

$$f(x) - f^* = f_1(x)(1 + x) + f_2(x)(1 - x), \quad x \in \mathbb{R},$$

for some s.o.s. polynomials $f_1, f_2 \in \Sigma[x]$ such that the degree of the summands is less than $2r - 1$. Again, for the problem

$$f^* = \min_x \left\{ f(x) \;:\; 1 - x \geq 0,\; 1 + x \geq 0 \right\},$$

the relaxation (3.14) with $i = r$ is exact.

3.3.4 Numerical experiments

In this section, we report on the performance of Algorithm 3.8 using the software package GloptiPoly[3] on a series of benchmark nonconvex continuous optimization problems. It is worth noticing that Algorithm 3.13 is implemented in GloptiPoly to extract global minimizers.

[3] GloptiPoly is a Matlab-based free software devoted to solve or approximate the Generalized Problem of Moments, whose polynomial optimization is only a particular instance; see [45]. It can be downloaded at `http://www.laas.fr/~henrion`.

In Table 3.1, we record the problem name, the source of the problem, the number of decision variables (var), the number of inequality or equality constraints (cstr), and the maximum degree arising in the polynomial expressions (deg), the CPU time in seconds (CPU) and the order of the relaxation (order).

At the time of the experiment, GloptiPoly was using the semidefinite optimization solver SeDuMi; see [123]. As indicated by the label "dim" in the rightmost column, the quadratic problems 2.8, 2.9 and 2.11 in [35] involve more than 19 variables and could not be handled by the current version of GloptiPoly. Except for problems 2.4 and 3.2, the computational load is moderate. In almost all reported instances, the global optimum was reached exactly by a semidefinite relaxation of small order.

3.3.5 Exploiting sparsity

Despite their nice properties, the size of the semidefinite relaxations (3.14) grows rapidly with the dimension n. Typically, the moment matrix $\mathbf{M}_i(\mathbf{y})$ is $s(i) \times s(i)$ with $s(i) = \binom{n+i}{n}$, and there are $\binom{n+2i}{n}$ variables $y_{\boldsymbol{\alpha}}$. This makes the applicability of Algorithm 3.8 limited to small or medium size problems only. Fortunately, in many practical applications of large size moment problems, some sparsity pattern is often present and may be exploited.

Suppose that there is no *coupling* between some subsets of variables in the polynomials g_k that define the set $\mathbf{K}$, and the polynomial f. That is, by no coupling between two sets of variables we mean that there is *no* monomial involving some variables of such subsets in any of the polynomials f, (g_k).

Recalling the notation of Section 1.7, let $I_0 = \{1, \ldots, n\}$ be the union $\cup_{k=1}^p I_k$ of p subsets I_k, $k = 1, \ldots, p$, with cardinal denoted n_k (with possible overlaps). For an arbitrary $J \subseteq I_0$, let $\mathbb{R}[\mathbf{x}(J)]$ denote the ring of polynomials in the variables $\mathbf{x}(J) = (x_i : i \in J)$, and so $\mathbb{R}[\mathbf{x}(I_0)] = \mathbb{R}[\mathbf{x}]$. So let Assumptions 1.28 and 1.29 hold and, as in Section 1.7, let $m' = m + p$ and $\{1, \ldots, m'\} = \cup_{i=1}^p J_i$ and $\mathbf{K}$ be as in (1.27) after having added in its definition the p redundant quadratic constraints (1.26).

With $k \in \{1, \ldots, p\}$ fixed, and $g \in \mathbb{R}[\mathbf{x}(I_k)]$, let $\mathbf{M}_i(\mathbf{y}, I_k)$ (resp. $\mathbf{M}_i(g\,\mathbf{y}, I_k)$) be the moment (resp. localizing) submatrix obtained from $\mathbf{M}_i(\mathbf{y})$ (resp. $\mathbf{M}_i(g\,\mathbf{y})$) by retaining only those rows (and columns) $\alpha \in \mathbb{N}^n$ of $\mathbf{M}_i(\mathbf{y})$ (resp. $\mathbf{M}_i(g\,\mathbf{y})$) such that $\alpha_i = 0$ whenever $i \notin I_k$. What we call the *sparse* semidefinite relaxation is the following semidefinite program:

$$
\begin{aligned}
\rho_i^{\mathrm{sparse}} = \inf_{\mathbf{y}} \ & L_{\mathbf{y}}(f) \\
\text{s.t. } & \mathbf{M}_i(\mathbf{y}, I_k) \succeq \mathbf{0}, \quad k = 1, \ldots, p, \\
& \mathbf{M}_{i-v_j}(g_j\mathbf{y}, I_k) \succeq \mathbf{0}, \quad j \in J_k,\ k = 1, \ldots, p, \\
& y_{\mathbf{0}} = 1,
\end{aligned}
\tag{3.26}
$$

Problem	var	cstr	deg	CPU	order
[63, Ex. 1]	2	0	4	0.13	2
[63, Ex. 2]	2	0	4	0.13	2
[63, Ex. 3]	2	0	6	1.13	8
[63, Ex. 5]	2	3	2	0.22	2
[35, Pb. 2.2]	5	11	2	11.8	3
[35, Pb. 2.3]	6	13	2	1.86	2
[35, Pb. 2.4]	13	35	2	1012	2
[35, Pb. 2.5]	6	15	2	1.58	2
[35, Pb. 2.6]	10	31	2	67.7	2
[35, Pb. 2.7]	10	25	2	75.3	2
[35, Pb. 2.8]	20	10	2	–	dim
[35, Pb. 2.9]	24	10	2	–	dim
[35, Pb. 2.10]	10	11	2	45.3	2
[35, Pb. 2.11]	20	10	2	–	dim
[35, Pb. 3.2]	8	22	2	3032	3
[35, Pb. 3.3]	5	16	2	1.20	2
[35, Pb. 3.4]	6	16	2	1.50	2
[35, Pb. 3.5]	3	8	2	2.42	4
[35, Pb. 4.2]	1	2	6	0.17	3
[35, Pb. 4.3]	1	2	50	0.94	25
[35, Pb. 4.4]	1	2	5	0.25	3
[35, Pb. 4.5]	1	2	4	0.14	2
[35, Pb. 4.6]	2	2	6	0.41	3
[35, Pb. 4.7]	1	2	6	0.20	3
[35, Pb. 4.8]	1	2	4	0.16	2
[35, Pb. 4.9]	2	5	4	0.31	2
[35, Pb. 4.10]	2	6	4	0.58	4

Table 3.1: Continuous optimization problems. CPU times and semidefinite relaxation orders required to reach global optimality.

whereas its dual is the semidefinite program

$$\begin{aligned}(\rho_i^{\text{sparse}})^* = &\sup_{\lambda,\sigma_{kj}} \lambda \\ &\text{s.t. } f - \lambda = \sum_{k=1}^{p} \left(\sigma_{k0} + \sum_{j \in J_k} \sigma_{kj}\, g_j \right), \\ &\quad \sigma_{k0}, \sigma_{kj} \in \Sigma[\mathbf{x}(I_k)], \quad k = 1, \ldots, p, \\ &\quad \deg \sigma_{k0},\ \deg \sigma_{kj} g_j \le 2i, \quad k = 1, \ldots, p.\end{aligned} \tag{3.27}$$

Theorem 3.16. *Let Assumptions 1.28 and 1.29 hold. Consider the sparse semidefinite relaxations (3.26) and (3.27). If the running intersection property (1.29) holds, then $\rho_i^{\text{sparse}} \uparrow f^*$ and $(\rho_i^{\text{sparse}})^* \uparrow f^*$ as $i \to \infty$.*

The proof is very similar to that of Theorem 3.9, except that one now uses the sparse version Theorem 1.30 of Theorem 1.17.

To see the gain in terms of number of variables and size of the moment and localizing matrices, let $\tau = \sup_k |I(k)|$. Then the semidefinite relaxation (3.26) has at most $O(p\tau^{2i})$ moment variables instead of $O(n^{2i})$, and p moment matrices of size at most $O(\tau^i)$ instead of a single one of size $O(n^i)$. Similarly, the localizing matrix $\mathbf{M}_{i-v_j}(g_j\mathbf{y}, I_k)$ is of size at most $O(\tau^{i-v_j})$ instead of $O(n^{i-v_j})$.

This yields big savings if τ is small compared to n. For instance, if τ is relatively small (say, e.g., 6, 7) then one may solve optimization problems with $n = 1000$ variables, whereas with $n = 1000$ one may not even implement the first standard semidefinite relaxation (3.14) with $i = 1$!

3.4 Constrained polynomial optimization: LP-relaxations

In this section we derive linear programming relaxations for problem (3.1).

3.4.1 The case of a basic semi-algebraic set

Let $\mathbf{K} \subset \mathbb{R}^n$ be as in (3.2) and let $\widehat{g}_j$ be the normalized version associated with g_j, $j = 1, \ldots, m$, defined in (1.17). Let $i \in \mathbb{N}$ be fixed and consider the following linear optimization problem:

$$L_i = \inf_{\mathbf{y}} \left\{ L_{\mathbf{y}}(f) \;:\; y_0 = 1;\; L_{\mathbf{y}}\big(\widehat{\mathbf{g}}^{\boldsymbol{\alpha}}(1 - \widehat{\mathbf{g}})^{\boldsymbol{\beta}}\big) \ge 0,\ \forall\, |\boldsymbol{\alpha} + \boldsymbol{\beta}| \le i \right\}, \tag{3.28}$$

with $L_{\mathbf{y}}$ being as in (2.3), and where the vector notation $\widehat{\mathbf{g}}$ has been introduced just before (1.18). Indeed, problem (3.28) is a linear optimization problem, because, since $|\boldsymbol{\alpha} + \boldsymbol{\beta}| \le i$, the conditions $L_{\mathbf{y}}(\widehat{\mathbf{g}}^{\boldsymbol{\alpha}}(1 - \widehat{\mathbf{g}})^{\boldsymbol{\beta}}) \ge 0$ yield finitely many linear inequality constraints on finitely many coefficients $y_{\boldsymbol{\alpha}}$ of the infinite sequence $\mathbf{y}$. The dual of problem (3.28) is

$$L_i^* = \sup_{\lambda,\mathbf{u}\geq 0} \left\{ \lambda \,:\, f - \lambda = \sum_{\substack{\boldsymbol{\alpha},\boldsymbol{\beta} \in \mathbb{N}^m \\ |\boldsymbol{\alpha}+\boldsymbol{\beta}|\leq i}} u_{\boldsymbol{\alpha},\boldsymbol{\beta}}\, \widehat{\mathbf{g}}^{\boldsymbol{\alpha}}\,(1-\widehat{\mathbf{g}})^{\boldsymbol{\beta}} \right\}. \tag{3.29}$$

Theorem 3.17. *The sequence $(L_i)_i$ is monotone nondecreasing and, under Assumption* 1.23,

$$\lim_{i\to\infty} L_i^* = \lim_{i\to\infty} L_i = f^*. \tag{3.30}$$

The proof is along the same lines as that of Theorem 3.9, except that one now uses Theorem 1.24 instead of Theorem 1.17.

So the relaxations (3.28)–(3.29) are linear programs, a good news because, in principle, with current LP packages, one is able to solve linear programs with million variables and constraints! However, as we will see, the linear programs (3.28)–(3.29) suffer from some serious drawbacks.

3.4.2 The case of a convex polytope

In the particular case where the basic semi-algebraic set $\mathbf{K} \subset \mathbb{R}^n$ is compact and all the polynomials g_i in the definition (3.2) of $\mathbf{K}$ are affine (and so $\mathbf{K}$ is a polytope), one may specialize the linear relaxation (3.28) to

$$L_i = \inf_{\mathbf{y}} \left\{ L_{\mathbf{y}}(f) \;:\; y_0 = 1;\; L_{\mathbf{y}}(\mathbf{g}^{\boldsymbol{\alpha}}) \geq 0,\; \forall\, |\boldsymbol{\alpha}| \leq i \right\}, \tag{3.31}$$

and its associated dual

$$L_i^* = \max_{\lambda,\mathbf{u}\geq 0} \left\{ \lambda \,:\, f - \lambda = \sum_{\substack{\boldsymbol{\alpha} \in \mathbb{N}^m \\ |\boldsymbol{\alpha}|\leq i}} u_{\boldsymbol{\alpha}}\, \mathbf{g}^{\boldsymbol{\alpha}} \right\}. \tag{3.32}$$

Theorem 3.18. *If all g_j's in* (3.2) *are affine and* $\mathbf{K}$ *is a convex polytope with nonempty interior, then*

$$\lim_{i\to\infty} L_i^* = \lim_{i\to\infty} L_i = f^*. \tag{3.33}$$

The proof is analogous to that of Theorem 3.17.

3.5 Contrasting LP and semidefinite relaxations

Theorem 3.18 implies that we can approach the global optimal value f^* as closely as desired by solving linear optimization problems (3.28) of increasing size. This should be interesting because very powerful linear optimization software packages are available, in contrast with semidefinite optimization software packages that have not yet reached the level of maturity of the linear optimization packages.

Unfortunately we next show that in general the LP-relaxations (3.28) cannot be exact, that is, the convergence in (3.30) (or in (3.33) when $\mathbf{K}$ is a convex polyope) is only asymptotic, not finite. Indeed, assume that the convergence is finite, i.e., $L_i^* = f^*$ for some $i \in \mathbb{N}$. Suppose that the interior of $\mathbf{K}$ ($\operatorname{int}\mathbf{K}$) is given by

$$\operatorname{int}\mathbf{K} = \{\mathbf{x} \in \mathbb{R}^n \,:\, g_j(\mathbf{x}) > 0,\ j = 1, \ldots, m\}.$$

Then, if there exists a global minimizer $\mathbf{x}^* \in \operatorname{int}\mathbf{K}$, and $1 - \widehat{g}_j(\mathbf{x}^*) > 0$ for all $j = 1, \ldots, m$, we would get the contradiction

$$0 = f(\mathbf{x}^*) - L_i^* = \sum_{\boldsymbol{\alpha},\boldsymbol{\beta} \in \mathbb{N}^m} u_{\boldsymbol{\alpha},\boldsymbol{\beta}}\, \widehat{\mathbf{g}}(\mathbf{x}^*)^{\boldsymbol{\alpha}} \big(1 - \widehat{\mathbf{g}}(\mathbf{x}^*)\big)^{\boldsymbol{\beta}} > 0.$$

Similarly, if $\mathbf{x}^* \in \partial\mathbf{K}$, let $J^* = \{j \,:\, g_j(\mathbf{x}^*) = 0\}$. For the same reasons, if there exists $\mathbf{x} \in \mathbf{K}$ with $f(\mathbf{x}) > f(\mathbf{x}^*)$ and $g_j(\mathbf{x}) = 0$ for all $j \in J^*$, then finite convergence cannot take place.

Example 3.19. Let $\mathbf{K} = \{x \in \mathbb{R} \,:\, 0 \leq x \leq 1\}$ and $x \mapsto f(x) = x(x-1)$. This is a convex optimization problem with global minimizer $x^* = 1/2$ in the interior of $\mathbf{K}$, and optimal value $f^* = -1/4$. The optimal values $(L_i)_i$ of the linear relaxations (3.28) are reported in Table 3.2. The example shows the rather slow monotone convergence of $L_i \to -0.25$, despite the original problem is convex.

i	2	4	6	10	15
L_i	−1/3	−1/3	−0.3	−0.27	−0.2695

Table 3.2: Slow convergence of the linear relaxation (3.28).

On the other hand, with $-f$ instead of f, the problem becomes a harder concave minimization problem. But this time the second relaxation is exact! Indeed, $f^* = 0$ and we have

$$f(x) - f^* = x - x^2 = x(1-x) = g_1(x)g_2(x),$$

with $g_1(x) = x$ and $g_2(x) = 1 - x$.

Example 3.19 illustrates that the convergence $L_i \uparrow f^*$ as $i \to \infty$ is in general asymptotic, not finite; as underlined, the global minimizer being in the interior of $\mathbf{K}$, convergence *cannot* be finite. In addition, Example 3.19 exhibits an annoying paradox, namely that LP-relaxations may perform better for the concave minimization problem than for the *a priori* easier convex minimization problem. Finally, notice that, for large values of i, the constraints of the LP-relaxations (3.29) should contain very large coefficients due to the presence of binomial coefficients in the terms $(1 - \widehat{g})^{\beta}$, a source of numerical instability and ill-conditioned problems.

3.6 Putinar versus Karush–Kuhn–Tucker

In this section, we derive global optimality conditions for polynomial optimization generalizing the local optimality conditions due to Karush–Kuhn–Tucker (KKT) for nonlinear optimization.

A vector $(\mathbf{x}^*, \boldsymbol{\lambda}^*) \in \mathbb{R}^n \times \mathbb{R}^m$ satisfies the KKT conditions associated with problem (3.1) (and is called a *KKT pair*) if

$$\begin{aligned} \nabla f(\mathbf{x}^*) &= \sum_{j=1}^m \boldsymbol{\lambda}_j^* \nabla g_j(\mathbf{x}^*), \\ \boldsymbol{\lambda}_j^* g_j(\mathbf{x}^*) &= 0, \quad j = 1, \ldots, m, \\ g_j(\mathbf{x}^*),\, \lambda_j^* &\geq 0, \quad j = 1, \ldots, m. \end{aligned} \tag{3.34}$$

- The nonnegative dual variables $\lambda \in \mathbb{R}^m$ are called *Karush–Kuhn–Tucker Lagrange multipliers*.
- In fact, most local optimization algorithms try to find a pair of vectors $(\mathbf{x}^*, \boldsymbol{\lambda}^*) \in \mathbb{R}^n \times \mathbb{R}_+^m$ that satisfies (3.34).
- In general, $\mathbf{x}^*$ is *not* a global minimizer of the *Lagrangian polynomial*

$$\mathbf{x} \longmapsto L_f(\mathbf{x}) = f(\mathbf{x}) - f^* - \sum_{j=1}^m \lambda_j^* g_j(\mathbf{x}). \tag{3.35}$$

- However, if f is convex, the g_i's are concave, and the interior of $\mathbf{K}$ is nonempty (i.e., Slater's condition holds), then (3.34) are necessary and sufficient optimality conditions for $\mathbf{x}^*$ to be an optimal solution of (3.1). Moreover, $\mathbf{x}^*$ is a global minimizer of the Lagrangian polynomial L_f which is nonnegative on $\mathbb{R}^n$, with $L_f(\mathbf{x}^*) = 0$.

The developments in the earlier section lead to the following global optimality conditions.

Theorem 3.20. *Let $\mathbf{x}^* \in \mathbf{K}$ be a global minimizer for problem* (3.1), *with global optimum f^*, and assume that $f - f^*$ has the representation* (3.13), *i.e.,*

$$f(\mathbf{x}) - f^* = \sigma_0(\mathbf{x}) + \sum_{j=1}^m \sigma_j(\mathbf{x})\, g_j(\mathbf{x}), \quad \mathbf{x} \in \mathbb{R}^n, \tag{3.36}$$

for some s.o.s. polynomials $(\sigma_j)_{j=0}^m \subset \Sigma[\mathbf{x}]$. Then:

(a) $\sigma_j(\mathbf{x}^*),\, g_j(\mathbf{x}^*) \geq 0$ *for all* $j = 1, \ldots, m$.

(b) $\sigma_j(\mathbf{x}^*) g_j(\mathbf{x}^*) = 0$ *for all* $j = 1, \ldots, m$.

(c) $\nabla f(\mathbf{x}^*) = \sum_{j=1}^m \sigma_j(\mathbf{x}^*) \nabla g_j(\mathbf{x}^*)$, *that is, $(\mathbf{x}^*, \boldsymbol{\lambda}^*)$ is a KKT pair, with* $\lambda_j^* = \sigma_j(\mathbf{x}^*)$, $\forall j = 1, \ldots, m$.

(d) $\mathbf{x}^*$ *is a global minimizer of the polynomial* $f - f^* - \sum_{j=1}^m \sigma_j g_j$.

Proof. Part (a) is obvious from $\mathbf{x}^* \in \mathbf{K}$, and the polynomials σ_j's are s.o.s.

(b) From (3.36) and the fact that $\mathbf{x}^*$ is a global minimizer, we obtain

$$f(\mathbf{x}^*) - f^* = 0 = \sigma_0(\mathbf{x}^*) + \sum_{j=1}^m \sigma_j(\mathbf{x}^*) g_j(\mathbf{x}^*),$$

which in turn implies part (b) because $g_j(\mathbf{x}^*) \geq 0$ for all $j = 1, \ldots, m$, and the polynomials (σ_j) are all s.o.s., hence nonnegative. This also implies $\sigma_0(\mathbf{x}^*) = 0$.

(c) Differentiating and using the fact that the polynomials (σ_j) are s.o.s., and using part (b), yields part (c).

(d) From (3.36) we obtain

$$f - f^* - \sum_{j=1}^m \sigma_j \, g_j = \sigma_0 \geq 0,$$

because $\sigma_0 \in \mathbb{R}[\mathbf{x}]$ is s.o.s., and, using property (b),

$$f(\mathbf{x}^*) - f^* - \sum_{j=1}^m \sigma_j(\mathbf{x}^*) \, g_j(\mathbf{x}^*) = 0,$$

which shows that $\mathbf{x}^*$ is a global minimizer of $f - f^* - \sum_j \sigma_j g_j$. □

Theorem 3.20 implies that if $f - f^*$ has a representation (3.36) then:

(a) (3.36) should be interpreted as a *global optimality condition.*

(b) The s.o.s. polynomial coefficients $\{\sigma_j\} \subset \mathbb{R}[\mathbf{x}]$ should be interpreted as generalized Karush–Kuhn–Tucker Lagrange multipliers.

(c) The polynomial $f - f^* - \sum_j \sigma_j g_j$ is a generalized Lagrangian polynomial, with s.o.s. polynomial multipliers instead of nonnegative scalars. It is a s.o.s. (hence nonnegative on $\mathbb{R}^n$), vanishes at every global minimizer $\mathbf{x}^* \in \mathbf{K}$, and so $\mathbf{x}^*$ is also a global minimizer of this Lagrangian.

Note that, in the local KKT optimality conditions (3.34), only the constraints $g_j(\mathbf{x}) \geq 0$ binding at $\mathbf{x}^*$ have a possibly nontrivial associated Lagrange (scalar) multiplier λ_j^*. In contrast, in the global optimality condition (3.36), every constraint $g_j(\mathbf{x}) \geq 0$ has a possibly nontrivial s.o.s. polynomial Lagrange multiplier $\mathbf{x} \mapsto \sigma_j(\mathbf{x})$. Note that, if $g_j(\mathbf{x}^*) > 0$, then necessarily $\sigma_j(\mathbf{x}^*) = 0 = \lambda_j^*$, as in the local KKT optimality conditions.

In nonconvex optimization, a constraint $g_j(\mathbf{x}) \geq 0$ that is not binding at a global minimizer $\mathbf{x}^* \in \mathbf{K}$ may still be important, i.e., if removed from the definition of $\mathbf{K}$, then the global minimum f^* may decrease strictly. In this case, and in contrast to local KKT optimality conditions (3.34), g_j is necessarily involved in the representation (3.13) of $f - f^*$ (when the latter exists), hence with a nontrivial s.o.s. multiplier σ_j which vanishes at $\mathbf{x}^*$.

To see this, consider the following trivial example.

Example 3.21. Let $n = 1$ and consider the following problem:

$$f^* = \min_x \{ -x \,:\, 1/2 - x \geq 0,\ x^2 - 1 = 0\},$$

with optimal value $f^* = 1$ and global minimizer $x^* = -1$. The constraint $1/2 - x \geq 0$ is not binding at $x^* = -1$, but, if removed, the global minimum jumps to -1 with new global minimizer $x^* = 1$. In fact, we have the representation

$$f(x) - f^* = -(x+1) = (1/2 - x)(x+1)^2 + (x^2 - 1)(x + 3/2),$$

which shows the important role of the constraint $1/2 - x \geq 0$ in the representation of $f - f^*$, via its nontrivial multiplier $x \mapsto \sigma_1(x) = (x+1)^2$. Note also that $\sigma_1(x^*) = 0 = \lambda_1$ and $\sigma_2(x^*) = x^* + 3/2 = -1/2 = \lambda_2$ are the KKT Lagrange multipliers $(\lambda_1, \lambda_2) \in \mathbb{R}_+ \times \mathbb{R}$ in the local optimality conditions (3.34). Notice that the Lagrange multiplier λ_2 is not constrained in sign because it is associated with an equality constraint (and not an inequality constraint as for λ_1); similarly, σ_2 is not required to be a s.o.s.

3.7 Discrete optimization

In this section, we consider problem (3.1) with $\mathbf{K}$ a real variety. More precisely, in the definition (3.2) of $\mathbf{K}$, the g_i's are such that $\mathbf{K}$ can be rewritten as

$$\mathbf{K} = \{\mathbf{x} \in \mathbb{R}^n \,:\, g_i(\mathbf{x}) = 0,\ i = 1, \ldots, m;\ h_j(\mathbf{x}) \geq 0,\ j = 1, \ldots, s\}. \qquad (3.37)$$

Indeed, (3.37) is a particular case of (3.2) where some inequality constraints $g_i(\mathbf{x}) \geq 0$ and $-g_i(\mathbf{x}) \geq 0$ are present.

The polynomials $(g_i)_{i=1}^m \subset \mathbb{R}[\mathbf{x}]$ define an ideal $I = \langle g_1, \ldots, g_m\rangle$ of $\mathbb{R}[\mathbf{x}]$, and we will consider the case where I is a zero-dimensional *radical ideal*, that is, I is radical and the variety

$$V_{\mathbb{C}}(I) = \{\mathbf{x} \in \mathbb{C}^n \,:\, g_i(\mathbf{x}) = 0,\ i = 1, \ldots, m\}$$

is a finite set; see Section 1.6.

This is an important special case, as it covers all basic 0/1 optimization problems. For instance, if we let $g_i(\mathbf{x}) = x_i^2 - x_i$, for all $i = 1, \ldots, n$, then we recover Boolean optimization, in which case $V_{\mathbb{C}}(I) = \{0,1\}^n$ and the ideal $I = \langle x_1^2 - x_1, \ldots, x_n^2 - x_n\rangle$ is radical. Similarly, given $(r_i)_{i=1}^n \subset \mathbb{N}$ and a finite set of points $(x_{ij})_{j=1}^{r_i} \subset \mathbb{R}$, $i = 1, \ldots, n$, let

$$g_i(\mathbf{x}) = \prod_{j=1}^{r_i} (x_i - x_{ij}), \quad i = 1, \ldots, n.$$

Then we recover (bounded) integer optimization problems, in which case $V_{\mathbb{C}}(I)$ is the *grid* $\{(x_{1j_1}, x_{2j_2}, \ldots, x_{nj_n})\}$, where $j_k \in \{1, \ldots, r_k\}$ for all $k = 1, \ldots, n$.

Theorem 3.22. *Let $f \in \mathbb{R}[\mathbf{x}]$, $\mathbf{K} \subset \mathbb{R}^n$ be as in* (3.37)*, and let the ideal $I = \langle g_1, \ldots, g_m \rangle$ be radical and zero-dimensional. Let (ρ_i) and (ρ_i^*) be the optimal values of the semidefinite relaxations defined in* (3.14) *and* (3.15)*. Then there is some $i_0 \in \mathbb{N}$ such that $\rho_i = \rho_i^* = f^*$ for all $i \geq i_0$. In addition, both* (3.14) *and* (3.15) *have an optimal solution.*

Proof. The polynomial $f - f^*$ is nonnegative on $\mathbf{K}$. Therefore, by Theorem 1.27, there exist polynomials $(\sigma_k)_{k=1}^m \subset \mathbb{R}[\mathbf{x}]$ and s.o.s. polynomials $(\sigma_0, (v_j)_{j=1}^s) \subset \Sigma[\mathbf{x}]$ such that

$$f - f^* = \sigma_0 + \sum_{k=1}^m \sigma_k \, g_k + \sum_{j=1}^s v_j \, h_j.$$

Let d_1, d_2, d_3 be the degree of σ_0 and the maximum degree of the polynomials $(\sigma_k \, g_k)_{i=1}^m$ and $(v_j \, h_j)_{j=1}^s$ respectively, and let $2i_0 \geq \max[d_1, d_2, d_3]$.

Then $(f^*, (\sigma_k), (v_j))$ is a feasible solution of the relaxation (3.15) for $i = i_0$, and with value f^*, so that $\rho_{i_0}^* \geq f^*$. As we also have $\rho_i^* \leq \rho_i \leq f^*$ whenever the semidefinite relaxations are well defined, we conclude that $\rho_i^* = \rho_i = f^*$ for all $i \geq i_0$. Finally, let μ be the Dirac probability measure at some global minimizer $\mathbf{x}^* \in \mathbf{K}$ of problem (3.1), and let $\mathbf{y} \subset \mathbb{R}$ be the vector of its moments. Then $\mathbf{y}$ is feasible for all the semidefinite relaxations (3.14), with value f^*, which completes the proof. □

In fact one may refine Theorem 3.22 and remove the assumption that the ideal is radical. But now the dual relaxation (3.15) is not guaranteed to have an optimal solution.

Theorem 3.23. *Let $f \in \mathbb{R}[\mathbf{x}]$, $\mathbf{K} \subset \mathbb{R}^n$ be as in* (3.37)*, and assume that the ideal $I = \langle g_1, \ldots, g_m \rangle$ is such that*

$$V_{\mathbb{R}}(I) = \{\mathbf{x} \in \mathbb{R}^n \; : \; g_j(\mathbf{x}) = 0, \; j = 1, \ldots, m\}$$

is finite. Let (ρ_i) be the optimal values of the semidefinite relaxations defined in (3.14)*. Then there is some $i_0 \in \mathbb{N}$ such that $\rho_i = f^*$ for all $i \geq i_0$.*

Notice that in Theorem 3.23 one may even tolerate that I is not zero-dimensional provided that its real associated variety $V_{\mathbb{R}}(I)$ is finite.

3.7.1 Boolean optimization

It is worth noting that, in the semidefinite relaxations (3.14), the constraints $\mathbf{M}_r(g_i\mathbf{y}) = 0$ translate into simplifications via elimination of variables in the moment matrix $\mathbf{M}_r(\mathbf{y})$ and the localizing matrices $\mathbf{M}_r(h_j\mathbf{y})$. Indeed, consider for instance the Boolean optimization case, i.e., when $g_i(\mathbf{x}) = x_i^2 - x_i$ for all $i = 1, \ldots, n$. Then the constraints $\mathbf{M}_r(g_i\mathbf{y}) = \mathbf{0}$ for all $i = 1, \ldots, n$ simply state that, whenever $|\boldsymbol{\alpha}| \leq 2r$, one replaces every variable $y_{\boldsymbol{\alpha}}$ with the variable $y_{\boldsymbol{\beta}}$, where, for $k = 1, \ldots, n$,

$$\beta_k = \begin{cases} 0, & \text{if } \alpha_k = 0; \\ 1, & \text{otherwise.} \end{cases}$$

Indeed, with $x_i^2 = x_i$ for all $i = 1, \ldots, n$, one has $\mathbf{x}^{\boldsymbol{\alpha}} = \mathbf{x}^{\boldsymbol{\beta}}$, with $\boldsymbol{\beta}$ as above. For instance, with $n = 2$, we obtain

$$\mathbf{M}_2(\mathbf{y}) = \begin{bmatrix} y_{00} & y_{10} & y_{01} & y_{10} & y_{11} & y_{01} \\ y_{10} & y_{10} & y_{11} & y_{10} & y_{11} & y_{11} \\ y_{01} & y_{11} & y_{01} & y_{11} & y_{11} & y_{01} \\ y_{10} & y_{10} & y_{11} & y_{10} & y_{11} & y_{11} \\ y_{11} & y_{11} & y_{11} & y_{11} & y_{11} & y_{11} \\ y_{01} & y_{11} & y_{01} & y_{11} & y_{11} & y_{01} \end{bmatrix}.$$

In addition, every column (row) of $\mathbf{M}_r(\mathbf{y})$ corresponding to a monomial $\mathbf{x}^{\boldsymbol{\alpha}}$, with $\alpha_k > 1$ for some $k \in \{1, \ldots, n\}$, is identical to the column corresponding to the monomial $\mathbf{x}^{\boldsymbol{\beta}}$, with $\boldsymbol{\beta}$ as above. Hence, the constraint $\mathbf{M}_r(\mathbf{y}) \succeq \mathbf{0}$ reduces to the new constraint $\widehat{\mathbf{M}}_r(\mathbf{y}) \succeq \mathbf{0}$, with the new simplified moment matrix

$$\widehat{\mathbf{M}}_2(\mathbf{y}) = \begin{bmatrix} y_{00} & y_{10} & y_{01} & y_{11} \\ y_{10} & y_{10} & y_{11} & y_{11} \\ y_{01} & y_{11} & y_{01} & y_{11} \\ y_{11} & y_{11} & y_{11} & y_{11} \end{bmatrix}.$$

Theorem 3.22 has little practical value. For instance, in the case of Boolean optimization, we may easily show that $\rho_i = \rho_{i_0}$ for all $i \geq i_0 = n$. But in this case the simplified matrix $\widehat{\mathbf{M}}_{i_0}(\mathbf{y})$ has size $2^n \times 2^n$, and solving problem (3.1) by simple enumeration would be as efficient! However, in general, one obtains the exact global optimum f^* at some earlier relaxation with value $\rho_i = f^*$, i.e., with $i \ll i_0$.

Numerical experiments

One also reports the performance of GloptiPoly on a series of small-size combinatorial optimization problems (in particular, the MAXCUT problem). In Table 3.3 we first let GloptiPoly converge to the global optimum, in general extracting several solutions. The number of extracted solutions is reported in the column labelled "sol", while "dim" indicates that the problem could not be solved because of excessive memory requirement. Then, we slightly perturbed the criterion to be optimized in order to destroy the problem symmetry. Proceeding this way, the optimum solution is generically unique and convergence to the global optimum is easier.

Of course, the size of the combinatorial problems is relatively small and GloptiPoly cannot compete with *ad hoc* heuristics, which may solve problems

Problem	var	cstr	deg	CPU	order	sol
QP [35, Pb. 13.2.1.1]	4	4	2	0.10	1	1
QP [35, Pb. 13.2.1.2]	10	0	2	3.61	2	1
Max-Cut P_1 [35]	10	0	2	38.1	3	10
Max-Cut P_2 [35]	10	0	2	2.7	2	2
Max-Cut P_3 [35]	10	0	2	2.6	2	2
Max-Cut P_4 [35]	10	0	2	2.6	2	2
Max-Cut P_5 [35]	10	0	2	–	4	dim
Max-Cut P_6 [35]	10	0	2	2.6	2	2
Max-Cut P_7 [35]	10	0	2	44.3	3	4
Max-Cut P_8 [35]	10	0	2	2.6	2	2
Max-Cut P_9 [35]	10	0	2	49.3	3	6
Max-Cut cycle C_5 [5]	5	0	2	0.19	3	10
Max-Cut complete K_5 [5]	5	0	2	0.19	4	20
Max-Cut 5-node [5]	5	0	2	0.24	3	6
Max-Cut antiweb AW_9^2 [5]	9	0	2	–	4	dim
Max-Cut 10-node Petersen [5]	10	0	2	39.6	3	10
Max-Cut 12-node [5]	12	0	2	–	3	dim

Table 3.3: Discrete optimization problems. CPU times and semidefinite relaxation order required to reach global optimum and extract several solutions.

with many more variables. But these numerical experiments are reported only to show the potential of the method, as in most cases the global optimum is reached at the second semidefinite relaxation in the hierarchy.

3.7.2 Back to unconstrained optimization

We have seen in Section 3.2 that for the unconstrained optimization problem (3.1), the semidefinite relaxations (3.6) reduce to a single one, and with a 0-1 answer, depending on whether or not the polynomial $f - f^*$ is a sum of squares. Therefore, in general, according to Theorem 3.2, the SDP (3.6) provides only a lower bound on f^*.

However, if one knows *a priori* some bound M on the Euclidean norm $\|\mathbf{x}^*\|$ of a global minimizer $\mathbf{x}^* \in \mathbb{R}^n$, then it suffices to replace the original unconstrained problem (3.1) (where $\mathbf{K} = \mathbb{R}^n$) with a constrained problem where $\mathbf{K}$ is the basic semi-algebraic set

$$\mathbf{K} = \{\mathbf{x} \in \mathbb{R}^n : M^2 - \|\mathbf{x}\|^2 \geq 0\}.$$

It is immediate to verify that Assumption 1.16 holds, and, therefore, the machinery described in Section 3.3 applies, and the semidefinite relaxations (3.14) with $\mathbf{K}$ as above converge to f^*.

Another approach which avoids the *a priori* knowledge of this bound M consists of taking

$$\mathbf{K} = \{\mathbf{x} \in \mathbb{R}^n : \nabla f(\mathbf{x}) = 0\}, \tag{3.38}$$

since, if a global minimizer $\mathbf{x}^* \in \mathbb{R}^n$ exists, then necessarily $\nabla f(\mathbf{x}^*) = 0$, and in addition $\mathbf{x}^*$ is also a global minimizer of f on $\mathbf{K}$ defined in (3.38). However, convergence of the relaxation (3.14) has been proved for a compact set $\mathbf{K}$. Fortunately, the set $\mathbf{K}$ in (3.38) has nice properties. Indeed:

Proposition 3.24. *For almost all polynomials $f \in \mathbb{R}[\mathbf{x}]_d$, the gradient ideal $I_f = \left\langle \frac{\partial f}{\partial x_1}, \ldots, \frac{\partial f}{\partial x_n} \right\rangle$ is zero-dimensional and radical.*

And so we get the following result:

Theorem 3.25. *With $f \in \mathbb{R}[\mathbf{x}]_d$, and $\mathbf{K} \subset \mathbb{R}^n$ as in (3.38), consider the semidefinite relaxation defined in (3.14) with optimal value ρ_i. Let*

$$\mathcal{F}_d = \{f \in \mathbb{R}[\mathbf{x}]_d : \exists \mathbf{x}^* \in \mathbb{R}^n \text{ s.t. } f(\mathbf{x}^*) = f^* = \min\{f(\mathbf{x}) : \mathbf{x} \in \mathbb{R}^n\}\}.$$

Then, for almost all $f \in \mathcal{F}_d$, $\rho_i = f^*$ for some index i, i.e., finite convergence takes place.

Theorem 3.25 is a direct consequence of Proposition 3.24 and Theorem 3.22.

3.8 Global minimization of a rational function

In this section, we consider the problem

$$\rho^* = \inf\left\{\frac{p(\mathbf{x})}{q(\mathbf{x})} : \mathbf{x} \in \mathbf{K}\right\}, \tag{3.39}$$

where $p, q \in \mathbb{R}[\mathbf{x}]$ and $\mathbf{K}$ is defined as in (3.2). Note that, if p, q have no common zero on $\mathbf{K}$, then $\rho^* > -\infty$ only if q does not change sign on $\mathbf{K}$. Therefore, we will assume that q is strictly positive on $\mathbf{K}$.

Proposition 3.26. *The optimal value ρ^* of problem (3.39) is the same as the optimal value of the optimization problem*

$$\rho = \inf_{\mu \in \mathcal{M}(\mathbf{K})_+} \left\{\int_{\mathbf{K}} p \, d\mu : \int_{\mathbf{K}} q \, d\mu = 1\right\}. \tag{3.40}$$

Proof. Assume first that $\rho^* > -\infty$, so that $p(\mathbf{x}) \geq \rho^* q(\mathbf{x})$ for all $\mathbf{x} \in \mathbf{K}$, and let $\mu \in \mathcal{M}(\mathbf{K})_+$ be a feasible measure for problem (3.40). Then $\int_{\mathbf{K}} p\, d\mu \geq \rho^* \int_{\mathbf{K}} q\, d\mu$, leading to $\rho \geq \rho^*$.

Conversely, let $\mathbf{x} \in \mathbf{K}$ be fixed arbitrarily, and let $\mu \in \mathcal{M}(\mathbf{K})_+$ be the measure $q(\mathbf{x})^{-1}\delta_{\mathbf{x}}$, where $\delta_{\mathbf{x}}$ is the Dirac measure at the point $\mathbf{x} \in \mathbf{K}$. Then $\int_{\mathbf{K}} q\, d\mu = 1$, so that μ is a feasible measure for problem (3.40). Moreover, its value satisfies $\int_{\mathbf{K}} p\, d\mu = p(\mathbf{x})/q(\mathbf{x})$. As $\mathbf{x} \in \mathbf{K}$ was arbitrary, $\rho \leq \rho^*$, and the result follows. Finally, if $\rho^* = -\infty$ then from what precedes we also have $\rho = -\infty$. □

Let $2v_k$ or $2v_k - 1$ be the degree of the polynomial $g_k \in \mathbb{R}[\mathbf{x}]$ in the definition (3.2) of $\mathbf{K}$, for all $k = 1, \ldots, m$. Proceeding as in Section 3.3, we obtain the following semidefinite relaxation for $i \geq \max\{\deg p, \deg q, \max_k v_k\}$, which is analogous to (3.14):

$$\begin{aligned} \rho_i = \inf_{\mathbf{y}} \ & L_{\mathbf{y}}(p) \\ \text{s.t.} \ & \mathbf{M}_i(\mathbf{y}), \mathbf{M}_{i-v_k}(g_k\, \mathbf{y}) \succeq \mathbf{0}, \quad k = 1, \ldots, m, \\ & L_{\mathbf{y}}(q) = 1. \end{aligned} \tag{3.41}$$

Note that, in contrast to (3.14), where $y_{\mathbf{0}} = 1$, in general $y_{\mathbf{0}} \neq 1$ in (3.41). In fact, the last constraint $L_{\mathbf{y}}(q) = 1$ in (3.41) yields $y_{\mathbf{0}} = 1$ whenever $q = 1$, that is, problem (3.40) reduces to problem (3.3). The dual of problem (3.41) reads:

$$\begin{aligned} \rho_i^* = \sup_{\sigma_k, \lambda} \ & \lambda \\ \text{s.t.} \ & p - \lambda q = \sigma_0 + \sum_{k=1}^m \sigma_k\, g_k; \\ & \sigma_k \in \Sigma[\mathbf{x}]; \quad \deg \sigma_k \leq i - v_k, \quad k = 0, 1, \ldots, m \end{aligned} \tag{3.42}$$

(with $v_0 = 1$). Recall that $q > 0$ on $\mathbf{K}$.

Theorem 3.27. *Let* $\mathbf{K}$ *be as in* (3.2), *and let Assumption* 1.16 *hold. Consider the semidefinite relaxations* (3.41) *and* (3.42). *Then,*

(a) ρ^* *is finite and* $\rho_i^* \uparrow \rho^*$, $\rho_i \uparrow \rho^*$ *as* $i \to \infty$.
(b) *In addition, if* $\mathbf{K} \subset \mathbb{R}^n$ *has nonempty interior then* $\rho_i^* = \rho_i$ *for every* i.
(c) *Let* $\mathbf{x}^* \in \mathbf{K}$ *be a global minimizer of* p/q *on* $\mathbf{K}$. *If the polynomial* $p - \rho^* q \in \mathbb{R}[\mathbf{x}]$, *nonnegative on* $\mathbf{K}$, *has the representation* (3.13), *then both problems* (3.41) *and* (3.42) *have an optimal solution, and* $\rho_i^* = \rho_i = \rho^*$ *for all* $i \geq i_0$, *for some* $i_0 \in \mathbb{N}$.

Proof. (a) As $q > 0$ on $\mathbf{K}$ and $\mathbf{K}$ is compact, the rational function p/q attains its minimum ρ^* on $\mathbf{K}$ at some point $\mathbf{x}^* \in \mathbf{K}$. Let $\mathbf{y}$ be the vector of moments (up to order $2i$) of the measure $q(\mathbf{x}^*)^{-1}\delta_{\mathbf{x}^*}$. Obviously $\mathbf{y}$ is feasible for (3.41) with value $\rho_i = L_{\mathbf{y}}(p) = p(\mathbf{x}^*)/q(\mathbf{x}^*) = \rho^*$. Hence $\rho_i \leq \rho^*$ for every i, and, by weak duality,

$\rho_i^* \leq \rho_i \leq \rho^*$ for every i. On the other hand, observe that the polynomial $p - \rho^* q$ is nonnegative on $\mathbf{K}$. So, with $\epsilon > 0$ fixed arbitrarily, $p - (\rho^* - \epsilon)q = p - \rho^* q + \epsilon q > 0$ on $\mathbf{K}$. Therefore, by Theorem 1.17,

$$p - (\rho^* - \epsilon)q = \sigma_0 + \sum_{k=1}^{m} \sigma_k g_k$$

for some s.o.s. polynomials $(\sigma_k) \subset \Sigma[\mathbf{x}]$. And so $(\rho^* - \epsilon, (\sigma_k))$ is a feasible solution of (3.42) provided that i is large enough. But then $\rho_i^* \geq \rho^* - \epsilon$, which, combined with the inequality $\rho_i^* \leq \rho^*$ and the fact that $\epsilon > 0$ was arbitrary, yields the desired result.

(b) The equality $\rho_i^* = \rho_i$ follows from Slater's condition, which is satisfied for problem (3.41). Indeed, let $\mu \in \mathcal{M}(\mathbf{K})_+$ be a measure with uniform distribution on $\mathbf{K}$ and scaled to ensure that $\int q\, d\mu = 1$. (As $\mathbf{K}$ is compact, this is always possible.) Then the vector $\mathbf{y}$ of its moments is a strictly feasible solution of problem (3.41) ($\mathbf{M}_i(\mathbf{y}) \succ \mathbf{0}$, and $\mathbf{K}$ having a nonempty interior implies that $\mathbf{M}_{i-v_k}(g_k\, \mathbf{y}) \succ \mathbf{0}$ for all $k = 1, \ldots, m$). Thus, there is no duality gap, i.e., $\rho_i^* = \rho_i$, and the result follows from (a).

(c) If $p - \rho^* q$ has the representation (3.13), then, as we did for part (a), we can find a feasible solution to problem (3.42) with value ρ^* for all $i \geq i_0$, for some $i_0 \in \mathbb{N}$. Hence,

$$\rho_i^* = \rho_i = \rho^*, \quad \forall i \geq i_0.$$

Finally, (3.41) has an optimal solution $\mathbf{y}$. It suffices to take for $\mathbf{y}$ the vector of moments of the measure $q(\mathbf{x}^*)^{-1}\delta_{\mathbf{x}^*}$, i.e., $y_{\boldsymbol{\alpha}} = q(\mathbf{x}^*)^{-1}(\mathbf{x}^*)^{\boldsymbol{\alpha}}$, for all $\boldsymbol{\alpha} \in \mathbb{N}^n$. □

3.9 Exploiting symmetry

In this section we briefly describe how symmetry can be exploited to replace a semidefinite program invariant under the action of some group of permutations with a much simpler one. In particular, it can be applied to the semidefinite relaxations (3.14) when $f \in \mathbb{R}[\mathbf{x}]$ and the polynomials $(g_j) \subset \mathbb{R}[\mathbf{x}]$ that define $\mathbf{K}$ are all invariant under some group of permutations.

Let S_n be the space of $n \times n$ real symmetric matrices and let $\mathrm{Aut}(S_n)$ be the group of automorphisms on S_n. Let $\mathcal{G}$ be a finite group acting on $\mathbb{R}^n$ via $\rho_0 \colon \mathcal{G} \to \mathrm{GL}(\mathbb{R}^n)$, which in turn induces an action $\rho \colon \mathcal{G} \to \mathrm{Aut}(S_n)$ on S_n by $\rho(g)(\mathbf{X}) = \rho_0(g)^T \mathbf{X}\, \rho_0(g)$ for all $g \in \mathcal{G}$ and $\mathbf{X} \in S_n$.

Assume that $\rho_0(g)$ is orthonormal for every $g \in \mathcal{G}$. A matrix $\mathbf{X} \in \mathbb{R}^{n \times n}$ is said to be *invariant* under the action of $\mathcal{G}$ if $\rho(g)(\mathbf{X}) = \mathbf{X}$ for every $g \in \mathcal{G}$, and $\mathbf{X}$ is invariant if and only if $\mathbf{X}$ is an element of the *commutant algebra*

$$A^{\mathcal{G}} = \{\mathbf{X} \in \mathbb{R}^{n \times n} \;:\; \rho_0(g)\,\mathbf{X} = \mathbf{X}\,\rho_0(g),\ \forall g \in \mathcal{G}\}. \tag{3.43}$$

Of particular interest is the case when $\mathcal{G}$ is a subgroup of the group $\mathcal{P}_n$ of permutations of $\{1,\ldots,n\}$, in which case $\rho_0(g)(\mathbf{x}) = (x_{g(i)})$ for every $\mathbf{x} \in \mathbb{R}^n$, and $\rho(g)(\mathbf{X})_{ij} = X_{g(i),g(j)}$ for all $1 \le i,j \le n$.

For every $(i,j) \in \{1,\ldots,n\} \times \{1,\ldots,n\}$, the orbit $O_{\mathcal{G}}(i,j)$ under the action of $\mathcal{G}$ is the set of couples $\{(g(i),g(j)) : g \in \mathcal{G}\}$. With ω the number of orbits, and $1 \le \ell \le \omega$, define the $n \times n$ matrix $\widetilde{\mathbf{D}}_\ell$ by $(\widetilde{\mathbf{D}}_\ell)_{ij} = 1$ if (i,j) belongs to orbit ℓ, and 0 otherwise. Normalize to $\mathbf{D}_\ell = \widetilde{\mathbf{D}}_\ell / \sqrt{\langle \widetilde{\mathbf{D}}_\ell, \widetilde{\mathbf{D}}_\ell \rangle}$, for all $1 \le \ell \le \omega$, and define:

- The multiplication table

$$\mathbf{D}_i\,\mathbf{D}_j = \sum_{\ell=1}^{\omega} \gamma_{ij}^{\ell}\,\mathbf{D}_\ell, \quad i,j = 1,\ldots,\omega,$$

  for some $(\gamma_{ij}^{\ell}) \subset \mathbb{R}$.
- The $\omega \times \omega$ matrices $\mathbf{L}_1,\ldots,\mathbf{L}_\omega$ by:

$$(\mathbf{L}_k)_{ij} = \gamma_{kj}^{i}, \quad i,j,k = 1,\ldots,\omega.$$

Then the commutant algebra (3.43) reads

$$A^{\mathcal{G}} = \left\{ \sum_{\ell=1}^{\omega} x_\ell\,\mathbf{D}_\ell \;:\; x_\ell \in \mathbb{R} \right\}$$

with dimension $\dim A^{\mathcal{G}} = \omega$.

Exploiting symmetry in a semidefinite program is possible thanks to the following crucial property of the matrices $(\mathbf{D}_l)$:

Theorem 3.28. *The mapping $\mathbf{D}_\ell \mapsto \mathbf{L}_\ell$ is a $\star$-isomorphism, called the regular $\star$-representation of $A^{\mathcal{G}}$, and, in particular,*

$$\sum_{\ell=1}^{\omega} x_\ell\,\mathbf{D}_\ell \succeq 0 \quad \Longleftrightarrow \quad \sum_{\ell=1}^{\omega} x_\ell\,\mathbf{L}_\ell \succeq 0. \tag{3.44}$$

Application to semidefinite programming

Consider the semidefinite program

$$\sup_{\mathbf{X} \succeq 0} \left\{ \langle \mathbf{C}, \mathbf{X} \rangle \;:\; \langle \mathbf{A}_k, \mathbf{X} \rangle = b_k,\; k = 1,\ldots,p \right\} \tag{3.45}$$

and assume that it is invariant under the action of $\mathcal{G}$, that is, $\mathbf{C} \in A^{\mathcal{G}}$, and the feasible region is globally invariant, meaning that if $\mathbf{X}$ feasible in (3.45) then so is $\rho(g)(\mathbf{X})$ for every $g \in \mathcal{G}$.

By convexity, for every feasible $\mathbf{X}$ of (3.45), the matrix

$$\mathbf{X}_0 = \frac{1}{|\mathcal{G}|} \sum_{g \in \mathcal{G}} \rho(g)(\mathbf{X})$$

is feasible, invariant under the action of $\mathcal{G}$, and with the same objective value as $\mathbf{X}$. Therefore, we can include in the semidefinite program (3.45) the additional linear constraint $\mathbf{X} \in A^{\mathcal{G}}$ without affecting the optimal value.

Therefore, writing $\mathbf{X} = \sum_{l=1}^{\omega} x_l \mathbf{D}_l$ and setting

$$c_l = \langle \mathbf{C}, \mathbf{D}_l \rangle, \quad a_{kl} = \langle \mathbf{A}_k, \mathbf{D}_l \rangle, \quad \forall l = 1, \dots, \omega;\ k = 1, \dots, p,$$

the semidefinite program (3.45) has the same optimal value as

$$\sup_{\mathbf{x} \in \mathbb{R}^{\omega}} \left\{ \mathbf{c}'\mathbf{x} \ :\ \mathbf{a}_k'\mathbf{x} = b_k,\ k = 1, \dots, p;\ \sum_{l=1}^{\omega} x_l\, \mathbf{L}_l \succeq 0 \right\}. \tag{3.46}$$

Observe that in (3.45) we have n variables and an $n \times n$ positive semidefinite matrix $\mathbf{X}$, whereas in (3.46) we only have ω variables and an $\omega \times \omega$ positive semidefinite matrix.

3.10 Notes and sources

For a survey on semidefinite programming and its multiple applications, the interested reader is referred to Vandenberghe and Boyd (1996) [126].

Most of the material in this chapter is from Lasserre (2000–2006) [62, 63, 64, 65, 66, 67, 69]. Shor (1987, 1998) [119, 120] was the first to prove that the global minimization of a univariate polynomial is a convex optimization problem. Later, Nesterov (2000) [92] defined exact semidefinite formulations for the univariate case, while converging semidefinite relaxations for the general multivariate case were treated in Lasserre (2000–2002) [62, 63, 65, 66] and Parrilo (2000, 2003) [97, 98]. De Klerk et al. (2006) [28] provided a polynomial time approximation scheme (PTAS) for minimizing polynomials of fixed degree on the simplex.

Building on Shor's ideas, Sherali and Adams (1990, 1999) [116, 117] and Sherali et al. (1992) [118] proposed their RLT (Reformulation-Linearization Technique), the earliest hierarchy of LP relaxations for polynomial optimization, and proved finite convergence for 0/1 problems. Other linear relaxations for 0/1 programs have also been proposed by Balas et al. (1993) [8] and Lovász and Schrijver (1991) [84], while a hierarchy of semidefinite relaxations for 0/1 problems were first proposed by Lovász and Schrijver (1991) [84], who also proved finite convergence.

A comparison between semidefinite and linear relaxations (in the spirit of Sherali and Adams) for general polynomial optimization problems is made in Lasserre (2002) [64], in the light of the results of Chapter 1 on the representation of positive polynomials. For 0/1 problems, Laurent (2003) [79] compares the linear relaxations of Sherali and Adams (1990) [116] and Lovász and Schrijver (1991) [84], and the semidefinite relaxations of Lovász and Schrijver (1991) [84] and Lasserre (2002) [65] within the common framework of the moment matrix, and proved that the latter semidefinite relaxations are the strongest. This has motivated research on integrality gaps for difficult combinatorial optimization problems. (The integrality gap measures the ratio between the optimal value of the relaxation and that of the problem to solve.) In particular, Chlamtac (2007) [22] and Chlamtac and Singh (2008) [23] showed that the hierarchy of semidefinite relaxations provides improved approximation algorithms for finding independent sets in graphs and for colouring problems. See also the related work of Schoenebeck (2008) [112].

Theorem 3.22 is from Parrilo (2003) [98], an extension to the general setting (3.37) of the grid case studied in Lasserre (2002) [65, 66]; see also Laurent (2007) [81] for refinements. Recent approaches to unconstrained optimization via optimizing on the gradient ideal appear in Hanzon and Jibetean (2003) [39] with matrix methods and in Jibetean and Laurent (2005) [50] with semidefinite programming. In both approaches, one slightly perturbates p (of degree say $2d$) by adding monomials $\{x_i^{2d+2}\}$ with a small coefficient ϵ and obtain a sequence of polynomials f_ϵ with the property that $V = \{\mathbf{x} : \nabla f_\epsilon(\mathbf{x}) = 0\}$ is finite and the minima f_ϵ^* converge to f^*. In particular, $\{\partial f_\epsilon / \partial x_i\}$ form a Gröbner basis of the ideal they generate. On the other hand, Proposition 3.24 is Proposition 1 in Nie et al. (2006) [94]. To handle the case where no global minimizer exists, Schweighofer (2006) [115] uses s.o.s. and the concept of gradient tentacles and Vui and Son (2009) [128] uses the truncated tangency variety.

Polynomials satisfying sparsity patterns were investigated in Kojima et al. (2005) [54] and the sparse semidefinite relaxations (3.26) were first proposed in Waki et al. (2006) [129] as a heuristic to solve global optimization problems with a large number of variables and which satisfy some sparsity pattern. Their convergence in Theorem 3.16 was proved in Lasserre (2006) [70] if the sparsity pattern satifies the running intersection property. The sparse relaxations have been implemented in the SparsePOP software of Waki et al. (2009) [130], and numerical experiments show that one may then solve global optimization problems with $n = 1000$ variables for which even the first nonsparse semidefinite relaxation of the hierarchy (3.14) cannot be implemented. Kim et al. (2009) [51] provide a nice application for sensor network localization. Section 3.8 is inspired from Jibetean and de Klerk (2006) [49], the first to prove Theorem 3.27.

Section 3.9 is inspired from Laurent (2008) [83], and Theorem 3.28 is from de Klerk et al. (2007) [29]. For exploiting symmetry in the context of sums of squares and semidefinite programming, see also the work of Gaterman and Parrilo (2004) [36] and Vallentin (2007) [125]. For instance, these ideas have been used sucessfully in coding theroy for large error correcting codes based on computing the stability number of some related graph; see, e.g., Laurent (2007) [82] and Schrijver (2005) [113].

Chapter 4

Convexity in Polynomial Optimization

If on the one hand practice seems to reveal that convergence of the semidefinite relaxations (3.14) is often fast and even finite, on the other hand we have seen that their size grows rapidly with the rank in the hierarchy. And so, if sparsity in the original problem data is not exploited, the approach is limited to small or to medium size problems only. On the other hand, it is well known that a large class of convex optimization problems can be solved efficiently. Therefore, as the moment approach is dedicated to solving difficult nonconvex (most of the time NP-hard) problems, it should have the highly desirable feature to somehow *recognize* "easy" problems like convex ones. That is, when applied to such easy problems it should show some significant improvement or a particular nice behavior not necessarily valid in the general case. This is the issue that we investigate in this chapter.

4.1 Convexity and polynomials

We first introduce some material which shows that in the presence of convexity some of the results presented in Chapter 1 have a nice specialization.

4.1.1 Algebraic certificates of convexity

We first consider the problem of detecting whether some given basic semi-algebraic set $\mathbf{K}$ as in (3.2) is convex. By detecting we mean that if $\mathbf{K}$ is convex then one may obtain a *certificate* (or a proof) of convexity by some algorithm. The geometric characterization of convexity,

$$\lambda \mathbf{x} + (1-\lambda)\mathbf{y} \in \mathbf{K}, \quad \forall \mathbf{x}, \mathbf{y} \in \mathbf{K}, \quad \lambda \in (0,1), \tag{4.1}$$

is not a certificate because it cannot be checked by an algorithm.

Given the basic closed semi-algebraic set $\mathbf{K}$ defined in (1.10), let $\widehat{\mathbf{K}} = \mathbf{K} \times \mathbf{K}$, and so $\widehat{\mathbf{K}}$ is the basic closed semi-algebraic set

$$\widehat{\mathbf{K}} = \big\{(\mathbf{x}, \mathbf{y}) \,:\, \widehat{g}_j(\mathbf{x}, \mathbf{y}) \geq 0,\ j = 1, \ldots, 2m\big\}, \tag{4.2}$$

where

$$(\mathbf{x},\mathbf{y}) \longmapsto \widehat{g}_j(\mathbf{x},\mathbf{y}) = g_j(\mathbf{x}), \quad \text{for } j = 1,\dots,m,$$
$$(\mathbf{x},\mathbf{y}) \longmapsto \widehat{g}_j(\mathbf{x},\mathbf{y}) = g_{j-m}(\mathbf{y}), \quad \text{for } j = m+1,\dots,2m,$$

and let $P(\widehat{g}) \subset \mathbb{R}[\mathbf{x},\mathbf{y}]$ be the preordering associated with the polynomials $\widehat{g}_j$ that define $\widehat{\mathbf{K}}$ in (4.2), i.e.,

$$P(\widehat{g}) = \left\{ \sum_{J\subseteq\{1,\dots,2m\}} \left(\prod_{k\in J} \widehat{g}_k \right) \sigma_J \,:\, \sigma_J \in \Sigma[\mathbf{x},\mathbf{y}] \right\}. \tag{4.3}$$

Our algebraic certificate of convexity is a follows.

Theorem 4.1. *Let $\mathbf{K} \subset \mathbb{R}^n$ be the basic closed semi-algebraic set defined in* (1.10). *Then $\mathbf{K}$ is convex if and only if, for every $j = 1,\dots,m$ and all $(\mathbf{x},\mathbf{y}) \in \mathbb{R}^n \times \mathbb{R}^n$,*

$$\theta_j(\mathbf{x},\mathbf{y})\, g_j\left(\frac{\mathbf{x}+\mathbf{y}}{2}\right) = g_j\left(\frac{\mathbf{x}+\mathbf{y}}{2}\right)^{2p_j} + h_j(\mathbf{x},\mathbf{y}) \tag{4.4}$$

for some polynomials $\theta_j, h_j \in P(\widehat{g})$ and some integer $p_j \in \mathbb{N}$.

Proof. The set $\mathbf{K}$ is convex if and only if (4.1) holds, or, equivalently, if and only if $(\mathbf{x}+\mathbf{y})/2 \in \mathbf{K}$ whenever $\mathbf{x},\mathbf{y} \in \mathbf{K}$. That is, if and only if, for every $j = 1,\dots,m$,

$$g_j\left(\frac{\mathbf{x}+\mathbf{y}}{2}\right) \geq 0, \quad \forall (\mathbf{x},\mathbf{y}) \in \widehat{\mathbf{K}}. \tag{4.5}$$

But then (4.4) is just an application of Stengle's Nichtnegativstellensatz, i.e., Theorem 1.14(a), applied to (4.5). □

The polynomials $\theta_j, h_j \in P(\widehat{g})$, $j = 1,\dots,m$, obtained in (4.4) indeed provide an obvious algebraic certificate of convexity for $\mathbf{K}$. This is because if (4.4) holds then for all $\mathbf{x},\mathbf{y} \in \mathbf{K}$ one has $\theta_j(\mathbf{x},\mathbf{y}) \geq 0$ and $h_j(\mathbf{x},\mathbf{y}) \geq 0$ because $\theta_j, h_j \in P(\widehat{g})$; and so $\theta_j(\mathbf{x},\mathbf{y})g_j((\mathbf{x}+\mathbf{y})/2) \geq 0$. Therefore, if $\theta_j(\mathbf{x},\mathbf{y}) > 0$ then $g_j((\mathbf{x}+\mathbf{y})/2) \geq 0$, whereas if $\theta_j(\mathbf{x},\mathbf{y}) = 0$ then $g_j((\mathbf{x}+\mathbf{y})/2)^{2p_j} = 0$, which in turn implies that $g_j((\mathbf{x}+\mathbf{y})/2) = 0$. Hence, for every $j = 1,\dots,m$, $g_j((\mathbf{x}+\mathbf{y})/2) \geq 0$ for all $\mathbf{x},\mathbf{y} \in \mathbf{K}$, which implies that $\mathbf{K}$ is convex.

In principle, the algebraic certificate can be obtained numerically. Indeed, there is a bound on $p_j \in \mathbb{N}$ and the degrees of the s.o.s. weights σ_J in the representation (4.3) of the polynomial certificates $\theta_j, h_j \in P(\widehat{g})$ in (4.4); see the discussion just after Theorem 1.14. Therefore, checking whether (4.4) holds reduces to checking whether some semidefinite program has a feasible solution. However, the bound is so large that for practical implementation one should proceed as follows. Fix an *a priori* bound M on $p_j \in \mathbb{N}$ and on the degrees of the s.o.s. polynomial weights σ_J that define $h_j, \theta_j \in P(\widehat{g})$; then check whether (4.4) holds true by solving the associated semidefinite program. If $\mathbf{K}$ is convex and the degrees of the certificates are small, then by increasing M one eventually finds a feasible solution. However, in practice such a certificate can be obtained only up to some machine precision, because of numerical inaccuracies inherent to semidefinite programming solvers.

We next provide another algebraic certificate of convexity for the class of basic closed semi-algebraic sets $\mathbf{K}$ whose defining polynomials (g_j) satisfy the following nondegeneracy assumption on the boundary $\partial\mathbf{K}$.

Assumption 4.2 (Nondegeneracy)**.** For a basic closed semi-algebraic set $\mathbf{K} \subset \mathbb{R}^n$ as in (1.10), the polynomials $(g_j) \subset \mathbb{R}[\mathbf{x}]$ satisfy the nondegeneracy property if, for every $j = 1, \ldots, m$, $\nabla g_j(\mathbf{x}) \neq 0$ whenever $\mathbf{x} \in \mathbf{K}$ and $g_j(\mathbf{x}) = 0$.

We first have the following characterization of convexity:

Lemma 4.3. *Let* $\mathbf{K}$ *be as in* (1.10) *and let Slater's condition*[1] *and Assumption* 4.2 *hold. Then* $\mathbf{K}$ *is convex if and only if, for every* $j = 1, \ldots, m$,

$$\langle \nabla g_j(\mathbf{y}), \mathbf{x} - \mathbf{y} \rangle \geq 0, \quad \forall \mathbf{x}, \mathbf{y} \in \mathbf{K} \text{ with } g_j(\mathbf{y}) = 0. \tag{4.6}$$

Proof. The *only if* part is obvious. Indeed, if $\langle \nabla g_j(\mathbf{y}), \mathbf{x} - \mathbf{y} \rangle < 0$ for some $\mathbf{x} \in \mathbf{K}$ and $\mathbf{y} \in \mathbf{K}$ with $g_j(\mathbf{y}) = 0$, then there is some $\bar{t} > 0$ such that $g_j(\mathbf{y} + t(\mathbf{x} - \mathbf{y})) < 0$ for all $t \in (0, \bar{t})$ and so the point $\mathbf{x}' = t\mathbf{x} + (1-t)\mathbf{y}$ does not belong to $\mathbf{K}$, which in turn implies that $\mathbf{K}$ is not convex.

For the *if* part, (4.6) implies that at every point of the boundary there exists a supporting hyperplane for $\mathbf{K}$. As $\mathbf{K}$ is closed with nonempty interior, the result follows from [111, Theorem 1.3.3]. □

As a consequence, we obtain the following algebraic certificate of convexity.

Corollary 4.4 (Algebraic certificate of convexity)**.** *Let* $\mathbf{K}$ *be as in* (1.10), *and let Slater's condition and Assumption* 4.2 *hold. Then* $\mathbf{K}$ *is convex if and only if, for every* $j = 1, \ldots, m$,

$$h_j(\mathbf{x}, \mathbf{y}) \, \langle \nabla g_j(\mathbf{y}), \mathbf{x} - \mathbf{y} \rangle = \langle \nabla g_j(\mathbf{y}), \mathbf{x} - \mathbf{y} \rangle^{2\ell} + \theta_j(\mathbf{x}, \mathbf{y}) + \varphi_j(\mathbf{x}, \mathbf{y}) \, g_j(\mathbf{y}) \tag{4.7}$$

for some integer $\ell \in \mathbb{N}$, *some polynomial* $\varphi_j \in \mathbb{R}[\mathbf{x}, \mathbf{y}]$ *and some polynomials* h_j, θ_j *in the preordering of* $\mathbb{R}[\mathbf{x}, \mathbf{y}]$ *generated by the family of polynomials* $(g_k(\mathbf{x}), g_p(\mathbf{y}))$, $k, p \in \{1, \ldots, m\}$, $p \neq j$.

Proof. By Lemma 4.3, $\mathbf{K}$ is convex if and only if, for every $j = 1, \ldots, m$, the polynomial $(\mathbf{x}, \mathbf{y}) \mapsto \langle \nabla g_j(\mathbf{y}), \mathbf{x} - \mathbf{y} \rangle$ is nonnegative on the set Ω_j defined by

$$\Omega_j = \{(\mathbf{x}, \mathbf{y}) \in \mathbf{K} \times \mathbf{K} \, : \, g_j(\mathbf{y}) = 0\}. \tag{4.8}$$

Then (4.7) follows from Theorem 1.14(a). □

---

[1] Recall that *Slater's condition* holds if there exists $\mathbf{x}_0 \in \mathbf{K}$ such that $g_j(\mathbf{x}_0) > 0$ for all $j = 1, \ldots, m$.

So if $\mathbf{K}$ is convex then $(\ell, h_j, \theta_j, \varphi_j)$ provides us with the desired certificate of convexity, which, in principle, can be obtained numerically as for the algebraic certificate (4.4). However, in practice, such a certificate can be obtained only up to some machine precision, because of numerical inaccuracies inherent to semidefinite programming solvers. See the discussion after Theorem 4.1.

Observe that, in Corollary 4.4, $\mathbf{K}$ is not necessarily compact. For compact basic semi-algebraic sets $\mathbf{K}$ that satisfy Assumption 1.16, one provides the following certificate.

Assumption 4.5 (Certificate of convexity). Slater's condition and Assumption 4.2 hold for the set $\mathbf{K} \subset \mathbb{R}^n$ in (1.10). In addition, for every $j = 1, \ldots, m$, the polynomial g_j satisfies

$$\begin{aligned}(\mathbf{x}, \mathbf{y}) \longmapsto \left\langle \nabla g_j \mathbf{y}), \mathbf{x} - \mathbf{y} \right\rangle = &\sum_{k=0}^{m} \sigma_{jk}\, g_k(\mathbf{x}) \\ &+ \sum_{k=0,\, k\neq j}^{m} \psi_{jk}\, g_k(\mathbf{y}) + \psi_j\, g_j(\mathbf{y}),\end{aligned} \tag{4.9}$$

for some s.o.s. polynomials (σ_{jk}), $(\psi_{jk})_{k\neq j} \subset \Sigma[\mathbf{x}, \mathbf{y}]$ and some $\psi_j \in \mathbb{R}[\mathbf{x}, \mathbf{y}]$.

If Assumption 4.5 holds, then $\mathbf{K}$ is convex because obviously (4.9) implies (4.6), which, in turn, by Lemma 4.3, implies that $\mathbf{K}$ is convex.

By fixing an *a priori* bound $2d_j$ on the polynomial $(\sigma_{jk} g_k, \psi_{jk} g_k, \psi_j g_j)$, checking whether (4.9) holds reduces to solving a semidefinite program, simpler than for checking whether (4.7) holds. For instance, for every $j = 1, \ldots, m$, it suffices to solve the semidefinite program (recall that $r_k = \lceil (\deg g_k)/2 \rceil$, $k = 1 \ldots, m$)

$$\begin{aligned}\rho_j = \min_{\mathbf{z}} \;& L_{\mathbf{z}}(\langle \nabla g_j(\mathbf{y}), \mathbf{x} - \mathbf{y} \rangle) \\ \text{s.t. } & \mathbf{M}_{d_j}(\mathbf{z}) \succeq 0, \\ & \mathbf{M}_{d_j - r_k}(g_k(\mathbf{x})\, \mathbf{z}) \succeq 0, \quad k = 1, \ldots, m, \\ & \mathbf{M}_{d_j - r_k}(g_k(\mathbf{y})\, \mathbf{z}) \succeq 0, \quad k = 1, \ldots, m;\; k \neq j, \\ & \mathbf{M}_{d_j - r_j}(g_j(\mathbf{y})\, \mathbf{z}) = 0, \\ & y_0 = 1.\end{aligned} \tag{4.10}$$

If $\rho_j = 0$ for every $j = 1, \ldots, m$, then Assumption 4.5 holds. However, again because of the numerical inaccuracies inherent to the SDP solvers, one would only get $\rho_j \approx 0$; and so, this certificate of convexity is valid only up to machine precision.

Example 4.6. Consider the following simple illustrative example in $\mathbb{R}^2$:

$$\mathbf{K} = \left\{\mathbf{x} \in \mathbb{R}^2 \;:\; x_1 x_2 - 1/4 \geq 0;\; 0.5 - (x_1 - 0.5)^2 - (x_2 - 0.5)^2 \geq 0\right\}. \tag{4.11}$$

Obviously $\mathbf{K}$ is convex but its defining polynomial $\mathbf{x} \mapsto g_1(\mathbf{x}) = x_1x_2 - 1/4$ is not concave.

With $d_1 = 3$, solving (4.10) using GloptiPoly 3 yields the optimal value $\rho_1 \approx -4.58.10^{-11}$, which, in view of the machine precision for the SDP solvers used in GloptiPoly, could be considered to be zero, but of course with no guarantee. For $j = 2$ there is no test to perform because $-g_2$ being quadratic and convex yields

$$\langle \nabla g_2(\mathbf{y}), \mathbf{x}-\mathbf{y}\rangle = g_2(\mathbf{x}) - g_2(\mathbf{y}) + \underbrace{(\mathbf{x}-\mathbf{y})'\big(-\nabla^2 g_2(\mathbf{y})\big)(\mathbf{x}-\mathbf{y})}_{\text{s.o.s.}}, \qquad (4.12)$$

which is in the form (4.9) with $d_2 = 1$, and so $\rho_2 = 0$.

4.1.2 Representation of convex polynomials

If $C \subseteq \mathbb{R}^n$ is a nonempty convex set, a function $f\colon C \to \mathbb{R}$ is convex on C if

$$f\big(\lambda x + (1-\lambda)y\big) \leq \lambda f(x) + \big(1-\lambda\big)f(y), \quad \forall \lambda \in (0,1),\ x, y \in C.$$

Similarly, f is strictly convex on C if the above inequality is strict for every $x, y \in C$, $x \neq y$, and all $\lambda \in (0,1)$.

If $C \subseteq \mathbb{R}^n$ is an open convex set and f is twice differentiable on C, then f is convex on C if and only if its Hessian[2] $\nabla^2 f$ is positive semidefinite on C (denoted $\nabla^2 f \succeq 0$ on C). Finally, if $\nabla^2 f$ is positive definite on C (denoted $\nabla^2 f \succ 0$ on C) then f is strictly convex on C.

Definition 4.7. A polynomial $f \in \mathbb{R}[\mathbf{x}]_{2d}$ is said to be *s.o.s.-convex* if its Hessian $\nabla^2 f$ is a s.o.s. matrix polynomial, that is, $\nabla^2 f = \mathbf{L}\mathbf{L}'$ for some real matrix polynomial $\mathbf{L} \in \mathbb{R}[\mathbf{x}]^{n\times s}$ (for some $s \in \mathbb{N}$).

Example 4.8. A polynomial $f \in \mathbb{R}[\mathbf{x}]$ is said to be *separable* if $f = \sum_{i=1}^n f_i$ for some univariate polynomials $f_i \in \mathbb{R}[x_i]$. Thus, every convex separable polynomial f is s.o.s.-convex because its Hessian $\nabla^2 f(\mathbf{x})$ is a positive semidefinite diagonal matrix, with all diagonal entries $(f_i''(x_i))_i$ nonnegative for all $\mathbf{x} \in \mathbb{R}^n$. Hence, being univariate and nonnegative, f_i'' is s.o.s. for every $i = 1, \ldots, n$, from which one easily concludes that f is s.o.s.-convex.

An important feature of s.o.s.-convexity is that it can be checked numerically by solving a semidefinite program. Of course every s.o.s.-convex polynomial is convex. Also, the s.o.s.-convex polynomials have the following interesting property.

Lemma 4.9. *If a symmetric matrix polynomial* $\mathbf{P} \in \mathbb{R}[\mathbf{x}]^{r\times r}$ *is s.o.s. then the double integral*

$$(\mathbf{x}, \mathbf{u}) \longmapsto \mathbf{F}(\mathbf{x}, \mathbf{u}) = \int_0^1 \int_0^t \mathbf{P}\big(\mathbf{u} + s(\mathbf{x}-\mathbf{u})\big)\, ds\, dt \qquad (4.13)$$

is also a symmetric s.o.s. matrix polynomial $\mathbf{F} \in \mathbb{R}[\mathbf{x}, \mathbf{u}]^{r\times r}$.

[2] Recall that the *Hessian* $\nabla^2 f(\mathbf{x})$ is the $n \times n$ symmetric matrix whose entry (i, j) is $\partial^2 f/\partial x_i \partial x_j$ evaluated at $\mathbf{x}$.

Proof. Writing $\mathbf{P} = (p_{ij})_{1\leq i,j\leq n}$ with $p_{ij} \in \mathbb{R}[\mathbf{x}]$ for every $1 \leq i,j \leq n$, let $d = \max_{i,j} \deg p_{ij}$. With $\mathbf{x},\mathbf{u}$ fixed, introduce the univariate matrix polynomial $s \mapsto \mathbf{Q}(s) = \mathbf{P}(\mathbf{u}+s(\mathbf{x}-\mathbf{u}))$ so that $\mathbf{Q} = (q_{ij})$ where $q_{ij} \in \mathbb{R}[s]$ has degree at most d for every $1 \leq i,j \leq n$. Observe that

$$\int_0^1\int_0^t \mathbf{P}\big(\mathbf{u}+s(\mathbf{x}-\mathbf{u})\big)\,ds\,dt = \int_0^1\int_0^t \mathbf{Q}(s)\,ds\,dt = \int_\Delta \mathbf{Q}(s)\,d\mu,$$

where μ is the measure uniformly distributed on the set $\Delta = \{(s,t) : 0 \leq t \leq 1;\ 0 \leq s \leq t\}$ with $\mu(\Delta) = 1/2$. By an extension of Tchakaloff's theorem ([124]), there exists a measure φ finitely supported on, say m, points $\{(s_k,t_k)\}_{k=1}^m \subset \Delta$ whose moments up to degree d match exactly those of μ. That is, there exist positive weights h_k, $k = 1,\dots,m$, such that

$$\int_\Delta f\,d\mu = \sum_{k=1}^m h_k\, f(s_k,t_k), \quad \forall f \in \mathbb{R}[s,t],\ \deg f \leq d.$$

So let $\mathbf{x},\mathbf{u} \in \mathbb{R}^n$ be fixed. If $\mathbf{P} = \mathbf{L}\mathbf{L}'$ for some $\mathbf{L} \in \mathbb{R}[\mathbf{x}]^{r\times q}$, one obtains:

$$\begin{aligned}\int_0^1\int_0^t \mathbf{P}\big(\mathbf{u}+s(\mathbf{x}-\mathbf{u})\big)\,ds\,dt &= \sum_{k=1}^m h_k\,\mathbf{Q}(s_k)\\ &= \sum_{k=1}^m h_k \mathbf{L}\big(\mathbf{u}+s_k(\mathbf{x}-\mathbf{u})\big)\,\mathbf{L}'\big(\mathbf{u}+s_k(\mathbf{x}-\mathbf{u})\big)\\ &= \mathbf{A}\mathbf{A}'\end{aligned}$$

for some $\mathbf{A} \in \mathbb{R}[\mathbf{x},\mathbf{u}]^{r\times q}$. □

And, as a consequence,

Lemma 4.10. *For a polynomial $f \in \mathbb{R}[\mathbf{x}]$ and all $\mathbf{x},\mathbf{u} \in \mathbb{R}^n$:*

$$\begin{aligned} f(\mathbf{x}) &= f(\mathbf{u}) + \nabla f(\mathbf{u})'(\mathbf{x}-\mathbf{u})\\ &\quad + (\mathbf{x}-\mathbf{u})'\underbrace{\int_0^1\int_0^t \nabla^2 f\big(\mathbf{u}+s(\mathbf{x}-\mathbf{u})\big)\,ds\,dt}_{\mathbf{F}(\mathbf{x},\mathbf{u})}(\mathbf{x}-\mathbf{u}).\end{aligned}$$

Hence if f is s.o.s.-convex and $f(\mathbf{u}) = 0, \nabla f(\mathbf{u}) = 0$, then f is a s.o.s. polynomial.

Let $\mathbf{K}$ be the basic semi-algebraic set in (1.10), $Q(g)$ be as in (1.12), and define $Q_c(g) \subset Q(g)$ to be the set

$$Q_c(g) = \left\{\sigma_0 + \sum_{j=1}^m \lambda_j\, g_j \;:\; \lambda \in \mathbb{R}_+^m;\ \sigma_0 \in \Sigma[\mathbf{x}],\ \sigma_0 \text{ convex}\right\}. \tag{4.14}$$

The set $Q_c(g)$ is a specialization of the quadratic module $Q(g)$ in the convex case, in that the weights associated with the g_j's are nonnegative scalars, i.e., s.o.s. polynomials of degree 0, and the s.o.s. polynomial σ_0 is convex. In particular, every $f \in Q_c(g)$ is nonnegative on $\mathbf{K}$. Let $\mathbf{F_K} \subset \mathbb{R}[\mathbf{x}]$ be the convex cone of convex polynomials nonnegative on $\mathbf{K}$. Recall that $\|f\|_1$ denotes the l_1-norm of the vector of coefficients, i.e., $\|f\|_1 = \sum_\alpha |f_\alpha|$.

Theorem 4.11. *Let* $\mathbf{K}$ *be as in* (1.10), *Slater's condition hold and* g_j *be concave for every* $j = 1, \ldots, m$. *Then, with* $Q_c(g)$ *as in* (4.14), *the set* $Q_c(g) \cap \mathbf{F_K}$ *is dense in* $\mathbf{F_K}$ *for the* l_1*-norm* $\|\cdot\|_1$. *In particular, if* $\mathbf{K} = \mathbb{R}^n$ *(so that* $\mathbf{F}_{\mathbb{R}^n} = \mathbf{F}$ *is now the set of nonnegative convex polynomials), then* $\Sigma[\mathbf{x}] \cap \mathbf{F}$ *is dense in* $\mathbf{F}$.

Theorem 4.11 states that if f is convex and nonnegative on $\mathbf{K}$ (including the case $\mathbf{K} \equiv \mathbb{R}^n$) then one may approximate f by a sequence $(f_{\epsilon r}) \subset Q_c(g) \cap \mathcal{F}_\mathbf{K}$ with $\|f - f_{\epsilon r}\|_1 \to 0$ as $\epsilon \to 0$ (and $r \to \infty$). For instance, with $r_0 = \lfloor (\deg f)/2 \rfloor + 1$, the polynomial $f_{\epsilon r}$ may be defined as $\mathbf{x} \mapsto f_{\epsilon r}(\mathbf{x}) = f + \epsilon(\theta_{r_0}(\mathbf{x}) + \theta_r(\mathbf{x}))$ with θ_r as in (1.6). Observe that Theorem 4.11 provides f with a *certificate* of nonnegativity on $\mathbf{K}$. Indeed, let $\mathbf{x} \in \mathbf{K}$ be fixed arbitrarily. Then, as $f_{\epsilon r} \in Q_c(g)$, one has $f_{\epsilon r}(\mathbf{x}) \geq 0$. Letting $\epsilon \downarrow 0$ yields $0 \leq \lim_{\epsilon \to 0} f_{\epsilon r}(\mathbf{x}) = f(\mathbf{x})$. And, as $\mathbf{x} \in \mathbf{K}$ was arbitrary, $f \geq 0$ on $\mathbf{K}$.

For the class of s.o.s.-convex polynomials, we have a more precise result:

Theorem 4.12. *Let* $\mathbf{K}$ *be as in* (1.10) *and let Slater's condition hold. Let* $f \in \mathbb{R}[\mathbf{x}]$ *be such that* $f^* = \min_\mathbf{x}\{f(\mathbf{x}) \,:\, \mathbf{x} \in \mathbf{K}\} = f(\mathbf{x}^*)$ *for some* $\mathbf{x}^* \in \mathbf{K}$. *If* f *and* $(-g_j)_{j=1}^m$ *are s.o.s.-convex, then* $f - f^* \in Q_c(g)$.

Proof. As Slater's condition holds, there exists a vector of KKT Lagrange multipliers $\boldsymbol{\lambda} \in \mathbb{R}^m_+$ such that the Lagrangian polynomial

$$L_f = f - f^* - \sum_{j=1}^m \lambda_j \, g_j \tag{4.15}$$

is nonnegative and satisfies $L_f(\mathbf{x}^*) = 0$ as well as $\nabla L_f(\mathbf{x}^*) = 0$. Moreover, as f and $(-g_j)$ are s.o.s.-convex, then so is L_f because $\nabla^2 L_f = \nabla^2 f - \sum_j \lambda_j \nabla^2 g_j$ with $\boldsymbol{\lambda} \geq 0$. Therefore, by Lemma 4.10, $L_f \in \Sigma[\mathbf{x}]$, i.e.,

$$f - f^* - \sum_{j=1}^m \lambda_j \, g_j = f_0,$$

for some $f_0 \in \Sigma[\mathbf{x}]$. As L_f is convex then so is f_0 and so $f - f^* \in Q_c(g)$. □

Hence the class of s.o.s.-convex polynomials is very interesting, because a nice representation result is available. Notice that the above representation holds with $f \geq 0$ on $\mathbf{K}$ and $\mathbf{K}$ is not required to be compact. Another interesting case is when the Hessian of f is positive definite on $\mathbf{K}$.

Theorem 4.13. *Let* $\mathbf{K}$ *be as in* (1.10) *and let Assumption* 1.16 *and Slater's condition both hold. Let* g_j *be concave,* $j = 1, \ldots, m$, *and let* $f \in \mathbb{R}[\mathbf{x}]$ *be convex and such that* $\nabla^2 f \succ 0$ *on* $\mathbf{K}$. *If* $f \geq 0$ *on* $\mathbf{K}$ *then* $f \in Q(g)$, *i.e.,*

$$f = f_0 + \sum_{j=1}^{m} f_j \, g_j, \tag{4.16}$$

for some s.o.s. polynomials $(f_j)_{j=0}^m \subset \Sigma[\mathbf{x}]$.

Proof. Under Assumption 1.16 the set $\mathbf{K}$ is compact. Hence, let $f^* = \min_{\mathbf{x} \in \mathbf{K}} f(\mathbf{x})$ and let $\mathbf{x}^* \in \mathbf{K}$ be a minimizer. As f and $-g_j$ are convex for $j = 1, \ldots, m$, and Slater's condition holds, there exists a vector of nonnegative multipliers $\boldsymbol{\lambda} \in \mathbb{R}^m_+$ such that the Lagrangian polynomial L_f in (4.15) is nonnegative on $\mathbb{R}^n$ with $L_f(\mathbf{x}^*) = 0$ and $\nabla L_f(\mathbf{x}^*) = 0$. Next, by definition of L_f, $\nabla^2 L_f \succeq \nabla^2 f$ because $-\nabla^2 g_j \succeq 0$ on $\mathbf{K}$ for every $j = 1, \ldots, m$, and so $\nabla^2 L_f \succ 0$ on $\mathbf{K}$ because $\nabla^2 f \succ 0$ on $\mathbf{K}$. Hence there is some $\delta > 0$ such that $\nabla^2 L_f(\mathbf{x}) \succeq \delta \, \mathbf{I}_n$ for all $\mathbf{x} \in \mathbf{K}$ (where $\mathbf{I}_n$ denotes the $n \times n$ identity matrix). By Lemma 4.10,

$$L_f(\mathbf{x}) = \big\langle (\mathbf{x} - \mathbf{x}^*), \mathbf{F}(\mathbf{x}, \mathbf{x}^*) (\mathbf{x} - \mathbf{x}^*) \big\rangle,$$

where $\mathbf{F}(\mathbf{x}, \mathbf{x}^*)$ is the matrix polynomial defined by

$$\mathbf{x} \longmapsto \mathbf{F}(\mathbf{x}, \mathbf{x}^*) = \int_0^1 \int_0^t \nabla^2 L_f(\mathbf{x}^* + s(\mathbf{x} - \mathbf{x}^*)) \, ds \, dt.$$

As $\mathbf{K}$ is convex, $\mathbf{x}^* + s(\mathbf{x} - \mathbf{x}^*) \in \mathbf{K}$ for all $s \in [0, 1]$, and so, for every $\xi \in \mathbb{R}^n$,

$$\xi' \mathbf{F}(\mathbf{x}, \mathbf{x}^*) \xi \geq \delta \int_0^1 \int_0^t \xi' \xi \, ds \, dt = \frac{\delta}{2} \xi' \xi.$$

Therefore, $\mathbf{F}(\mathbf{x}, \mathbf{x}^*) \succeq \frac{\delta}{2} \mathbf{I}_n$ for every $\mathbf{x} \in \mathbf{K}$, and, by Theorem 1.22,

$$\mathbf{F}(\mathbf{x}, \mathbf{x}^*) = \mathbf{F}_0(\mathbf{x}) + \sum_{j=1}^{n} \mathbf{F}_j(\mathbf{x}) \, \mathbf{g}_j(\mathbf{x}),$$

for some real symmetric s.o.s. matrix polynomials $(\mathbf{F}_j) \subset \mathbb{R}[\mathbf{x}]^{n \times n}$. Therefore,

$$\begin{aligned}
L_f(\mathbf{x}) &= \big\langle \mathbf{x} - \mathbf{x}^*, \mathbf{F}(\mathbf{x}, \mathbf{x}^*) (\mathbf{x} - \mathbf{x}^*) \big\rangle \\
&= \big\langle \mathbf{x} - \mathbf{x}^*, \mathbf{F}_0(\mathbf{x}) (\mathbf{x} - \mathbf{x}^*) \big\rangle + \sum_{j=1}^{n} g_j(\mathbf{x}) \big\langle \mathbf{x} - \mathbf{x}^*, \mathbf{F}_j(\mathbf{x}) (\mathbf{x} - \mathbf{x}^*) \big\rangle \\
&= \sigma_0(\mathbf{x}) + \sum_{j=1}^{n} \sigma_j(\mathbf{x}) \, g_j(\mathbf{x}),
\end{aligned}$$

for some s.o.s. polynomials $(\sigma_j)_{j=0}^m \subset \Sigma[\mathbf{x}]$. Hence, recalling the definition of L_f,

$$\begin{aligned} f(\mathbf{x}) &= f^* + L_f(\mathbf{x}) + \sum_{j=1}^m \lambda_j\, g_j(\mathbf{x}) \\ &= \big(f^* + \sigma_0(\mathbf{x})\big) + \sum_{j=1}^m \big(\lambda_j + \sigma_j(\mathbf{x})\big)\, g_j(\mathbf{x}), \quad \mathbf{x} \in \mathbb{R}^n, \end{aligned}$$

that is, $f \in Q(g)$ because $f^* \geq 0$ and $\boldsymbol{\lambda} \geq 0$. □

So, for the class of compact basic semi-algebraic sets $\mathbf{K}$ defined by concave polynomials, one obtains a refinement of Putinar's Positivstellensatz Theorem 1.17, for the class of convex functions whose Hessian is positive definite on $\mathbf{K}$ and that are only nonnegative on $\mathbf{K}$.

4.1.3 An extension of Jensen's inequality

Recall that, if μ is a probability measure on $\mathbb{R}^n$ with $E_\mu(\mathbf{x}) < \infty$, Jensen's inequality states that if $f \in L_1(\mu)$ and f is convex, then

$$E_\mu(f) \left(= \int_{\mathbb{R}^n} f\, d\mu \right) \,\geq\, f(E_\mu(\mathbf{x})),$$

a very useful property in many applications. We next provide an extension of Jensen's inequality to a larger class of linear functionals when one restricts its application to the class of s.o.s.-convex polynomials (as opposed to all convex functions). Recall that $\mathbb{N}^n_{2d} = \{\boldsymbol{\alpha} \in \mathbb{N}^n \,:\, |\boldsymbol{\alpha}| \leq 2d\}$.

Theorem 4.14. *Let $f \in \mathbb{R}[\mathbf{x}]_{2d}$ be s.o.s.-convex and let $\mathbf{y} = (y_\alpha)_{\alpha \in \mathbb{N}^n_{2d}}$ satisfy $y_0 = 1$ and $\mathbf{M}_d(\mathbf{y}) \succeq 0$. Then*

$$L_{\mathbf{y}}(f) \,\geq\, f(L_{\mathbf{y}}(\mathbf{x})), \tag{4.17}$$

where $L_{\mathbf{y}}(\mathbf{x}) = (L_{\mathbf{y}}(x_1), \ldots, L_{\mathbf{y}}(x_n))$.

Proof. Let $\mathbf{z} \in \mathbb{R}^n$ be fixed, arbitrary, and consider the polynomial $\mathbf{x} \mapsto f(\mathbf{x}) - f(\mathbf{z})$. Then, from Lemma 4.10,

$$f(\mathbf{x}) - f(\mathbf{z}) \,=\, \big\langle \nabla f(\mathbf{z}), \mathbf{x} - \mathbf{z} \big\rangle + \big\langle (\mathbf{x} - \mathbf{z}), \mathbf{F}(\mathbf{x})(\mathbf{x} - \mathbf{z}) \big\rangle, \tag{4.18}$$

with $\mathbf{F}\colon \mathbb{R}^n \to \mathbb{R}[\mathbf{x}]^{n \times n}$ being the matrix polynomial

$$\mathbf{x} \longmapsto \mathbf{F}(\mathbf{x}) = \int_0^1 \int_0^t \nabla^2 f\big(\mathbf{z} + s(\mathbf{x} - \mathbf{z})\big)\, ds\, dt.$$

As f is s.o.s.-convex, by Lemma 4.9, $\mathbf{F}$ is a s.o.s. matrix polynomial and so the polynomial $\mathbf{x} \mapsto g(\mathbf{x}) = \langle (\mathbf{x} - \mathbf{z}), \mathbf{F}(\mathbf{x})(\mathbf{x} - \mathbf{z}) \rangle$ is s.o.s., i.e., $g \in \Sigma[\mathbf{x}]$. Then, applying $L_{\mathbf{y}}$ to the polynomial $\mathbf{x} \mapsto f(\mathbf{x}) - f(\mathbf{z})$ and using (4.18) yields (recall that $y_{\mathbf{0}} = 1$)

$$\begin{aligned} L_{\mathbf{y}}(f) - f(\mathbf{z}) &= \langle \nabla f(\mathbf{z}), L_{\mathbf{y}}(\mathbf{x}) - \mathbf{z} \rangle + L_{\mathbf{y}}(g) \\ &\geq \langle \nabla f(\mathbf{z}), L_{\mathbf{y}}(\mathbf{x}) - \mathbf{z} \rangle \end{aligned}$$

because $L_{\mathbf{y}}(g) \geq 0$. As $\mathbf{z} \in \mathbb{R}^n$ was arbitrary, $\mathbf{z} = L_{\mathbf{y}}(\mathbf{x})$ $(= (L_{\mathbf{y}}(x_1), \ldots, L_{\mathbf{y}}(x_n)))$ yields the desired result. □

Hence, (4.17) is Jensen's inequality extended to linear functionals

$$L_{\mathbf{y}} : \mathbb{R}[\mathbf{x}]_{2d} \longrightarrow \mathbb{R}$$

in the dual cone of $\Sigma[\mathbf{x}]_d$, that is, vectors $\mathbf{y} = (y_{\boldsymbol{\alpha}})$ such that $\mathbf{M}_d(\mathbf{y}) \succeq 0$ and $y_{\mathbf{0}} = L_{\mathbf{y}}(1) = 1$; hence $\mathbf{y}$ is *not* necessarily the (truncated) moment sequence of some probability measure μ. As a consequence, we also get:

Corollary 4.15. *Let f be a convex univariate polynomial, $g \in \mathbb{R}[\mathbf{x}]$ (and so $f \circ g \in \mathbb{R}[\mathbf{x}]$). Let $d = \lceil (\deg f \circ g)/2 \rceil$, and let $\mathbf{y} = (y_\alpha)_{\alpha \in \mathbb{N}^n_{2d}}$ be such that $y_{\mathbf{0}} = 1$ and $\mathbf{M}_d(\mathbf{y}) \succeq 0$. Then*

$$L_{\mathbf{y}}(f \circ g) \geq f(L_{\mathbf{y}}(g)). \tag{4.19}$$

4.2 Convex polynomial programs

In this section we show that, for problem (3.1), the moment-s.o.s approach based on the semidefinite relaxations (3.14) *recognizes* a large class of convex "easy" problems. That is, when applied to such easy problems as (3.1), finite convergence takes place, i.e., the optimal value f^* is obtained by solving a particular semidefinite program of the hierarchy. Recall that this is not the case in general for the moment approach based on the LP-relaxations (3.28).

4.2.1 The s.o.s.-convex case

With $f \in \mathbb{R}[\mathbf{x}]_d$ and $2i \geq \max[\deg f, \max_j \deg g_j]$, consider the semidefinite program

$$\rho_i = \inf_{\mathbf{y}} \{ L_{\mathbf{y}}(f) \,:\, \mathbf{M}_i(\mathbf{y}) \succeq 0;\ L_{\mathbf{y}}(g_j\,\mathbf{y}) \geq 0,\ j = 1, \ldots, m;\ y_{\mathbf{0}} = 1 \} \tag{4.20}$$

and its dual

$$\begin{aligned} \rho_i^* = \sup_{\boldsymbol{\lambda}, \gamma, \sigma_0} \ & \gamma \\ \text{s.t.}\ & f - \gamma = \sigma_0 + \sum_{j=1}^m \lambda_j\, g_j, \\ & \gamma \in \mathbb{R},\ \lambda \in \mathbb{R}^m_+,\ \sigma_0 \in \Sigma[\mathbf{x}]_i. \end{aligned} \tag{4.21}$$

Theorem 4.16. *Let* $\mathbf{K}$ *be as in* (3.2) *and Slater's condition hold. Let* $f \in \mathbb{R}[\mathbf{x}]$ *be such that* $f^* = \min_{\mathbf{x}}\{f(\mathbf{x}) : \mathbf{x} \in \mathbf{K}\} = f(\mathbf{z})$ *for some* $\mathbf{z} \in \mathbf{K}$. *Assume that* f *and* $-g_j$ *are s.o.s.-convex,* $j = 1, \ldots, m$. *Then* $f^* = \rho_i = \rho_i^*$. *Moreover, if* $\mathbf{y}$ *is an optimal solution of* (4.20) *then* $\mathbf{x}^* = (L_{\mathbf{y}}(x_i))_i \in \mathbf{K}$ *is a global minimizer of* f *on* $\mathbf{K}$.

Proof. Recall the definition of $Q_c(g)$ in (4.14). By Theorem 4.12, $f - f^* \in Q_c(g)$, i.e., $f - f^* = \sigma_0 + \sum_j \lambda_j g_j$ for some $\boldsymbol{\lambda} \in \mathbb{R}^m_+$ and some $\sigma_0 \in \Sigma[\mathbf{x}]$. Therefore, $(f^*, \boldsymbol{\lambda}, \sigma_0)$ is a feasible solution of (4.21), which yields $\rho_i^* \geq f^*$, and which, combined with $\rho_i^* \leq \rho_i \leq f^*$, yields $\rho_i^* = \rho_i = f^*$. Obviously (4.20) is solvable; thus, let $\mathbf{y}$ be an optimal solution. By Theorem 4.14, $f^* = \rho_i^* = L_{\mathbf{y}}(f) \geq f(L_{\mathbf{y}}(\mathbf{x}))$, and, similarly, $0 \geq L_{\mathbf{y}}(-g_j) \geq -g_j(L_{\mathbf{y}}(\mathbf{x}))$, $j = 1, \ldots, m$, which shows that $\mathbf{x}^* = L_{\mathbf{y}}(\mathbf{x})\,(= L_{\mathbf{y}}(x_i)) \in \mathbf{K}$ is a global minimizer of f on $\mathbf{K}$. □

Therefore, when the polynomials f and $(-g_j)$ are s.o.s.-convex, the first semidefinite program in the hierarchy of semidefinite programs (3.14) is exact as it is either identical to (4.20) or more constrained, hence with optimal value $\rho_i = f^*$. In other words, the moment-s.o.s. approach recognizes s.o.s.-convexity.

4.2.2 The strictly convex case

If f or some of the $-g_j$'s is not s.o.s.-convex but $\nabla^2 f \succ 0$ (so that f is strictly convex) and $-g_j$ is convex for every $j = 1, \ldots, m$, then one obtains the following result.

Theorem 4.17. *Let* $\mathbf{K}$ *be as in* (1.10) *with* g_j *concave for* $j = 1, \ldots, m$, *and let Assumption* 1.16 *and Slater's condition hold. Assume that* f *is convex and* $\nabla^2 f(\mathbf{x}) \succ 0$ *for all* $\mathbf{x} \in \mathbf{K}$. *Then the hierarchy of semidefinite relaxations* (3.14)–(3.15) *has finite convergence. That is,* $f^* = \rho_i^* = \rho_i$ *for some index* i. *In addition, both primal and dual relaxations are solvable.*

Proof. Let $f^* = \min_{\mathbf{x} \in \mathbf{K}} f(\mathbf{x})$. By Theorem 4.13,

$$f - f^* = \sigma_0 + \sum_{j=1}^{m} \sigma_j \, g_j,$$

for some s.o.s. polynomials $(\sigma_j)_j \subset \Sigma[\mathbf{x}]$. Let $2i_0 \geq \max_j \deg \sigma_j + \deg g_j$ (with $g_0 = 1$). Then $(f^*, (\sigma_j))$ is a feasible solution of the semidefinite program (3.15). Hence $f^* \leq \rho_{i_0}^* \leq \rho_{i_0} \leq f^*$, which yields the desired result $f^* = \rho_{i_0} = \rho_{i_0}^*$. □

When compared to Theorem 4.16 for the s.o.s.-convex case, in the strictly convex case the simplified semidefinite relaxation (4.20) is not guaranteed to be exact. However, finite convergence still occurs for the standard semidefinite relaxations (3.14). Hence the hierarchy of semidefinite relaxations exhibits a particularly nice behavior for convex problems, a highly desirable property since convex optimization problems are easier to solve.

4.3 Notes and sources

Most of this chapter is from Lasserre (2009) [75] and Lasserre (2010) [76]. Lemmas 4.9 and 4.10 are from Helton and Nie (2009) [43], who introduced the very interesting class of s.o.s.-convex polynomials. Theorems 4.11 and 4.12 are from Lasserre (2008) [74]. Some properties of convex polynomials with applications to polynomial programming have been investigated in Andronov et al. (1982) [4] and Belousov and Klatte (2002) [11].

Chapter 5

Parametric Polynomial Optimization

5.1 Introduction

Roughly speaking, given a set of parameters $\mathbf{Y}$ and an optimization problem whose description depends on $\mathbf{y} \in \mathbf{Y}$ (call it $\mathbf{P_y}$), *parametric optimization* is concerned with the behavior and properties of the optimal value as well as primal (and possibly dual) optimal solutions of $\mathbf{P_y}$, when $\mathbf{y}$ varies in $\mathbf{Y}$. This is a quite challenging problem for which in general only local information around some nominal value $\mathbf{y}_0$ of the parameter can be obtained. Sometimes, in the context of optimization with data uncertainty, some probability distribution on the parameter set $\mathbf{Y}$ is available and in this context one is also interested in, e.g., the distributions of the optimal value and optimal solutions, all viewed as random variables. The purpose of this chapter is to show that if one restricts to the case of *polynomial* parametric optimization then all this important information can be obtained, or at least approximated, as closely as desired.

Hence we here restrict our attention to parametric polynomial optimization, that is, when $\mathbf{P_y}$ is described by *polynomial* equality and inequality constraints on both the parameter vector $\mathbf{y}$ and the optimization variables $\mathbf{x}$. Moreover, the set $\mathbf{Y}$ is restricted to be a compact basic semi-algebraic set of $\mathbb{R}^p$, and preferably a set sufficiently simple so that one may obtain the moments of some probability measure φ on $\mathbf{Y}$, absolutely continuous with respect to the Lebesgue measure. For instance, if $\mathbf{Y}$ is a simple set (like a simplex, a box) one may choose φ to be the probability measure uniformly distributed on $\mathbf{Y}$; typical $\mathbf{Y}$ candidates are polyhedra. Or, sometimes, in the context of optimization with data uncertainty, φ is already specified. We also suppose that $\mathbf{P_y}$ has a unique optimal solution for φ-almost all values of the parameter $\mathbf{y} \in \mathbf{Y}$. In this specific context one may get insightful information on the set of all global optimal solutions of $\mathbf{P_y}$, via what we call the *"joint + marginal"* approach, that we briefly describe below.

The "joint + marginal" approach

We are given a probability distribution φ on $\mathbf{Y}$, and we assume that all its moments are known and available. Call $J(\mathbf{y})$ (resp. $\mathbf{X}^*_{\mathbf{y}} \in \mathbb{R}^n$) the optimal value (resp. the solution set) of $\mathbf{P}_{\mathbf{y}}$ for the value $\mathbf{y} \in \mathbf{Y}$ of the parameter.

- One first defines an infinite-dimensional optimization problem $\mathbf{P}$ whose optimal value is just $\rho = \int_{\mathbf{Y}} J(\mathbf{y})\, d\varphi(\mathbf{y})$, i.e., the φ-mean of the global optimum. Any optimal solution of $\mathbf{P}_{\mathbf{y}}$ is a probability measure μ^* on $\mathbb{R}^n \times \mathbb{R}^p$ with marginal φ on $\mathbf{Y} \subset \mathbb{R}^p$. It turns out that μ^* encodes all information on the set of global minimizers $\mathbf{X}^*_{\mathbf{y}}$, $\mathbf{y} \in \mathbf{Y}$. Whence the name *"joint+ marginal"* as μ^* is a *joint* distribution of $\mathbf{x}$ and $\mathbf{y}$, and φ is the *marginal* of μ^* on $\mathbb{R}^p$.
- One then provides a hierarchy of semidefinite relaxations of $\mathbf{P}$ with associated sequence of optimal values $(\rho_i)_i$. An optimal solution $\mathbf{z}^i$ of the ith semidefinite relaxation is a sequence $\mathbf{z}^i = (z^i_{\alpha\beta})$ indexed in the monomial basis $(\mathbf{x}^\alpha \mathbf{y}^\beta)$ of the subspace $\mathbb{R}[\mathbf{x}, \mathbf{y}]_{2i}$ of polynomials of degree at most $2i$. If for φ-almost all $\mathbf{y} \in \mathbf{Y}$, $\mathbf{P}_{\mathbf{y}}$ has a unique *global* optimal solution $\mathbf{x}^*(\mathbf{y}) \in \mathbb{R}^n$, then as $i \to \infty$, $\mathbf{z}^i$ converges pointwise to the sequence of moments of μ^* defined above. In particular, one obtains the distribution of the global minimizer $\mathbf{x}^*(\mathbf{y})$, and, therefore, one may approximate as closely as desired any polynomial functional of $\mathbf{x}^*(\mathbf{y})$, like, e.g., the φ-mean or variance of $\mathbf{x}^*(\mathbf{y})$.
- From the hierarchy of semidefinite relaxations one may approximate as closely as desired any moment of the measure $d\psi(\mathbf{y}) = x^*_k(\mathbf{y})\, d\varphi(\mathbf{y})$ on $\mathbf{Y}$. And so, the kth coordinate function $\mathbf{y} \mapsto x^*_k(\mathbf{y})$ of the global minimizer of $\mathbf{P}_{\mathbf{y}}$, $\mathbf{y} \in \mathbf{Y}$, can be estimated, e.g., by maximum entropy methods. Of course the latter estimation is not pointwise but it still provides useful information on optimal solutions, e.g., the shape of the function $\mathbf{y} \mapsto x^*_k(\mathbf{y})$.
- At last but not least, from an optimal solution of the dual of the ith semidefinite relaxation, one obtains a piecewise polynomial approximation of the optimal value function $\mathbf{y} \mapsto J(\mathbf{y})$, that converges φ-almost uniformly to J.

Finally, the computational complexity of the above methodology is roughly the same as the moment-s.o.s. approach described in earlier chapters for an optimization problem with $n + p$ variables, since we consider the joint distribution of the n variables $\mathbf{x}$ and the p parameters $\mathbf{y}$. Hence, the approach is particularly interesting when the number of parameters is small, say 1 or 2. But this computational price may not seem that high in view of the ambitious goal of the approach. After all, keep in mind that by applying the moment-s.o.s. approach to a single $(n+p)$-variable problem, one obtains information on global minimizers of an n-variable problem that depends on p parameters, that is, one approximates n *functions* of p variables!

5.2 A related linear program

Recall that $\mathbb{R}[\mathbf{x},\mathbf{y}]$ denotes the ring of polynomials in the variables $\mathbf{x} = (x_1,\ldots,x_n)$ and the variables $\mathbf{y} = (y_1,\ldots,y_p)$, whereas $\mathbb{R}[\mathbf{x},\mathbf{y}]_k$ denotes its subspace of polynomials of degree at most k. Let $\Sigma[\mathbf{x},\mathbf{y}] \subset \mathbb{R}[\mathbf{x},\mathbf{y}]$ denote the subset of polynomials that are sums of squares (in short s.o.s.).

The parametric optimization problem

Let $\mathbf{Y} \subset \mathbb{R}^p$ be a compact set, called the *parameter set*, and let $f, g_j : \mathbb{R}^n \times \mathbb{R}^p \to \mathbb{R}$, $j = 1,\ldots,m$, be continuous. For each $\mathbf{y} \in \mathbf{Y}$ fixed, consider the following optimization problem:

$$J(\mathbf{y}) = \inf_{\mathbf{x}} \left\{ f_{\mathbf{y}}(\mathbf{x}) \;:\; g_{\mathbf{y}j}(\mathbf{x}) \geq 0,\; j = 1,\ldots,m \right\}, \tag{5.1}$$

where the functions $f_{\mathbf{y}}, g_{\mathbf{y}j} : \mathbb{R}^n \to \mathbb{R}$ are defined via

$$\left.\begin{aligned} \mathbf{x} &\longmapsto f_{\mathbf{y}}(\mathbf{x}) = f(\mathbf{x},\mathbf{y}) \\ \mathbf{x} &\longmapsto g_{\mathbf{y}j}(\mathbf{x}) = g_j(\mathbf{x},\mathbf{y}),\; j = 1,\ldots,m, \end{aligned}\right\} \quad \forall \mathbf{x} \in \mathbb{R}^n,\; \mathbf{y} \in \mathbb{R}^p.$$

Next, let $\mathbf{K} \subset \mathbb{R}^n \times \mathbb{R}^p$ be the set

$$\mathbf{K} = \left\{ (\mathbf{x},\mathbf{y}) \;:\; \mathbf{y} \in \mathbf{Y};\; g_j(\mathbf{x},\mathbf{y}) \geq 0,\; j = 1,\ldots,m \right\}, \tag{5.2}$$

and, for each $\mathbf{y} \in \mathbf{Y}$, let

$$\mathbf{K}_{\mathbf{y}} = \left\{ \mathbf{x} \in \mathbb{R}^n \;:\; g_{\mathbf{y}j}(\mathbf{x}) \geq 0,\; j = 1,\ldots,m \right\}. \tag{5.3}$$

The interpretation is as follows: $\mathbf{Y}$ is a set of parameters and, for each instance $\mathbf{y} \in \mathbf{Y}$ of the parameter, one wishes to compute an optimal *decision* vector $\mathbf{x}^*(\mathbf{y})$ that solves problem (5.1). Let φ be a Borel probability measure on $\mathbf{Y}$ with a positive density with respect to the Lebesgue measure on $\mathbb{R}^p$. For instance, choose for φ the probability measure

$$\varphi(B) = \left(\int_{\mathbf{Y}} d\mathbf{y} \right)^{-1} \int_B d\mathbf{y}, \quad \forall B \in \mathcal{B}(\mathbf{Y}),$$

uniformly distributed on $\mathbf{Y}$. Sometimes, e.g., in the context of optimization with data uncertainty, φ is already specified. We will use φ (or more precisely its moments) to get information on the distribution of optimal solutions $\mathbf{x}^*(\mathbf{y})$ of $\mathbf{P}_{\mathbf{y}}$, viewed as random variables.

In the rest of the chapter we assume that, for every $\mathbf{y} \in \mathbf{Y}$, the set $\mathbf{K}_{\mathbf{y}}$ in (5.3) is nonempty.

5.2.1 A related infinite-dimensional linear program

Recall that $\mathcal{M}(\mathbf{K})_+$ is the set of finite Borel measures on $\mathbf{K}$, and consider the following infinite-dimensional linear program:

$$\mathbf{P}: \quad \rho = \inf_{\mu \in \mathcal{M}(\mathbf{K})_+} \left\{ \int_{\mathbf{K}} f \, d\mu \ : \ \pi\mu = \varphi \right\}, \tag{5.4}$$

where $\pi\mu$ denotes the marginal of μ on $\mathbb{R}^p$, that is, $\pi\mu$ is a probability measure on $\mathbb{R}^p$ defined by

$$\pi\mu(B) = \mu\big(\mathbb{R}^n \times B\big), \quad \forall B \in \mathcal{B}(\mathbb{R}^p).$$

Notice that $\mu(\mathbf{K}) = 1$ for any feasible solution μ of $\mathbf{P}$. Indeed, as φ is a probability measure and $\pi\mu = \varphi$, one has $1 = \varphi(\mathbf{Y}) = \mu(\mathbb{R}^n \times \mathbb{R}^p) = \mu(\mathbf{K})$.

Recall that, for two Borel spaces X, Y, the graph $\mathrm{Gr}\,\psi \subset X \times Y$ of a set-valued mapping $\psi\colon X \to Y$ is the set

$$\mathrm{Gr}\,\psi = \big\{(\mathbf{x}, \mathbf{y}) \ : \ \mathbf{x} \in X;\ \mathbf{y} \in \psi(\mathbf{x})\big\}.$$

If ψ is measurable then any measurable function $h\colon X \to Y$ with $h(\mathbf{x}) \in \psi(\mathbf{x})$ for every $\mathbf{x} \in X$ is called a (measurable) *selector*.

Lemma 5.1. *Let both $\mathbf{Y} \subset \mathbb{R}^n$ and $\mathbf{K}$ in (5.2) be compact. Then the set-valued mapping $\mathbf{y} \mapsto \mathbf{K}_{\mathbf{y}}$ is Borel-measurable. In addition:*

(a) *The mapping $\mathbf{y} \mapsto J(\mathbf{y})$ is measurable.*

(b) *There exists a measurable selector $g\colon Y \to \mathbf{K}_{\mathbf{y}}$ such that $J(\mathbf{y}) = f(g(\mathbf{y}), \mathbf{y})$ for every $\mathbf{y} \in \mathbf{Y}$.*

Proof. As $\mathbf{K}$ and $\mathbf{Y}$ are both compact, the set-valued mapping $\mathbf{y} \mapsto \mathbf{K}_{\mathbf{y}} \subset \mathbb{R}^n$ is compact-valued. Moreover, the graph of $\mathbf{K}_{\mathbf{y}}$ is by definition the set $\mathbf{K}$, which is a Borel subset of $\mathbb{R}^n \times \mathbb{R}^p$. Hence, by [46, Proposition D.4], $\mathbf{K}_{\mathbf{y}}$ is a measurable function from $\mathbf{Y}$ to the space of nonempty compact subsets of $\mathbb{R}^n$, topologized by the Hausdorff metric. Next, since $\mathbf{x} \mapsto f_{\mathbf{y}}(\mathbf{x})$ is continuous for every $\mathbf{y} \in \mathbf{Y}$, (a) and (b) follow from [46, Proposition D.5]. □

Theorem 5.2. *Let both $\mathbf{Y} \subset \mathbb{R}^p$ and $\mathbf{K}$ in (5.2) be compact and assume that, for every $\mathbf{y} \in \mathbf{Y}$, the set $\mathbf{K}_{\mathbf{y}} \subset \mathbb{R}^n$ in (5.3) is nonempty. Let $\mathbf{P}$ be the optimization problem (5.4) and let $\mathbf{X}^*_{\mathbf{y}} = \{\mathbf{x} \in \mathbb{R}^n : f(\mathbf{x}, \mathbf{y}) = J(\mathbf{y})\}$, $y \in \mathbf{Y}$. Then:*

(a) *$\mathbf{P}$ has an optimal solution with optimal value $\rho = \int_{\mathbf{Y}} J(\mathbf{y})\, d\varphi(\mathbf{y})$.*

(b) *For every optimal solution μ^* of $\mathbf{P}$, and for φ-almost all $\mathbf{y} \in \mathbf{Y}$, there is a probability measure $\psi^*(\,\cdot \mid \mathbf{y})$ on $\mathbb{R}^n$, supported on $\mathbf{X}^*_{\mathbf{y}}$, such that:*

$$\mu^*(C \times B) = \int_B \psi^*(C \mid \mathbf{y})\, d\varphi(\mathbf{y}), \quad \forall B \in \mathcal{B}(\mathbf{Y}),\ C \in \mathcal{B}(\mathbb{R}^n). \tag{5.5}$$

(c) *Assume that, for φ-almost all $\mathbf{y} \in \mathbf{Y}$, the set of minimizers of $\mathbf{X}^*_{\mathbf{y}}$ is the singleton $\{\mathbf{x}^*(\mathbf{y})\}$ for some $\mathbf{x}^*(\mathbf{y}) \in \mathbf{K}_y$. Then there is a measurable mapping $g\colon \mathbf{Y} \to \mathbf{K}_{\mathbf{y}}$ such that*

$$g(\mathbf{y}) = \mathbf{x}^*(\mathbf{y}) \text{ for every } \mathbf{y} \in \mathbf{Y}; \quad \rho = \int_{\mathbf{Y}} f(g(\mathbf{y}), \mathbf{y})\, d\varphi(\mathbf{y}), \tag{5.6}$$

and, for every $\alpha \in \mathbb{N}^n$ and $\beta \in \mathbb{N}^p$,

$$\int_{\mathbf{K}} \mathbf{x}^\alpha \mathbf{y}^\beta\, d\mu^*(\mathbf{x}, \mathbf{y}) = \int_{\mathbf{Y}} \mathbf{y}^\beta g(\mathbf{y})^\alpha\, d\varphi(\mathbf{y}). \tag{5.7}$$

Proof. (a) As $\mathbf{K}$ is compact then so is $\mathbf{K}_{\mathbf{y}}$ for every $\mathbf{y} \in \mathbf{Y}$. Next, as $\mathbf{K}_{\mathbf{y}} \neq \emptyset$ for every $\mathbf{y} \in \mathbf{Y}$ and f is continuous, the set $\mathbf{X}^*_{\mathbf{y}} = \{\mathbf{x} \in \mathbb{R}^n : f(\mathbf{x}, \mathbf{y}) = J(\mathbf{y})\}$ is nonempty for every $\mathbf{y} \in \mathbf{Y}$. Let μ be any feasible solution of $\mathbf{P}$ and so, by definition, its marginal on $\mathbb{R}^p$ is just φ. Since $\mathbf{X}^*_{\mathbf{y}} \neq \emptyset$ for all $\mathbf{y} \in \mathbf{Y}$, one has $f_{\mathbf{y}}(\mathbf{x}) \geq J(\mathbf{y})$ for all $\mathbf{x} \in \mathbf{K}_{\mathbf{y}}$ and all $\mathbf{y} \in \mathbf{Y}$. So, $f(\mathbf{x}, \mathbf{y}) \geq J(\mathbf{y})$ for all $(\mathbf{x}, \mathbf{y}) \in \mathbf{K}$ and therefore

$$\int_{\mathbf{K}} f d\mu \geq \int_{\mathbf{K}} J(\mathbf{y})\, d\mu(\mathbf{y}) = \int_{\mathbf{Y}} J(\mathbf{y})\, d\varphi(\mathbf{y}),$$

which proves that $\rho \geq \displaystyle\int_{\mathbf{Y}} J(\mathbf{y})\, d\varphi(\mathbf{y})$.

On the other hand, recall that $\mathbf{K}_{\mathbf{y}} \neq \emptyset$ for all $\mathbf{y} \in \mathbf{Y}$. Consider the set-valued mapping $\mathbf{y} \mapsto \mathbf{X}^*_{\mathbf{y}} \subset \mathbf{K}_{\mathbf{y}}$. As f is continuous and $\mathbf{K}$ is compact, $\mathbf{X}^*_{\mathbf{y}}$ is compact-valued. In addition, as $f_{\mathbf{y}}$ is continuous, by [46, D6] (or [106]) there exists a measurable selector $g\colon \mathbf{Y} \to \mathbf{X}^*_y$ (and so $f(g(\mathbf{y}), \mathbf{y}) = J(\mathbf{y})$). Therefore, for every $\mathbf{y} \in \mathbf{Y}$, let $\delta_{g(\mathbf{y})}$ be the Dirac probability measure with support on the singleton $g(\mathbf{y}) \in \mathbf{X}^*_{\mathbf{y}}$, and let μ be the probability measure defined by

$$\mu(C, B) = \int_B 1_C(g(\mathbf{y}))\, d\varphi(\mathbf{y}), \quad \forall B \in \mathcal{B}(\mathbf{Y}),\ C \in \mathcal{B}(\mathbb{R}^n).$$

It follows that

$$\begin{aligned} \int_{\mathbf{K}} f\, d\mu &= \int_{\mathbf{Y}} \left(\int_{\mathbf{K}_{\mathbf{y}}} f(\mathbf{x}, \mathbf{y})\, d\delta_{g(\mathbf{y})} \right) d\varphi(\mathbf{y}) \\ &= \int_{\mathbf{Y}} f(g(\mathbf{y}), \mathbf{y})\, d\varphi(\mathbf{y}) = \int_{\mathbf{Y}} J(\mathbf{y})\, d\varphi(\mathbf{y}) \leq \rho, \end{aligned}$$

which shows that μ is an optimal solution of $\mathbf{P}$.

(b) Let μ^* be an arbitrary optimal solution of $\mathbf{P}$, hence supported on $\mathbf{K}_{\mathbf{y}} \times \mathbf{Y}$. Therefore, as $\mathbf{K}$ is contained in the Cartesian product $\mathbb{R}^p \times \mathbb{R}^n$, the probability measure μ^* can be disintegrated as

$$\mu^*(C, B) = \int_B \psi^*(C \,|\, \mathbf{y})\, d\varphi(\mathbf{y}), \quad \forall B \in \mathcal{B}(\mathbf{Y}),\ C \in \mathcal{B}(\mathbb{R}^n),$$

where, for all $\mathbf{y} \in \mathbf{Y}$, $\psi^*(\cdot \,|\, \mathbf{y})$ is a probability measure[1] on $\mathbf{K}_{\mathbf{y}}$. Hence, from (a)

[1]The object $\psi^*(\cdot \,|\, \cdot)$ is called a *stochastic kernel*; see, e.g., [32, pp. 88 and 89] or [46, D8].

we get

$$\rho = \int_{\mathbf{Y}} J(\mathbf{y})\, d\varphi(\mathbf{y}) = \int_{\mathbf{K}} f(\mathbf{x},\mathbf{y})\, d\mu^*(\mathbf{x},\mathbf{y})$$
$$= \int_{\mathbf{Y}} \left(\int_{\mathbf{K}_{\mathbf{y}}} f(\mathbf{x},\mathbf{y})\, \psi^*(d\mathbf{x} \,|\, \mathbf{y}) \right) d\varphi(\mathbf{y}).$$

Therefore, using $f(\mathbf{x},\mathbf{y}) \geq J(\mathbf{y})$ on $\mathbf{K}$,

$$0 = \int_{\mathbf{Y}} \left(\int_{\mathbf{K}_{\mathbf{y}}} \underbrace{J(\mathbf{y}) - f(\mathbf{x},\mathbf{y})}_{\leq 0}\, \psi^*(d\mathbf{x} \,|\, \mathbf{y}) \right) d\varphi(\mathbf{y}),$$

which implies $\psi^*(\mathbf{X}^*(\mathbf{y}) \,|\, \mathbf{y}) = 1$ for φ-almost all $\mathbf{y} \in \mathbf{Y}$.

(c) Let $g\colon \mathbf{Y} \to \mathbf{K}_{\mathbf{y}}$ be the measurable mapping of Lemma 5.1(b). As $J(\mathbf{y}) = f(g(\mathbf{y}),\mathbf{y})$ and $(g(\mathbf{y}),\mathbf{y}) \in \mathbf{K}$, necessarily $g(\mathbf{y}) \in \mathbf{X}^*_{\mathbf{y}}$ for every $\mathbf{y} \in \mathbf{Y}$. Next, let μ^* be an optimal solution of $\mathbf{P}$, and let $\alpha \in \mathbb{N}^n$, $\beta \in \mathbb{N}^p$. Then

$$\int_{\mathbf{K}} \mathbf{x}^\alpha \mathbf{y}^\beta\, d\mu^*(\mathbf{x},\mathbf{y}) = \int_{\mathbf{Y}} \mathbf{y}^\beta \left(\int_{\mathbf{X}^*_{\mathbf{y}}} \mathbf{x}^\alpha\, \psi^*(d\mathbf{x} | \, \mathbf{y}) \right) d\varphi(\mathbf{y})$$
$$= \int_{\mathbf{Y}} \mathbf{y}^\beta\, g(\mathbf{y})^\alpha\, d\varphi(\mathbf{y}),$$

the desired result. □

An optimal solution μ^* of $\mathbf{P}$ encodes *all* information on the optimal solutions $\mathbf{x}^*(\mathbf{y})$ of $\mathbf{P}_{\mathbf{y}}$. For instance, let $\mathbf{B}$ be a given Borel set of $\mathbb{R}^n$. Then, from Theorem 5.2,

$$\text{Prob}\left(\mathbf{x}^*(\mathbf{y}) \in \mathbf{B}\right) = \mu^*\left(\mathbf{B} \times \mathbb{R}^p\right) = \int_{\mathbf{Y}} \psi^*(\mathbf{B} \,|\, \mathbf{y})\, d\varphi(\mathbf{y}),$$

with ψ^* as in Theorem 5.2(b).

Consequently, if one knows an optimal solution μ^* of $\mathbf{P}$, then one may evaluate functionals on the solutions of $\mathbf{P}_{\mathbf{y}}$, $\mathbf{y} \in \mathbf{Y}$. That is, assuming that for φ-almost all $\mathbf{y} \in \mathbf{Y}$, the problem $\mathbf{P}_{\mathbf{y}}$ has a unique optimal solution $\mathbf{x}^*(\mathbf{y})$, and, given a measurable mapping $h\colon \mathbb{R}^n \to \mathbb{R}^q$, one may evaluate the functional

$$\int_{\mathbf{Y}} h(\mathbf{x}^*(\mathbf{y}))\, d\varphi(\mathbf{y}).$$

For instance, with $\mathbf{x} \mapsto h(\mathbf{x}) = \mathbf{x}$ one obtains the *mean* vector $\mathrm{E}_\varphi(\mathbf{x}^*(\mathbf{y})) = \int_{\mathbf{Y}} \mathbf{x}^*(\mathbf{y})\, d\varphi(\mathbf{y})$ of optimal solutions $\mathbf{x}^*(\mathbf{y})$, $\mathbf{y} \in \mathbf{Y}$.

Corollary 5.3. *Let both $\mathbf{Y} \subset \mathbb{R}^p$ and $\mathbf{K}$ in* (5.2) *be compact and assume that, for every $\mathbf{y} \in \mathbf{Y}$, the set $\mathbf{K}_{\mathbf{y}} \subset \mathbb{R}^n$ in* (5.3) *is nonempty. Let $\mathbf{P}$ be the optimization* *problem (5.4) and assume that, for φ-almost all $\mathbf{y} \in \mathbf{Y}$, the set $\mathbf{X}_y^* = \{\mathbf{x} \in \mathbf{K_y} : J(\mathbf{y}) = f(\mathbf{x}, \mathbf{y})\}$ is the singleton $\{\mathbf{x}^*(\mathbf{y})\}$. Then, for every measurable mapping $h \colon \mathbb{R}^n \to \mathbb{R}^q$,*

$$\int_{\mathbf{Y}} h(\mathbf{x}^*(\mathbf{y}))\, d\varphi(\mathbf{y}) = \int_{\mathbf{K}} h(\mathbf{x})\, d\mu^*(\mathbf{x}, \mathbf{y}), \tag{5.8}$$

where μ^* is an optimal solution of $\mathbf{P}$.

Proof. By Theorem 5.2(c),

$$\int_{\mathbf{K}} h(\mathbf{x})\, d\mu^*(\mathbf{x}, \mathbf{y}) = \int_{\mathbf{Y}} \left[\int_{\mathbf{X}_\mathbf{y}^*} h(\mathbf{x}) \psi^*(d\mathbf{x} \,|\, \mathbf{y}) \right] d\varphi(\mathbf{y}) = \int_{\mathbf{Y}} h(\mathbf{x}^*(\mathbf{y}))\, d\varphi(\mathbf{y}). \qquad \square$$

5.2.2 Duality

Consider the following infinite-dimensional linear program:

$$\mathbf{P}^* : \quad \rho^* = \sup_{p \in \mathbb{R}[\mathbf{y}]} \left\{ \int_{\mathbf{Y}} p\, d\varphi \; : \; f(\mathbf{x}, \mathbf{y}) - p(\mathbf{y}) \geq 0, \; \forall (\mathbf{x}, \mathbf{y}) \in \mathbf{K} \right\}. \tag{5.9}$$

Then $\mathbf{P}^*$ is a *dual* of $\mathbf{P}$.

Lemma 5.4. *Let both $\mathbf{Y} \subset \mathbb{R}^p$ and $\mathbf{K}$ in (5.2) be compact and let $\mathbf{P}$ and $\mathbf{P}^*$ be as in (5.4) and (5.9) respectively. Then there is no duality gap, i.e., $\rho = \rho^*$.*

Proof. For a topological space $\mathcal{X}$, denote by $C(\mathcal{X})$ the space of bounded continuous functions on $\mathcal{X}$. Recall that $\mathcal{M}(\mathbf{K})$ is the vector space of finite signed Borel measures on $\mathbf{K}$ (and so $\mathcal{M}(\mathbf{K})_+$ is its positive cone). Let $\pi \colon \mathcal{M}(\mathbf{K}) \to \mathcal{M}(\mathbf{Y})$ be defined by $(\pi\mu)(B) = \mu((\mathbb{R}^n \times B) \cap \mathbf{K})$ for all $B \in \mathcal{B}(\mathbf{Y})$, with adjoint mapping $\pi^* \colon C(\mathbf{Y}) \to C(\mathbf{K})$ defined as

$$(\mathbf{x}, \mathbf{y}) \longmapsto (\pi^* h)(\mathbf{x}, \mathbf{y}) = h(\mathbf{y}), \quad \forall h \in C(\mathbf{Y}).$$

Put (5.4) in the framework of infinite-dimensional linear programs on vector spaces, as described in, e.g., [3]. That is,

$$\rho = \inf_{\mu \in \mathcal{M}(\mathbf{K})} \{\langle f, \mu \rangle \; : \; \pi\mu = \varphi, \; \mu \geq 0\},$$

with dual

$$\tilde{\rho} = \sup_{h \in C(\mathbf{Y})} \{\langle h, \varphi \rangle \; : \; f - \pi^* h \geq 0 \text{ on } \mathbf{K}\}.$$

One first proves that $\rho = \tilde{\rho}$ and then $\tilde{\rho} = \rho^*$. Recall that, by Theorem 5.2, ρ is finite. Therefore, by [3, Theorem 3.10], it suffices to prove that the set $D = \{(\pi\mu, \langle f, \mu \rangle) : \mu \in \mathcal{M}(\mathbf{K})_+\}$ is closed for the respective weak $\star$ topologies $\sigma(\mathcal{M}(\mathbf{Y}) \times \mathbb{R}, C(\mathbf{Y}) \times \mathbb{R})$ and $\sigma(\mathcal{M}(\mathbf{K}), C(\mathbf{K}))$. Therefore, consider a converging sequence $\pi\mu_n \to a$ with $\mu_n \in \mathcal{M}(\mathbf{K})_+$. The sequence (μ_n) is uniformly bounded because

$$\mu_n(\mathbf{K}) = (\pi\mu_n)(\mathbf{Y}) = \langle 1, \pi\mu_n \rangle \longrightarrow \langle 1, a \rangle = a(\mathbf{K}).$$

But, by the Banach–Alaoglu Theorem (see, e.g., [7]), the bounded closed sets of $\mathcal{M}(\mathbf{K})_+$ are compact in the weak $\star$ topology, and so $\mu_{n_k} \to \mu \in \mathcal{M}(\mathbf{K})_+$ for some $\mu \in \mathcal{M}(\mathbf{K})_+$ and some subsequence (n_k). Hence $\pi\mu_{n_k} \to a = \pi\mu$ as the projection π is continuous. Similarly, $\langle f, \mu_{n_k}\rangle \to \langle f, \mu\rangle$ because $f \in C(\mathbf{K})$. This proves that D is closed and the desired result $\rho = \tilde{\rho}$.

So, given $\epsilon > 0$ fixed arbitrary, there is a function $h_\epsilon \in C(\mathbf{Y})$ such that $f - h_\epsilon \geq 0$ on $\mathbf{K}$ and $\int h_\epsilon \, d\varphi \geq \tilde{\rho} - \epsilon$. By compactness of $\mathbf{Y}$ and the Stone–Weierstrass theorem, there is $p_\epsilon \in \mathbb{R}[\mathbf{y}]$ such that $\sup_{\mathbf{y}\in\mathbf{Y}} |h_\epsilon - p_\epsilon| \leq \epsilon$. Hence the polynomial $\tilde{p}_\epsilon = p_\epsilon - \epsilon$ is feasible with value $\int_{\mathbf{Y}} \tilde{p}_\epsilon \, d\varphi \geq \tilde{\rho} - 3\epsilon$, and, as ϵ was arbitrary, the result $\tilde{\rho} = \rho^*$ follows. □

As next shown, optimal or nearly optimal solutions of $\mathbf{P}^*$ provide us with polynomial lower approximations of $J(\,\cdot\,)$ that converge to the optimal value function $\mathbf{y} \mapsto J(\mathbf{y})$ on $\mathbf{Y}$ for the $L_1(\varphi)$-norm. Moreover, one may also obtain a piecewise polynomial approximation that converges to $J(\,\cdot\,)$, φ-almost uniformly[2].

Corollary 5.5. *Let both $\mathbf{Y} \subset \mathbb{R}^p$ and $\mathbf{K}$ in (5.2) be compact and assume that, for every $\mathbf{y} \in \mathbf{Y}$, the set $\mathbf{K}_\mathbf{y}$ is nonempty. Let $\mathbf{P}^*$ be as in (5.9). If $(p_i)_{i\in\mathbb{N}} \subset \mathbb{R}[\mathbf{y}]$ is a maximizing sequence of (5.9), then*

$$\int_{\mathbf{Y}} |\, J(\mathbf{y}) - p_i(\mathbf{y}) \,|\, d\varphi(\mathbf{y}) \longrightarrow 0 \quad \text{as } i \to \infty. \tag{5.10}$$

Moreover, define

$$\tilde{p}_0 = p_0, \quad \mathbf{y} \longmapsto \tilde{p}_i(\mathbf{y}) = \max\,[\,\tilde{p}_{i-1}(\mathbf{y}), p_i(\mathbf{y})], \quad i = 1, \ldots$$

Then $\tilde{p}_i \to J(\,\cdot\,)$ φ-almost uniformly.

Proof. By Lemma 5.4, we already know that $\rho^* = \rho$ and so

$$\int_{\mathbf{Y}} p_i(\mathbf{y})\, d\varphi(\mathbf{y}) \uparrow \rho^* = \rho = \int_{\mathbf{Y}} J(\mathbf{y})\, d\varphi(\mathbf{y}).$$

Next, by feasibility of p_i in (5.9),

$$f(\mathbf{x}, \mathbf{y}) \geq p_i(\mathbf{y}) \quad \forall (\mathbf{x}, \mathbf{y}) \in \mathbf{K} \Longrightarrow \inf_{\mathbf{x}\in\mathbf{K}_\mathbf{y}} f(\mathbf{x}, \mathbf{y}) = J(\mathbf{y}) \geq p_i(\mathbf{y}), \quad \forall \mathbf{y} \in \mathbf{Y}.$$

Hence (5.10) follows from the inequality $p_i(\mathbf{y}) \leq J(\mathbf{y})$ on $\mathbf{Y}$.

Next, with $\mathbf{y} \in \mathbf{Y}$ fixed, the sequence $(\tilde{p}_i(\mathbf{y}))_i$ is obviously monotone nondecreasing and bounded above by $J(\mathbf{y})$, hence with a limit $p^*(\mathbf{y}) \leq J(\mathbf{y})$. Therefore, $\tilde{p}_i$ has the pointwise limit $\mathbf{y} \mapsto p^*(\mathbf{y}) \leq J(\mathbf{y})$, and, since $\tilde{p}_i$ is also bounded below uniformly, by the monotone convergence theorem (or also the bounded convergence

[2]Recall that a sequence of measurable functions (g_n) on a measure space $(\mathbf{Y}, \mathcal{B}(\mathbf{Y}), \varphi)$ converges *φ-almost uniformly* to g if and only if, for every $\epsilon > 0$, there is a set $A \in \mathcal{B}(\mathbf{Y})$ such that $\varphi(A) < \epsilon$ and $g_n \to g$ uniformly on $\mathbf{Y} \setminus A$.

theorem), $\int_{\mathbf{Y}} \tilde{p}_i(\mathbf{y})\, d\varphi(\mathbf{y}) \to \int_{\mathbf{Y}} p^*(\mathbf{y})\, d\varphi(\mathbf{y})$. This latter fact, combined with (5.10) and $p_i(\mathbf{y}) \leq \tilde{p}_i(\mathbf{y}) \leq J(\mathbf{y})$, yields

$$0 = \int \left(J(\mathbf{y}) - p^*(\mathbf{y})\right) d\varphi(\mathbf{y}),$$

which in turn implies that $p^*(\mathbf{y}) = J(\mathbf{y})$ for φ-almost all $\mathbf{y} \in \mathbf{Y}$. Therefore, $\tilde{p}_i(\mathbf{y}) \to J(\mathbf{y})$ for φ-almost all $\mathbf{y} \in \mathbf{Y}$. And so, by Egoroff's Theorem [7, Theorem 2.5.5], $\tilde{p}_i \to J(\cdot)$, φ-almost uniformly. □

5.3 A hierarchy of semidefinite relaxations

In general, solving the infinite-dimensional problem $\mathbf{P}$ and getting an optimal solution μ^* is out of reach. One possibility is to use numerical discretization schemes on a box containing $\mathbf{K}$; see for instance [47]. But in the present context of parametric optimization, if one selects finitely many *grid* points $(\mathbf{x}, \mathbf{y}) \in \mathbf{K}$, one is implicitly considering solving (or rather approximating) $\mathbf{P}_{\mathbf{y}}$ for finitely many points $\mathbf{y}$ in a grid of $\mathbf{Y}$, which we want to avoid. To avoid this numerical discretization scheme, we will use specific features of $\mathbf{P}$ when its data f (resp. $\mathbf{K}$) is a polynomial (resp. a compact basic semi-algebraic set).

Therefore, in this section we are now considering a *polynomial* parametric optimization problem, a special case of (5.1) as we assume the following:

- $f \in \mathbb{R}[\mathbf{x}, \mathbf{y}]$ and $g_j \in \mathbb{R}[\mathbf{x}, \mathbf{y}]$, for every $j = 1, \ldots, m$.
- $\mathbf{K}$ in (5.2) is compact and $\mathbf{Y} \subset \mathbb{R}^p$ is a compact basic semi-algebraic set.

For instance, $\mathbf{Y} \subset \mathbb{R}^p$ is the compact semi-algebraic set defined by

$$\mathbf{Y} = \left\{\mathbf{y} \in \mathbb{R}^p \,:\, g_k(\mathbf{y}) \geq 0,\ k = m+1, \ldots, t\right\} \tag{5.11}$$

for some polynomials $(g_k)_{k=m+1}^t \in \mathbb{R}[\mathbf{y}]$. Hence, the set $\mathbf{K} \subset \mathbb{R}^n \times \mathbb{R}^p$ in (5.2) is a compact basic semi-algebraic set.

We also assume that there is a probability measure φ on $\mathbf{Y}$, absolutely continuous with respect to the Lebesgue measure, whose moments $\gamma = (\gamma_\beta)$, $\beta \in \mathbb{N}^p$, are available. As already mentioned, if $\mathbf{Y}$ is a simple set (like, e.g., a simplex or a box) then one may choose φ to be the probability measure uniformly distributed on $\mathbf{Y}$, for which all moments can be computed easily. Sometimes, in the context of optimization with data uncertainty, the probability measure φ is already specified and in this case we assume that its moments $\gamma = (\gamma_\beta)$, $\beta \in \mathbb{N}^p$, can be computed effectively. Of course, if the set $\mathbf{Y}$ is discrete we choose a probability distribution with positive value at each point of $\mathbf{Y}$.

5.3.1 Semidefinite relaxations

We now describe what we call a *"joint + marginal" hierarchy* of semidefinite relaxations to compute (or at least approximate) the optimal value ρ of problem $\mathbf{P}$

in (5.4). With $\mathbf{K} \subset \mathbb{R}^n \times \mathbb{R}^p$ as in (5.2), let $v_k = \lceil (\deg g_k)/2 \rceil]$ for every $k = 1, \ldots, t$. Next, let $\gamma = (\gamma_\beta)$ with

$$\gamma_\beta = \int_{\mathbf{Y}} \mathbf{y}^\beta \, d\varphi(\mathbf{y}), \quad \forall \beta \in \mathbb{N}^p,$$

be the moments of a probability measure φ on $\mathbf{Y}$, absolutely continuous with respect to the Lebesgue measure, and let $i_0 = \max[\lceil (\deg f)/2 \rceil, \max_k v_k]$. For $i \geq i_0$, consider the following semidefinite relaxations:

$$\begin{aligned} \rho_i = \inf_{\mathbf{z}} \; & L_{\mathbf{z}}(f) \\ \text{s.t.} \; & \mathbf{M}_i(\mathbf{z}) \succeq 0, \\ & \mathbf{M}_{i-v_j}(g_j \, \mathbf{z}) \succeq 0, \quad j = 1, \ldots, t, \\ & L_{\mathbf{z}}(\mathbf{y}^\beta) = \gamma_\beta, \quad \forall \beta \in \mathbb{N}^p_i, \end{aligned} \tag{5.12}$$

where the sequence $\mathbf{z} = (z_{\alpha\beta})$, $\alpha \in \mathbb{N}^n, \beta \in \mathbb{N}^p$, is indexed in the canonical basis of $\mathbb{R}[\mathbf{x}, \mathbf{y}]$.

Theorem 5.6. *Let $\mathbf{K}, \mathbf{Y}$ be as in* (5.2) *and* (5.11) *respectively, and let $(g_k)_{k=1}^t$ satisfy Assumption* 1.16. *Assume that for every $\mathbf{y} \in \mathbf{Y}$ the set $\mathbf{K}_\mathbf{y}$ is nonempty, and, for φ-almost all $\mathbf{y} \in \mathbf{Y}$, $J(\mathbf{y})$ is attained at a unique optimal solution $\mathbf{x}^*(\mathbf{y})$. Consider the semidefinite relaxations* (5.12). *Then:*

(a) *$\rho_i \uparrow \rho$ as $i \to \infty$.*

(b) *Let $\mathbf{z}^i$ be a nearly optimal solution of* (5.12), *e.g., such that $L_{\mathbf{z}^i}(f) \leq \rho_i + 1/i$, and let $g \colon \mathbf{Y} \to \mathbf{K}_\mathbf{y}$ be the measurable mapping in Theorem* 5.2(c). *Then*

$$\lim_{i \to \infty} z^i_{\alpha\beta} = \int_{\mathbf{Y}} \mathbf{y}^\beta \, g(\mathbf{y})^\alpha \, d\varphi(\mathbf{y}), \quad \forall \alpha \in \mathbb{N}^n, \; \beta \in \mathbb{N}^p. \tag{5.13}$$

In particular, for every $k = 1, \ldots, n$,

$$\lim_{i \to \infty} z^i_{e(k)\beta} = \int_{\mathbf{Y}} \mathbf{y}^\beta \, g_k(\mathbf{y}) \, d\varphi(\mathbf{y}), \quad \forall \beta \in \mathbb{N}^p, \tag{5.14}$$

where $e(k) = (\delta_{j=k})_j \in \mathbb{N}^n$.

The proof is postponed to Section 5.5.

5.3.2 The dual semidefinite relaxations

The dual of the semidefinite relaxation (5.12) reads:

$$\begin{aligned}\rho_i^* = \sup_{p,(\sigma_i)} & \int_{\mathbf{Y}} p\, d\varphi \\ \text{s.t. } & f - p = \sigma_0 + \sum_{j=1}^{t} \sigma_j\, g_j, \\ & p \in \mathbb{R}[\mathbf{y}];\ \sigma_j \subset \Sigma[\mathbf{x},\mathbf{y}], \quad j = 0,1,\ldots,t, \\ & \deg p \leq 2i,\ \deg \sigma_0 \leq 2i,\ \deg \sigma_j g_j \leq 2i, \quad j = 1,\ldots,t.\end{aligned} \tag{5.15}$$

Observe that (5.15) is a strenghtening of (5.9), as one restricts to polynomials $p \in \mathbb{R}[\mathbf{y}]$ of degree at most $2i$ and the nonnegativity of $f - p$ in (5.9) is replaced with a stronger requirement in (5.15). Therefore, $\rho_i^* \leq \rho^*$ for every i.

Theorem 5.7. *Let $\mathbf{K}, \mathbf{Y}$ be as in* (5.2) *and* (5.11) *respectively, and let $(g_k)_{k=1}^t$ satisfy Assumption* 1.16. *Assume that for every $\mathbf{y} \in \mathbf{Y}$ the set $\mathbf{K_y}$ is nonempty, and consider the semidefinite relaxations* (5.15). *Then:*

(a) $\rho_i^* \uparrow \rho$ *as* $i \to \infty$.

(b) *Let $(p_i, (\sigma_j^i))$ be a nearly optimal solution of* (5.15), *e.g., such that $\int_{\mathbf{Y}} p_i d\varphi \geq \rho_i^* - 1/i$. Then $p_i \leq J(\cdot)$ and*

$$\lim_{i\to\infty} \int_{\mathbf{Y}} (J(\mathbf{y}) - p_i(\mathbf{y}))\, d\varphi(\mathbf{y}) = 0. \tag{5.16}$$

Moreover, if one defines

$$\tilde{p}_0 = p_0, \quad \mathbf{y} \longmapsto \tilde{p}_i(\mathbf{y}) = \max\left[\tilde{p}_{i-1}(\mathbf{y}), p_i(\mathbf{y})\right], \quad i = 1,\ldots,$$

then $\tilde{p}_i \to J(\cdot)$, φ-almost uniformly on $\mathbf{Y}$.

Proof. Recall that, by Lemma 5.4, $\rho = \rho^*$. Moreover, let $(p_k) \subset \mathbb{R}[\mathbf{y}]$ be a maximizing sequence of (5.9) as in Corollary 5.5 with value $s_k = \int_{\mathbf{Y}} p_k\, d\varphi$, and let $p'_k = p_k - 1/k$ for every k, so that $f - p'_k > 1/k$ on $\mathbf{K}$. By Theorem 1.17, there exist s.o.s. polynomials $(\sigma_j^k) \subset \Sigma[\mathbf{x},\mathbf{y}]$ such that $f - p'_k = \sigma_0^k + \sum_j \sigma_j^k g_j$. Letting d_k be the maximum degree of σ_0 and $\sigma_j g_j$, $j = 1,\ldots,t$, it follows that $(s_k - 1/k, (\sigma_j^k))$ is a feasible solution of (5.15) with $i = d_k$. Hence, $\rho^* \geq \rho_{d_k}^* \geq s_k - 1/k$ and the result (a) follows because $s_k \to \rho^*$ and the sequence ρ_i^* is monotone. On the other hand, (b) follows from Corollary 5.5. □

Thus, Theorem 5.7 provides a lower polynomial approximation $p_i \in \mathbb{R}[\mathbf{y}]$ of the optimal value function $J(\cdot)$. Its degree is bounded by $2i$, the order of the moments (γ_β) of φ taken into account in the semidefinite relaxation (5.15). Moreover, one may even define a piecewise polynomial lower approximation $\tilde{p}_i$ that converges to $J(\cdot)$ on $\mathbf{Y}$, φ-almost uniformly.

Functionals of the optimal solutions

Theorem 5.6 provides a way of approximating any polynomial functional on the optimal solutions of $\mathbf{P_y}$, $\mathbf{y} \in \mathbf{Y}$.

Corollary 5.8. *Let* $\mathbf{K}, \mathbf{Y}$ *be as in* (5.2) *and* (5.11) *respectively, and let* $(g_k)_{k=1}^t$ *satisfy Assumption* 1.16. *Assume that for every* $\mathbf{y} \in \mathbf{Y}$ *the set* $\mathbf{K_y}$ *is nonempty and, for* φ*-almost all* $\mathbf{y} \in \mathbf{Y}$, $J(\mathbf{y})$ *is attained at a unique optimal solution* $\mathbf{x}^*(\mathbf{y}) \in \mathbf{X}^*_{\mathbf{y}}$. *Let* $h \in \mathbb{R}[\mathbf{x}]$ *be of the form*

$$\mathbf{x} \longmapsto h(\mathbf{x}) = \sum_{\alpha\in\mathbb{N}^n} h_\alpha \, \mathbf{x}^\alpha,$$

and let $\mathbf{z}^i$ *be an optimal solution of the semidefinite relaxations* (5.12).

Then, for i *sufficiently large,*

$$\int_{\mathbf{Y}} h(\mathbf{x}^*(\mathbf{y}))\, d\varphi(\mathbf{y}) \approx \sum_{\alpha\in\mathbb{N}^n} h_\alpha \, \mathbf{z}^i_{\alpha 0}.$$

Proof. Follows from Theorem 5.6 and Corollary 5.3. □

5.3.3 Persistency for Boolean variables

One interesting and potentially useful application is in Boolean optimization. Indeed, suppose that, for some subset $I \subseteq \{1, \dots, n\}$, the variables (x_i), $i \in I$, are Boolean, that is, the definition of $\mathbf{K}$ in (5.2) includes the constraints $x_i^2 - x_i = 0$ for every $i \in I$.

Then, for instance, one might be interested in determining whether in an optimal solution $\mathbf{x}^*(\mathbf{y})$ of $\mathbf{P_y}$, and for some index $i \in I$, one has $x_i^*(\mathbf{y}) = 1$ (or $x_i^*(\mathbf{y}) = 0$) for every value of the parameter $\mathbf{y} \in \mathbf{Y}$. In [14], the probability that $x_k^*(\mathbf{y})$ is 1 is called the *persistency* of the Boolean variable $x_k^*(\mathbf{y})$.

Corollary 5.9. *Let* $\mathbf{K}, \mathbf{Y}$ *be as in* (5.2) *and* (5.11) *respectively. Let Assumption* 1.16 *hold. Assume that for every* $\mathbf{y} \in \mathbf{Y}$ *the set* $\mathbf{K_y}$ *is nonempty. Let* $\mathbf{z}^i$ *be a nearly optimal solution of the semidefinite relaxations* (5.12). *Then, for* $k \in I$ *fixed,*

(a) $x_k^*(\mathbf{y}) = 1$ *for* φ*-almost all* $\mathbf{y} \in \mathbf{Y}$ *only if* $\lim_{i\to\infty} z^i_{e(k)0} = 1$.

(b) $x_k^*(\mathbf{y}) = 0$ *for* φ*-almost all* $\mathbf{y} \in \mathbf{Y}$ *only if* $\lim_{i\to\infty} z^i_{e(k)0} = 0$.

Assume that, for φ*-almost all* $\mathbf{y} \in \mathbf{Y}$, $J(\mathbf{y})$ *is attained at a unique optimal solution* $\mathbf{x}^*(\mathbf{y}) \in \mathbf{X}^*_{\mathbf{y}}$. *Then* $\mathrm{Prob}(x_k^*(\mathbf{y}) = 1) = \lim_{i\to\infty} z^i_{e(k)0}$, *and so:*

(c) $x_k^*(\mathbf{y}) = 1$ *for* φ*-almost all* $\mathbf{y} \in \mathbf{Y}$ *if and only if* $\lim_{i\to\infty} z^i_{e(k)0} = 1$.

(d) $x_k^*(\mathbf{y}) = 0$ *for* φ*-almost all* $\mathbf{y} \in \mathbf{Y}$ *if and only if* $\lim_{i\to\infty} z^i_{e(k)0} = 0$.

Proof. (a) The *only if* part. Let $\alpha = e(k) \in \mathbb{N}^n$. From the proof of Theorem 5.6, there is a subsequence $(i_\ell)_\ell \subset (i)_i$ such that

$$\lim_{\ell\to\infty} z^{i_\ell}_{e(k)0} = \int_{\mathbf{K}} x_k \, d\mu^*,$$

where μ^* is an optimal solution of $\mathbf{P}$. Hence, by Theorem 5.2(b), μ^* can be disintegrated into $\psi^*(d\mathbf{x} \,|\, \mathbf{y}) d\varphi(\mathbf{y})$, where $\psi^*(\,\cdot\,|\,\mathbf{y})$ is a probability measure on $\mathbf{X}^*_{\mathbf{y}}$ for every $\mathbf{y} \in \mathbf{Y}$. Therefore,

$$\begin{aligned}
\lim_{\ell\to\infty} z^{i_\ell}_{e(k)0} &= \int_{\mathbf{Y}} \left(\int_{\mathbf{X}^*_{\mathbf{y}}} x_k \psi^*(d\mathbf{x} \,|\, \mathbf{y}) \right) d\varphi(y), \\
&= \int_{\mathbf{Y}} \psi^*(\mathbf{X}^*_{\mathbf{y}} \,|\, \mathbf{y}) \, d\varphi(y) \quad [\text{because } x^*_k = 1 \text{ in } \mathbf{X}^*_{\mathbf{y}}] \\
&= \int_{\mathbf{Y}} d\varphi(y) = 1,
\end{aligned}$$

and as the subsequence $(i_\ell)_\ell$ was arbitrary, the whole sequence $(z^i_{e(k)0})$ converges to 1, the desired result. The proof of (b), being exactly the same, is omitted.

Next, if for every $\mathbf{y} \in \mathbf{Y}$, $J(\mathbf{y})$ is attained at a singleton, by Theorem 5.6(b),

$$\begin{aligned}
\lim_{i\to\infty} z^i_{e(k)0} &= \int_{\mathbf{Y}} x^*_k(\mathbf{y}) \, d\varphi(y) = \varphi(\{\mathbf{y} : x^*_k(\mathbf{y}) = 1\}) \\
&= \mathrm{Prob}(x^*_k(\mathbf{y}) = 1),
\end{aligned}$$

from which (c) and (d) follow. □

5.3.4 Estimating the density $g(\mathbf{y})$

By Corollary 5.8, one may approximate any polynomial functional of the optimal solutions, like for instance the mean, variance, etc. (with respect to the probability measure φ). However, one may also wish to approximate (in some sense) the "curve" $\mathbf{y} \mapsto x^*_k(\mathbf{y})$, that is, the curve described by the kth coordinate $x^*_k(\mathbf{y})$ of the optimal solutions $\mathbf{x}^*(\mathbf{y})$ when $\mathbf{y}$ varies in $\mathbf{Y}$ (and $\mathbf{X}^*_{\mathbf{y}}$ is a singleton for φ-almost all $\mathbf{y} \in \mathbf{Y}$).

So let $g\colon \mathbf{Y} \to \mathbb{R}^n$ be the measurable mapping in Theorem 5.6 (i.e., $g(\mathbf{y}) = \mathbf{x}^*(\mathbf{y})$) and suppose that one knows some lower bound vector $\mathbf{a} = (a_i) \in \mathbb{R}^n$, where:

$$a_i \le \inf \{\mathbf{x}_i : (\mathbf{x}, \mathbf{y}) \in \mathbf{K}\}, \quad i = 1, \ldots, n.$$

Then, for every $k = 1, \ldots, n$, the measurable function $\widehat{g}_k\colon \mathbf{Y} \to \mathbb{R}^n$ defined by

$$\mathbf{y} \longmapsto \widehat{g}_k(\mathbf{y}) = x^*_k(\mathbf{y}) - a_k, \quad \mathbf{y} \in \mathbf{Y}, \tag{5.17}$$

is nonnegative and integrable with respect to φ.

Hence, for every $k = 1, \ldots, n$, one may consider $d\lambda = \widehat{g}_k\, dx$ as a Borel measure on $\mathbf{Y}$ with unknown density $\widehat{g}_k$ with respect to φ, but with known moments $\mathbf{u} = (u_\beta)$. Indeed, using (5.14),

$$u_\beta = \int_{\mathbf{Y}} \mathbf{y}^\beta\, d\lambda(\mathbf{y}) = -a_k \int_{\mathbf{Y}} \mathbf{y}^\beta\, d\varphi(\mathbf{y}) + \int_{\mathbf{Y}} \mathbf{y}^\beta\, x_k^*(\mathbf{y})\, d\varphi(\mathbf{y})$$
$$= -a_k \gamma_\beta + z_{e(k)\beta}, \quad \forall \beta \in \mathbb{N}^p, \tag{5.18}$$

where, for every $k = 1, \ldots, n$,

$$z_{e(k)\beta} = \lim_{i\to\infty} z^i_{e(k)\beta}, \quad \forall \beta \in \mathbb{N}^p.$$

Thus we are now faced with a so-called density estimation problem, that is, given the sequence of moments $u_\beta = \int_{\mathbf{Y}} \mathbf{y}^\beta \widehat{g}_k(\mathbf{y})\, d\varphi(\mathbf{y})$, $\beta \in \mathbb{N}^p$, for an unknown nonnegative measurable function $\widehat{g}_k$ on $\mathbf{Y}$, "estimate" $\widehat{g}_k$. One possibility is the so-called *maximum entropy* approach, briefly described in the next section.

Maximum-entropy estimation

We briefly describe the *maximum entropy* estimation technique in the univariate case. Let $g \in L_1([0,1])$[3] be a nonnegative function only known via the first $2d+1$ moments $\mathbf{u} = (u_j)_{j=0}^{2d}$ of its associated measure $d\varphi = g\, d\mathbf{x}$ on $[0,1]$. (In the context of the previous section, the function g to estimate is $\mathbf{y} \mapsto \widehat{g}_k(\mathbf{y})$ in (5.17) from the sequence $\mathbf{u}$ in (5.18) of its (multivariate) moments.)

From that partial knowledge one wishes (a) to provide an estimate h_d of g such that the first $2d+1$ moments of $h_d\, d\mathbf{x}$ match those of $g\, d\mathbf{x}$, and (b) analyze the asymptotic behavior of h_d when $d \to \infty$. This problem has important applications in various areas of physics, engineering, and signal processing in particular.

An elegant methodology is to search for h_d in a (finitely) parametrized family $\{h_d(\lambda, x)\}$ of functions, and optimize over the unknown parameters λ via a suitable criterion. For instance, one may wish to select an estimate h_d that maximizes some appropriate *entropy*. Several choices of entropy functional are possible as long as one obtains a convex optimization problem in the finitely many coefficients λ_i's. For more details, the interested reader is referred to, e.g., [18, 19] and the many references therein.

We here choose the Boltzmann–Shannon entropy $\mathcal{H}\colon L_1([0,1]) \to \mathbb{R} \cup \{-\infty\}$,

$$h \longmapsto \mathcal{H}[h] = -\int_0^1 h(x) \ln h(x)\, dx, \tag{5.19}$$

a strictly concave functional. Therefore, the problem reduces to

$$\sup_h \left\{ \mathcal{H}[h] \, : \, \int_0^1 x^j\, h(x)\, dx = u_j, \; j = 0, \ldots, 2d \right\}. \tag{5.20}$$

[3] $L_1([0,1])$ denotes the Banach space of integrable functions on the interval $[0,1]$ of the real line, equipped with the norm $\|g\| = \int_0^1 |b(\mathbf{x})|\, d\mathbf{x}$.

The structure of this infinite-dimensional convex optimization problem permits to search for an optimal solution h_d^* of the form

$$x \longmapsto h_d^*(x) = \exp \sum_{j=0}^{2d} \lambda_j^* \, x^j, \tag{5.21}$$

and so λ^* is an optimal solution of the finite-dimensional unconstrained convex problem

$$\sup_{\lambda} \, \langle \mathbf{u}, \lambda \rangle - \int_0^1 \exp\left(\sum_{j=0}^{2d} \lambda_j^* x^j \right) dx.$$

An optimal solution can be calculated by applying first-order methods, in which case the gradient $\nabla v(\lambda)$ of the function

$$\lambda \longmapsto v(\lambda) = \langle \mathbf{u}, \lambda \rangle - \int_0^1 \exp\left(\sum_{j=0}^{2d} \lambda_j^* x^j \right) dx,$$

at current iterate λ, is provided by

$$\frac{\partial v(\lambda)}{\partial \lambda_k} = u_k - \int_0^1 x^k \exp\left(\sum_{j=0}^{2d} \lambda_j^* x^j \right) d\mathbf{x}, \quad k = 0, \ldots, 2d+1.$$

If one applies Newton's method, then computing the Hessian $\nabla^2 v(\lambda)$ at current iterate λ requires to compute

$$\frac{\partial^2 v(\lambda)}{\partial \lambda_k \partial \lambda_j} = - \int_0^1 x^{k+j} \exp\left(\sum_{j=0}^{2d} \lambda_j^* x^j \right) dx, \quad k, j = 0, \ldots, 2d+1.$$

In such simple cases like the box $[0, 1]$ (or $[a, b]^n$ in the multivariate case), such quantities are approximated via cubature formulas which behave very well for exponentials of polynomials; see, e.g., [88].

Of course, the maximum-entropy estimate that we obtain is not a pointwise estimate, and so at some points of $[0, 1]$ the maximum-entropy density h^* and the density f to estimate may differ significantly. However, in many cases, the *shape* of h^* gives a good idea of f.

5.3.5 Illustrative examples

Example 5.10. For illustration purposes, consider the example where $\mathbf{Y} = [0, 1]$,

$$\mathbf{K} = \left\{ (x, y) \,:\, 1 - x^2 + y^2 \geq 0, \, y \in \mathbf{Y} \right\} \subset \mathbb{R}^2, \; (x, y) \longmapsto f(x, y) = -x^2 y.$$

Hence, for each value of the parameter $y \in \mathbf{Y}$, the unique optimal solution is $x^*(y) = \sqrt{1-y^2}$; and so, in Theorem 5.6(b), $y \mapsto g(y) = \sqrt{1-y^2}$.

Let φ be the probability measure uniformly distributed on $[0,1]$. Therefore,

$$\rho = \int_0^1 J(y)\,d\varphi(y) = \int_0^1 y\big(1-y^2\big)\,dy = 1/4.$$

Solving (5.12) with $i = 4$, that is, with moments up to order 8, one obtains the optimal value 0.25001786 and the moment sequence

$$\begin{aligned}\mathbf{z} = (&1, 0.7812, 0.5, 0.6604, 0.3334, 0.3333, 0.5813, 0.25, 0.1964, 0.25, 0.5244,\\ &0.2, 0.1333, 0.1334, 0.2, 0.4810, 0.1667, 0.0981, 0.0833, 0.0983, 0.1667).\end{aligned}$$

Observe that

$$z_{1k} - \int_0^1 y^k\sqrt{1-y^2}\,dy \approx O\big(10^{-6}\big), \quad k = 0,\dots,4,$$

$$z_{1k} - \int_0^1 y^k\sqrt{1-y^2}\,dy \approx O\big(10^{-5}\big), \quad k = 5,6,7.$$

Using a maximum-entropy approach to approximate the density $y \mapsto g(y)$ on $[0,1]$ with the first 5 moments z_{1k}, $k = 0,\dots,4$, we find that the optimal function h_4^* in (5.21) is obtained with

$$\lambda^* = (-0.1564, 2.5316, -12.2194, 20.3835, -12.1867).$$

Both curves of g and h_4^* are displayed in Figure 5.1. Observe that, with only 5 moments, the maximum-entropy solution h_4^* approximates g relatively well, even if it differs significantly at some points. Indeed, the shape of h_4^* resembles that of g.

Example 5.11. Again with $\mathbf{Y} = [0,1]$, let

$$\mathbf{K} = \big\{(\mathbf{x},y)\ :\ 1 - x_1^2 - x_2^2 \geq 0\big\} \subset \mathbb{R}^2, \quad (\mathbf{x},y) \longmapsto f\big(\mathbf{x},y\big) = yx_1 + \big(1-y\big)x_2.$$

For each value of the parameter $y \in \mathbf{Y}$, the unique optimal solution $\mathbf{x}^* \in \mathbf{K}$ satisfies

$$\big(x_1^*\big)^2 + \big(x_2^*\big)^2 = 1; \quad \big(x_1^*\big)^2 = \frac{y^2}{y^2+(1-y)^2}, \quad \big(x_2^*\big)^2 = \frac{(1-y)^2}{y^2+(1-y)^2},$$

with optimal value

$$J(y) = -\frac{y^2}{\sqrt{y^2+(1-y)^2}} - \frac{(1-y)^2}{\sqrt{y^2+(1-y)^2}} = -\sqrt{y^2+(1-y)^2}.$$

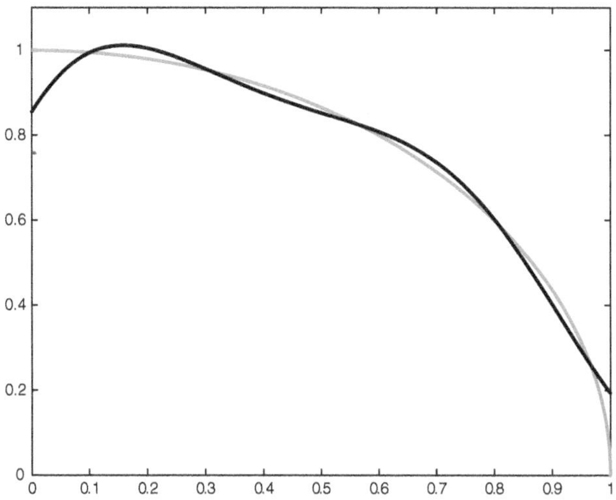

Figure 5.1: Example 5.10; $g(y) = \sqrt{1-y^2}$ versus $h_4^*(y)$.

So, in Theorem 5.6(b),

$$y \longmapsto g_1(y) = \frac{-y}{\sqrt{y^2+(1-y)^2}}, \quad y \longmapsto g_2(y) = \frac{y-1}{\sqrt{y^2+(1-y)^2}},$$

and with φ being the probability measure uniformly distributed on $[0,1]$,

$$\rho = \int_0^1 J(y)\, d\varphi(y) = -\int_0^1 \sqrt{y^2+(1-y)^2}\, dy \approx -0.81162.$$

Solving (5.12) with $i = 4$, that is, with moments up to order 8, one obtains $\rho_i \approx -0.81162$ with $\rho_i - \rho \approx O(10^{-6})$, and the moment sequence (z_{k10}), $k = 0, 1, 2, 3, 4$:

$$z_{k10} = (-0.6232,\ 0.4058,\ -0.2971,\ 0.2328,\ -0.1907),$$

and

$$z_{k10} - \int_0^1 y^k\, g_1(y)\, dy \approx O(10^{-5}), \quad k = 0, \dots, 4.$$

Using a maximum-entropy approach to approximate the density $y \mapsto -g_1(y)$ on $[0,1]$, with the first 5 moments z_{1k}, $k = 0, \dots, 4$, we find that the optimal function h_4^* in (5.21) is obtained with

$$\lambda^* = (-3.61284,\ 15.66153266,\ -29.430901,\ 27.326347,\ -9.9884452),$$

and we find that

$$z_{k10} + \int_0^1 y^k\, h_4^*(y)\, dy \approx O(10^{-11}), \quad k = 0, \dots, 4.$$

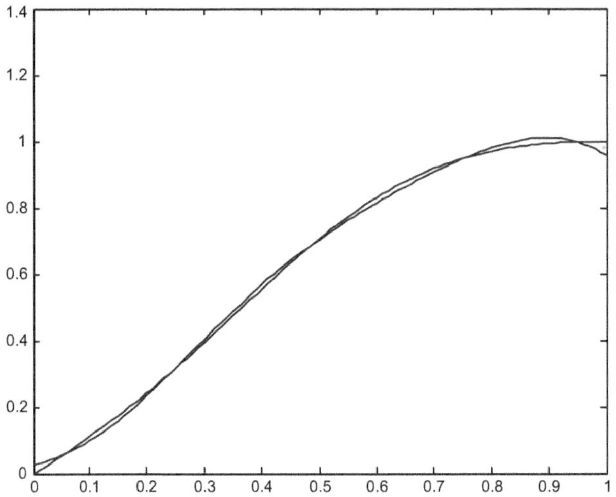

Figure 5.2: Example 5.11; $h_4^*(y)$ versus $-g_1(y) = y/\sqrt{y^2 + (1-y)^2}$.

Figure 5.2 displays the two functions $-g_1$ and h_4^*, and one observes a very good concordance.

We end up this section with the case where the density g_k to estimate is a step function which would be the case in an optimization problem $\mathbf{P_y}$ with Boolean variables (e.g., the variable x_k takes values in $\{0, 1\}$).

Example 5.12. Assume that, with a single parameter $y \in [0, 1]$, the density g_k to estimate is the step function

$$y \longmapsto g_k(y) = \begin{cases} 1 & \text{if } y \in [0, 1/3] \cup [2/3, 1], \\ 0 & \text{otherwise.} \end{cases}$$

The maximum-entropy estimate h_4^* in (5.21) with 5 moments is obtained with

$$\lambda^* = (-0.65473672,\ 19.170724,\ -115.39354,\ 192.4493171655,\ -96.226948865),$$

and we have

$$\int_0^1 y^k h_4^*(y)\,dy - \int_0^1 y^k\,dg_k(y) \approx O(10^{-8}), \quad k = 0, \dots, 4.$$

In particular, the persistency $\int_0^1 g_k(y)\,dy = 2/3$ of the variable $x_k^*(y)$ is very well approximated (up to a precision of 10^{-8}) by $\int h_4^*(y)\,dy$, with only 5 moments. Of course, in this case and with only 5 moments, the density h_4^* is not a good pointwise approximation of the step function g_k; however, as shown in Figure 5.3, its "shape" reveals the two steps of value 1 separated by a step of value 0. A better pointwise approximation would require many more moments.

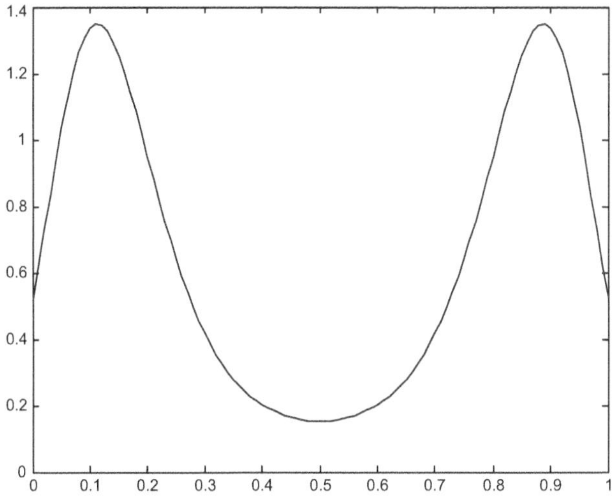

Figure 5.3: Example 5.12; $g_k(y) = 1_{[0,1/3]\cup[2/3,1]}$ versus $h_4^*(y)$.

5.4 A joint + marginal algorithm for polynomial optimization

We next show how to use the above results to define what we call the *joint + marginal algorithm* ((J + M)-algorithm) to provide an approximate solution for the polynomial optimization problem (3.1), i.e.,

$$\mathbf{P}: \quad f^* = \min\left\{f(\mathbf{x}) \,:\, \mathbf{x} \in \mathbf{K}\right\}, \tag{5.22}$$

where $f \in \mathbb{R}[\mathbf{x}]$ is a real-valued polynomial and $\mathbf{K} \subset \mathbb{R}^n$ is the basic closed semi-algebraic set defined by

$$\mathbf{K} = \left\{\mathbf{x} \in \mathbb{R}^n \,:\, g_i(\mathbf{x}) \geq 0,\ i = 1, \ldots, m\right\}, \tag{5.23}$$

for some polynomials $g_i \in \mathbb{R}[\mathbf{x}]$, $i = 1, \ldots, m$. The underlying idea behind the (J + M)-algorithm is very simple and can be summarized as follows.

For every $k = 1, \ldots, n$, let the compact interval $\mathbf{Y}_k = [\underline{x}_k, \overline{x}_k] \subset \mathbb{R}$ be contained in the projection of $\mathbf{K}$ into the x_k-coordinate axis.

(a) Treat x_1 as a parameter in the compact interval $\mathbf{Y}_1 = [\underline{x}_1, \overline{x}_1]$ with associated probability distribution φ_1 uniformly distributed on $\mathbf{Y}_1$.

(b) With $i \in \mathbb{N}$ fixed, solve the ith semidefinite relaxation (5.12) of the (J + M)-hierarchy applied to problem $\mathbf{P}(x_1)$ with $n - 1$ variables $(x_2, \ldots, x_n)$ and parameter $y = x_1$, which is problem $\mathbf{P}$ in (5.22) with the additional constraint that the variable $x_1 \in \mathbf{Y}_1$ is fixed. The dual (5.15) provides a univariate polynomial $x_1 \mapsto J_i^1(x_1)$ which, if i would increase, would converge to $J^1(x_1)$

in the $L_1(\varphi_1)$-norm. (The map $v \mapsto J^1(v)$ denotes the optimal value function of $\mathbf{P}(v)$, i.e., the optimal value of $\mathbf{P}$ in (5.22), given that the variable x_1 is fixed at the value v.) Next, compute $\tilde{x}_1 \in \mathbf{Y}_1$, a global minimizer of the univariate polynomial J_i^1 on $\mathbf{Y}_1$ (e.g., this can be done by solving a single semidefinite program). Ideally, when i is large enough, $\tilde{x}_1$ should be close to the first coordinate x_1^* of a global minimizer $\mathbf{x}^* = (x_1^*, \dots, x_n^*)$ of $\mathbf{P}$.

(c) Go back to step (b) with now $x_2 \in \mathbf{Y}_2 \subset \mathbb{R}$ instead of x_1, and with φ_2 being the probability measure uniformly distributed on $\mathbf{Y}_2$. With the same method, compute a global minimizer $\tilde{x}_2 \in \mathbf{Y}_2$ of the univariate polynomial $x_2 \mapsto J_i^2(x_2)$ on the interval $\mathbf{Y}_2$. Again, if i would increase, J_i^2 would converge in the $L_1(\varphi_2)$-norm to the optimal value function $v \mapsto J^2(v)$ of $\mathbf{P}(x_2)$ (i.e., the optimal value of $\mathbf{P}$ given that the variable x_2 is fixed at the value v). Iterate until one has obtained $\tilde{x}_n \in \mathbf{Y}_n \subset \mathbb{R}$.

One ends up with a point $\tilde{\mathbf{x}} \in \prod_{k=1}^n \mathbf{Y}_k$ and in general $\tilde{\mathbf{x}} \notin \mathbf{K}$. But then this point $\tilde{\mathbf{x}}$ may serve as initial guess in any local optimization procedure to find a local minimizer $\hat{\mathbf{x}} \in \mathbf{K}$ of $\mathbf{P}$. The rationale behind this algorithm is that if i is large enough then, by Theorem 5.7(b), the univariate polynomial $x_1 \mapsto J_i^1(x_1)$ provides a good approximation of $J^1(x_1)$. And so there is reasonable hope that a minimizer $\tilde{x}_1$ of J_i on $\mathbf{Y}_1$ is "close" to the coordinate x_1^* of some optimal solution $\mathbf{x}^*$ of $\mathbf{P}$. And similarly for $\tilde{x}_2, \dots, \tilde{x}_n$.

The computational complexity before the local optimization procedure is less than solving n times the ith semidefinite relaxation in the (J + M)-hierarchy, which is itself of same order as the ith semidefinite relaxation (3.14) in the hierarchy defined in Chapter 3, i.e., a polynomial in the input size of $\mathbf{P}$.

When the feasible set $\mathbf{K}$ is convex, one may define the following variant to obtain a *feasible* point $\tilde{\mathbf{x}} \in \mathbf{K}$. Again, let $\mathbf{Y}_1$ be the projection of $\mathbf{K}_1$ into the x_1-coordinate axis. Once $\tilde{x}_1 \in \mathbf{Y}_1$ is obtained in step (b), consider the new optimization problem $\mathbf{P}(\tilde{x}_1)$ in the $n-1$ variables $(x_2, \dots, x_n)$, obtained from $\mathbf{P}$ by fixing the variable $x_1 \in \mathbf{Y}_1$ at the value $\tilde{x}_1$. Its feasible set is the convex set $\mathbf{K}_1 = \mathbf{K} \cap \{\mathbf{x} : x_1 = \tilde{x}_1\}$. Let $\mathbf{Y}_2$ be the projection of $\mathbf{K}_1$ into the x_2-coordinate axis. Then go back to step (b) with now $x_2 \in \mathbf{Y}_2$ as parameter and $(x_3, \dots, x_n)$ as variables, to obtain a point $\tilde{x}_2 \in \mathbf{Y}_2$, etc. until a point $\tilde{\mathbf{x}} \in \prod_{k=1}^n \mathbf{Y}_k$ is obtained. Notice that now $\tilde{\mathbf{x}} \in \mathbf{K}$ because $\mathbf{K}$ is convex. Then proceed as before with $\tilde{\mathbf{x}}$ being the initial guess of a local minimization algorithm to obtain a local minimizer $\hat{\mathbf{x}} \in \mathbf{K}$ of $\mathbf{P}$.

The above variant can also be applied to 0/1 optimization problems where feasibility is easy to detect (like, e.g., the MAXCUT, knapsack or k-cluster problems). In this case $\mathbf{Y} = \{0, 1\}$ and $\varphi(\{0\}) = 1 - \varphi(\{1\}) = p$ for some $p \in (0, 1)$. The even more sophisticated *max-gap* variant consists of running n times the ith semidefinite relaxation of the (J + M)-hierarchy with x_1 as a parameter, then with x_2 as a parameter... and finally with x_n as a parameter, i.e., the basic procedure when $\mathbf{K}$ is not convex. Then select the index k for which the gap $|J_i^k(0) - J_i^k(1)|$ is maximum, and set $x_k = 0$ if $J_i^k(0) < J_i^k(1)$ and $x_k = 1$ otherwise. Then repeat the procedure with the $n-2$ variables $(x_j)_{j\neq k}$, etc. The rationale behind this more computationally demanding variant is that the larger the gap $|J_i^k(0) - J_i^k(1)|$, the more likely the resulting decision on x_k is right.

Some practical aspects

In step 1 one has to make sure that for every $y \in \mathbf{Y}$ one has $J(y) < \infty$, that is, the parametrized optimization problem has at least a feasible solution. If the set $\{x_1 : J(x_1) < \infty\}$ is connected (i.e., some interval $[a, b]$, as $\mathbf{K}$ is assumed to be compact) then one may obtain an approximation of a and b as follows.

Solve the first or second semidefinite relaxation of the hierarchy (3.14) applied to $\mathbf{P}$ with $\mathbf{x} \mapsto f(\mathbf{x}) = x_1$, and with $\mathbf{x} \mapsto f(\mathbf{x}) = -x_1$, yielding respective optimal values $a_1 \leq a$ and $b \leq -b_1$. If $\mathbf{K}$ is convex, one may obtain the exact value if $\mathbf{K}$ is a polytope or a very good approximation otherwise. In general, assuming that the approximation is good and choosing some scalar $0 < \rho < 1$ not too far from 1, one may reasonably expect that $a \leq \rho a_1 \leq -\rho b_1 \leq b$, and choose the parameter set $\mathbf{Y} = [\rho a_1, -\rho b_1]$.

Observe that, in 0/1-optimization problems, $\mathbf{Y} = \{0, 1\}$ and so if $a_1 > 0$ (resp. if $-b_1 < 1$) then one may directly decide whether $x_1^* = 1$ (resp. $x_1^* = 0$) in any optimal solution of $\mathbf{P}$. If $a_1 \leq 0$ and $b_1 \leq -1$ then one may safely choose $\mathbf{Y} = \{0, 1\}$.

In step 3 one has to find the global minimum of the univariate polynomial $x_1 \mapsto J_i^1(x_1)$ on some interval $\mathbf{Y} = [a, b]$ (that can be chosen to be $[0, 1]$ possibly after some change of variable). By Theorem 3.15, it suffices to solve a single semidefinite program. In the case of 0/1-optimization, one simply has to check whether $J_k(0) < J_k(1)$ to decide whether $\tilde{x}_1 = 0$ or 1.

Preliminary computational results

We have tested the (J + M)-algorithm on three NP-hard discrete optimization problems, namely the MAXCUT, k-cluster, and 0/1-knapsack problems. With n about 40 to 80, and according to the current status of public semidefinite solvers, we can only solve the first semidefinite relaxation (5.12), i.e., with $i = 1$.

The MAXCUT problem. The celebrated MAXCUT problem consists of solving

$$\mathbf{P}: \quad \min_{\mathbf{x}} \{\mathbf{x}'Q\mathbf{x} \,:\, \mathbf{x} \in \{-1, 1\}^n\},$$

for some real symmetric matrix $Q = (Q_{ij}) \in \mathbb{R}^{n\times n}$. In our sample of randomly generated problems, the entry Q_{ij} of the real symmetric matrix Q is set to zero with probability 1/2, and, when different from zero, Q_{ij} is randomly (and independently) generated according to the uniform probability distribution on the interval $[0, 10]$.

We have tested the max-gap variant of the (J + M)-algorithm with $i = 1$ and $p = 1/2$, on random graphs with $n = 20$, 30 and 40 variables. The semidefinite relaxation (5.12) reads:

$$\rho_1 = \min_{\mathbf{x},\mathbf{X}} \ \mathrm{trace}\,(Q\,\mathbf{X})$$
$$\text{s.t.} \quad \begin{bmatrix} 1 & \mathbf{x}' \\ \mathbf{x} & \mathbf{X} \end{bmatrix} \succeq 0, \quad \mathbf{X}' = \mathbf{X} \in \mathbb{R}^{n\times n},$$
$$X_{jj} = 1, \quad j = 1,\dots,n,$$
$$x_1 = 1 - 2p.$$

Without the marginal constraint $x_1 = 1 - 2p$, this is just the first semidefinite relaxation in the hierarchy (3.14) applied to $\mathbf{P}$, the celebrated Shor's relaxation with famous Goemans and Williamson's 0.878 performance guarantee (whose optimal value is denoted by $\mathbf{Q}_1$). For each value of n, we have generated 50 problems and 100 for $n = 40$. Let $\mathbf{P}_1$ denote the cost of the solution $\mathbf{x} \in \{-1, 1\}^n$ generated by the (J + M)-algorithm[4]. In Table 5.1 below, we have reported the average relative error $(\mathbf{P}_1 - \mathbf{Q}_1)/|\mathbf{Q}_1|$, which as one may see, is comparable with the Goemans and Williamson (GW) ratio.

n	20	30	40
$(\mathbf{P}_1 - \mathbf{Q}_1)/\lvert\mathbf{Q}_1\rvert$	10.3%	12.3%	12.5%

Table 5.1: Relative error for MAXCUT.

In fact, if one evaluates the true MAXCUT criterion $\frac{1}{2}\max\sum_{i,j} Q_{ij}(1-x_ix_j)$ then the relative error in Table 5.1 becomes around 4%, like for the solution obtained via the Goemans-Williamson randomized rounding procedure.

The k-cluster problem. The k-cluster problem consists of solving

$$\mathbf{P}: \quad \min_{\mathbf{x}} \left\{ \mathbf{x}'Q\mathbf{x} \ : \ \mathbf{x} \in \{0,1\}^n; \ \sum_{\ell=1}^{n} x_\ell = k \right\}, \tag{5.24}$$

again for some real symmetric matrix $Q = (Q_{ij}) \in \mathbb{R}^{n\times n}$, and some fixed integer $k \in \mathbb{N}$, $1 \le k < n$. Observe that one may add the n constraints $x_j(k - \sum_\ell x_\ell) = 0$, $j = 1,\dots,n$, in the definition (5.24) of $\mathbf{P}$ because they are redundant. However these constraints make the ith semidefinite relaxation (5.12) more constrained. As for MAXCUT with $i = 1$, the semidefinite relaxation (5.12), which reads

$$\rho_1 = \min_{\mathbf{x},\mathbf{X}} \ \mathrm{trace}\,(Q\,\mathbf{X})$$

[4] $\mathbf{Q}_1$ and $\mathbf{P}_1$ were computed with the GloptiPoly software of [45], dedicated to solving the generalized problem of moments.

$$\begin{aligned}
\text{s.t.}\quad & \begin{bmatrix} 1 & \mathbf{x}' \\ \mathbf{x} & \mathbf{X} \end{bmatrix} \succeq 0, \quad \mathbf{X}' = \mathbf{X} \in \mathbb{R}^{n\times n}, \\
& X_{jj} = x_j, \quad j = 1,\dots,n, \\
& k\,x_j - \sum_{\ell=1}^{n} X_{j\ell} \geq 0, \quad j = 1,\dots,n, \\
& x_1 = 1 - p,
\end{aligned}$$

is the same as the first in the hierarchy of semidefinite relaxations (3.14) to solve problem $\mathbf{P}$ (with optimal value denoted by $\mathbf{Q}_1$) but with the additional moment constraint $x_1 = 1-p$. Let $\mathbf{P}_1$ denote the cost of the solution $\mathbf{x} \in \{0,1\}^n$ generated by the (J + M)-algorithm. We have tested the (J + M)-algorithm on problems randomly generated as for MAXCUT, and with $k = n/2$. The average relative error $(\mathbf{P}_1 - \mathbf{Q}_1)/\mathbf{Q}_1$ was:

- 5.7% on 4 randomly generated problems with $n = 60$ variables.
- 4.5% and 5.6% on 2 randomly generated problems with $n = 70$ variables. The "max-gap" variant was a little better ($\approx$ 4% and $\approx$ 4.5% respectively).
- 5.7% on a problem with $n = 80$ variables.

The 0/1 knapsack problem. The 0/1 knapsack problem consists of solving

$$\mathbf{P}: \quad \max_{\mathbf{x}} \left\{ \mathbf{c}'\mathbf{x} \,:\, \mathbf{x} \in \{0,1\}^n;\ \sum_{\ell=1}^{n} a_\ell\, x_\ell \leq b \right\}, \tag{5.25}$$

for some real vector $\mathbf{c} \in \mathbb{R}^n$ and $\mathbf{a} \in \mathbb{N}^n$, $b \in \mathbb{N}$.

As for the k-cluster problem, one may add the n redundant constraints $x_j(b - \sum_\ell a_\ell x_\ell) \geq 0$ and $(1 - x_j)(b - \sum_\ell a_\ell x_\ell) \geq 0$, $j = 1,\dots,n$, in the definition (5.25) of $\mathbf{P}$. Again, with $i = 1$, the semidefinite relaxation (5.12), which reads

$$\begin{aligned}
\rho_i = \max_{\mathbf{x},\mathbf{X}}\ & \mathbf{c}'\mathbf{x} \\
\text{s.t.}\quad & \begin{bmatrix} 1 & \mathbf{x}' \\ \mathbf{x} & \mathbf{X} \end{bmatrix} \succeq 0, \quad \mathbf{X}' = \mathbf{X} \in \mathbb{R}^{n\times n}, \\
& X_{jj} = x_j, \quad j = 1,\dots,n, \\
& b\,x_j - \sum_{\ell=1}^{n} a_\ell X_{j\ell} \geq 0, \quad j = 1,\dots,n, \\
& b - b\,x_j - \sum_{\ell=1}^{n} a_\ell (x_\ell - X_{j\ell}) \geq 0, \quad j = 1,\dots,n, \\
& x_1 = 1 - p,
\end{aligned}$$

is the same as the first in the hierarchy of semidefinite relaxations (3.14) to solve problem $\mathbf{P}$ (with optimal value denoted $\mathbf{Q}_1$) but with the additional moment constraint $x_1 = 1-p$. Let $\mathbf{P}_1$ denote the cost of the solution $\mathbf{x} \in \{0,1\}^n$ generated by the (J + M)-algorithm. We have tested the max-gap variant of the (J + M)-algorithm on a sample of 16 problems with $n = 50$ variables and 3 problems with $n = 60$ variables, where $b = \sum_\ell a_\ell/2$ and the integers a_ℓ's are generated uniformly in $[10, 100]$. The vector $\mathbf{c}$ is generated by $c_\ell = s\epsilon + a_\ell$ with $s = 0.1$ and ϵ is a random variable uniformly distributed in $[0, 1]$. From the results reported in Table 5.2, one may see that very good relative errors are obtained.

n	50	60
$(\mathbf{Q}_1 - \mathbf{P}_1)/\mathbf{Q}_1$	2.1%	0.62%

Table 5.2: Relative error for 0/1 knapsack.

5.5 Appendix

Proof of Theorem 5.6. We already know that $\rho_i \leq \rho$ for all $i \geq i_0$. We also need to prove that $\rho_i > -\infty$ for sufficiently large i. Let $Q \subset \mathbb{R}[\mathbf{x},\mathbf{y}]$ be the *quadratic module* generated by the polynomials $(g_j) \subset \mathbb{R}[\mathbf{x},\mathbf{y}]$ that define $\mathbf{K}$, i.e.,

$$Q = \left\{\sigma \in \mathbb{R}[\mathbf{x},\mathbf{y}] \;:\; \sigma = \sigma_0 + \sum_{j=1}^{t} \sigma_j\, g_j \text{ with } (\sigma_j)_{j=0}^{t} \subset \Sigma[\mathbf{x},\mathbf{y}]\right\}.$$

In addition, let $Q(\ell) \subset Q$ be the set of elements $\sigma \in Q$ which have a representation $\sigma_0 + \sum_{j=0}^{t} \sigma_j\, g_j$ for some s.o.s. family $(\sigma_j) \subset \Sigma[\mathbf{x},\mathbf{y}]$ with $\deg \sigma_0 \leq 2\ell$ and $\deg \sigma_j g_j \leq 2\ell$ for all $j = 1, \ldots, t$.

Let $i \in \mathbb{N}$ be fixed. As $\mathbf{K}$ is compact, there exists N such that $N \pm \mathbf{x}^\alpha \mathbf{y}^\beta > 0$ on $\mathbf{K}$, for all $\alpha \in \mathbb{N}^n$ and $\beta \in \mathbb{N}^p$, with $|\alpha + \beta| \leq 2i$. Therefore, under Assumption 1.16, the polynomial $N \pm \mathbf{x}^\alpha \mathbf{y}^\beta$ belongs to Q; see Theorem 1.18. But there is even some $\ell(i)$ such that $N \pm \mathbf{x}^\alpha \mathbf{y}^\beta \in Q(\ell(i))$ for every $|\alpha + \beta| \leq 2i$. Of course we also have $N \pm \mathbf{x}^\alpha \mathbf{y}^\beta \in Q(\ell)$ for every $|\alpha + \beta| \leq 2i$, whenever $\ell \geq \ell(i)$. Therefore, let us take $\ell(i) \geq i_0$. For every feasible solution $\mathbf{z}$ of $\mathbf{Q}_{l(i)}$ one has

$$|z_{\alpha\beta}| = \left| L_{\mathbf{z}}(\mathbf{x}^\alpha \mathbf{y}^\beta) \right| \leq N, \quad \forall |\alpha + \beta| \leq 2i.$$

This follows from $z_0 = 1$, $\mathbf{M}_{\ell(i)}(\mathbf{z}) \succeq 0$ and $\mathbf{M}_{\ell(i)-v_j}(g_j\, \mathbf{z}) \succeq 0$, which implies

$$N z_0 \pm z_{\alpha\beta} = L_{\mathbf{z}}(N \pm \mathbf{x}^\alpha \mathbf{y}^\beta) = L_{\mathbf{z}}(\sigma_0) + \sum_{j=1}^{t} L_{\mathbf{z}}(\sigma_j\, g_j) \geq 0$$

(for some $(\sigma_j) \subset \Sigma[\mathbf{x},\mathbf{y}]$ with $\deg \sigma_j\, g_j \leq 2\ell(i)$). Therefore, in particular, $L_{\mathbf{z}}(f) \geq -N\sum_{\alpha,\beta}|f_{\alpha\beta}|$, which proves that $\rho_{\ell(i)} > -\infty$, and so $\rho_i > -\infty$ for all sufficiently large i.

From what precedes, and with $k \in \mathbb{N}$ arbitrary, let $\ell(k) \geq k$ and N_k be such that

$$N_k \pm \mathbf{x}^\alpha \mathbf{y}^\beta \in Q\big(\ell(k)\big), \quad \forall\, \alpha \in \mathbb{N}^n,\, \beta \in \mathbb{N}^p \text{ with } |\alpha+\beta| \leq 2k. \tag{5.26}$$

Let $i \geq \ell(i_0)$, and let $\mathbf{z}^i$ be a nearly optimal solution of (5.12) with value

$$\rho_i \leq L_{\mathbf{z}^i}(f) \leq \rho_i + \frac{1}{i} \quad \left(\leq \rho + \frac{1}{i}\right). \tag{5.27}$$

Fix $k \in \mathbb{N}$. Notice that, from (5.26), for every $i \geq \ell(k)$ one has

$$\big|\, L_{\mathbf{z}^i}\big(\mathbf{x}^\alpha\mathbf{y}^\beta\big)\,\big| \leq N_k z_0 = N_k, \quad \forall \alpha \in \mathbb{N}^n,\ \beta \in \mathbb{N}^p \text{ with } |\alpha+\beta| \leq 2k.$$

Therefore, for all $i \geq \ell(i_0)$,

$$\big|z^i_{\alpha\beta}\big| = \big|\, L_{\mathbf{z}^i}\big(\mathbf{x}^\alpha\mathbf{y}^\beta\big)\,\big| \leq N'_k, \quad \forall \alpha \in \mathbb{N}^n,\ \beta \in \mathbb{N}^p \text{ with } \big|\alpha+\beta\big| \leq 2k, \tag{5.28}$$

where $N'_k = \max[N_k, V_k]$, with

$$V_k = \max_{\alpha,\beta,i} \left\{ \big|z^i_{\alpha\beta}\big| \;:\; \big|\alpha+\beta\big| \leq 2k;\ \ell(i_0) \leq i \leq \ell(k) \right\}.$$

Complete each vector $\mathbf{z}^i$ with zeros to make it an infinite bounded sequence in ℓ_∞, indexed by the canonical basis $(\mathbf{x}^\alpha\mathbf{y}^\beta)$ of $\mathbb{R}[\mathbf{x},\mathbf{y}]$. In view of (5.28),

$$\big|z^i_{\alpha\beta}\big| \leq N'_k, \quad \forall \alpha \in \mathbb{N}^n,\ \beta \in \mathbb{N}^p \text{ with } 2k-1 \leq \big|\alpha+\beta\big| \leq 2k, \tag{5.29}$$

and for all $k = 1, 2, \ldots$

Hence, let $\widehat{\mathbf{z}}^i \in \ell_\infty$ be the new sequence defined by

$$\widehat{z}^i_{\alpha\beta} = \frac{z^i_{\alpha\beta}}{N'_k}, \quad \forall \alpha \in \mathbb{N}^n, \beta \in \mathbb{N}^p \text{ with } 2k-1 \leq |\alpha+\beta| \leq 2k, \quad \forall\, k = 1, 2, \ldots,$$

and in ℓ_∞ consider the sequence $\{\widehat{\mathbf{z}}^i\}_i$ as $i \to \infty$.

Obviously, the sequence $\widehat{\mathbf{z}}^i$ is in the unit ball B_1 of ℓ_∞, and so, by the Banach–Alaoglu theorem (see, e.g., Ash [7]), there exist $\widehat{\mathbf{z}} \in B_1$ and a subsequence $\{i_\ell\}$ such that $\widehat{\mathbf{z}}^{i_\ell} \to \widehat{\mathbf{z}}$ as $\ell \to \infty$ for the weak $\star$ topology $\sigma(\ell_\infty, \ell_1)$ of ℓ_∞. In particular, pointwise convergence holds, that is,

$$\lim_{\ell\to\infty} \widehat{z}^{i_\ell}_{\alpha\beta} \longrightarrow \widehat{z}_{\alpha\beta}, \quad \forall \alpha \in \mathbb{N}^n,\ \beta \in \mathbb{N}^p.$$

Next, define

$$z_{\alpha\beta} = \widehat{z}_{\alpha\beta} \times N'_k, \quad \forall \alpha \in \mathbb{N}^n,\ \beta \in \mathbb{N}^p \text{ with } 2k-1 \leq |\alpha+\beta| \leq 2k, \quad \forall k = 1, 2, \ldots$$

The pointwise convergence $\widehat{z}^{i_\ell} \to \widehat{y}$ implies the pointwise convergence $\mathbf{z}^{i_\ell} \to \mathbf{z}$, i.e.,

$$\lim_{\ell\to\infty} z^{i_\ell}_{\alpha\beta} \longrightarrow z_{\alpha\beta}, \quad \forall \alpha \in \mathbb{N}^n,\ \beta \in \mathbb{N}^p. \tag{5.30}$$

Next, let $s \in \mathbb{N}$ be fixed. From the pointwise convergence (5.30) we deduce that

$$\lim_{\ell\to\infty} \mathbf{M}_s(z^{i_\ell}) = \mathbf{M}_s(\mathbf{z}) \succeq 0.$$

Similarly,

$$\lim_{\ell\to\infty} \mathbf{M}_s(g_j\, \mathbf{z}^{i_\ell}) = \mathbf{M}_s(g_j\, \mathbf{z}) \succeq 0, \quad j = 1,\dots,t.$$

As s was arbitrary, we obtain

$$\mathbf{M}_s(\mathbf{y}) \succeq 0; \quad \mathbf{M}_s(g_j\, \mathbf{z}) \succeq 0, \quad j = 1,\dots,t, \quad s = 0,1,2,\dots \tag{5.31}$$

By Theorem 1.17, (5.31) implies that $\mathbf{z}$ is the sequence of moments of some finite measure μ^* with support contained in $\mathbf{K}$. Moreover, the pointwise convergence (5.30) also implies that

$$\gamma_\beta = \lim_{\ell\to\infty} z^{i_\ell}_{0\beta} = z_{0\beta} = \int_{\mathbf{K}} \mathbf{y}^\beta\, d\mu^* = \int_{\mathbf{Y}} \mathbf{y}^\beta\, d\varphi(\mathbf{y}). \tag{5.32}$$

As measures on compacts sets are determinate, (5.32) implies that φ is the marginal of μ^* on $\mathbf{Y}$, and so μ^* is feasible for $\mathbf{P}$. Finally, combining the pointwise convergence (5.30) with (5.27) yields

$$\rho \ge \lim_{\ell\to\infty} \rho_{i_\ell} = \lim_{\ell\to\infty} L_{\mathbf{z}^{i_\ell}}(f) = L_{\mathbf{z}}(f) = \int f\, d\mu^*,$$

which in turn yields that μ^* is an optimal solution of $\mathbf{P}$. And so $\rho_{i_\ell} \to \rho$ as $\ell \to \infty$. As the sequence (ρ_i) is monotone, this yields the desired result (a).

(b) Next, let $\alpha \in \mathbb{N}^n$ and $\beta \in \mathbb{N}^p$ be fixed, arbitrary. From (5.30), we have:

$$\lim_{\ell\to\infty} z^{i_\ell}_{\alpha\beta} = z_{\alpha\beta} = \int_{\mathbf{K}} \mathbf{x}^\alpha\, \mathbf{y}^\beta\, d\mu^*,$$

and, by Theorem 5.2(c),

$$\lim_{\ell\to\infty} z^{i_\ell}_{\alpha\beta} = \int_{\mathbf{K}} \mathbf{x}^\alpha\, \mathbf{y}^\beta\, d\mu^* = \int_{\mathbf{Y}} \mathbf{y}^\beta\, g(\mathbf{y})^\alpha\, d\varphi(\mathbf{y}),$$

and, as the converging subsequence was arbitrary, the above convergence holds for the whole sequence $(z^i_{\alpha\beta})$. □

5.6 Notes and sources

The material of this chapter is from Lasserre (2010) [77]. There is a vast and rich literature on the topic and, for a detailed treatment, the interested reader is referred to, e.g., Bonnans and Shapiro (2000) [17] and the many references therein. Sometimes, in the context of optimization with data uncertainty, some probability distribution φ on the parameter set $\mathbf{Y}$ is available and in this context one is also interested in, e.g., the distribution of the optimal value, optimal solutions, all viewed as random variables. In fact, very often only some moments of φ are known (typically first- and second-order moments) and the goal of this more realistic moment-based approach is to obtain optimal bounds over all distributions that share this moment information. In particular, for discrete optimization problems where coefficients of the cost vector (of the objective function to optimize) are random variables with joint distribution φ, some bounds on the expected optimal value have been obtained. More recently, Natarajan et al. (2009) [90] extended the earlier work in Bertsimas et al. (2005) [14] to even provide a convex optimization problem for computing the so-called *persistency* values[5] of (discrete) variables, for a particular distribution φ^* in a certain set Φ of distributions that share some common moment information. However, the resulting second-order cone program requires knowledge of the convex hull of some discrete set of points, which is possible when the number of vertices is small. The approach is nicely illustrated on a discrete choice problem and a stochastic knapsack problem. For more details on persistency in discrete optimization, the interested reader is referred to Natarajan et al. (2009) [90] and the references therein. Later, Natarajan et al. (2009) [91] have considered mixed zero-one linear programs with uncertainty in the objective function and first- and second-order moment information. They show that computing the supremum of the expected optimal value (where the supremum is over all distributions sharing this moment information) reduces to solving a completely positive program.

Moment-based approaches also appear in robust optimization and stochastic programming under data uncertainty, where the goal is different, since, in this context, decisions of interest must be taken *before* the uncertain data is known. For these *min-max* type problems, a popular approach initiated in Arrow et al. (1958) [6] is to optimize decisions against the worst possible distribution θ (on uncertain data) taken in some set Φ of candidate distributions that share some common moment information (in general first- and second-order moments). Recently, Delage and Ye (2009) [30] have even considered the case where only a confidence interval is known for first- and second-order moments. For a nice discussion, the interested reader is referred to Delage and Ye (2009) [30] and the many references therein.

In the context of solving systems of polynomial equations whose coefficients are themselves polynomials of some parameter $\mathbf{y} \in \mathbf{Y}$, specific "parametric" meth-

[5]Given a 0-1 optimization problem $\max\{\mathbf{C}'\mathbf{x} : \mathbf{x} \in \mathcal{X} \cap \{0,1\}^n\}$ and a distribution φ on $\mathbf{C}$, the persistency value of the variable x_i is $\mathrm{Prob}_\varphi(x_i^* = 1)$ at an optimal solution $\mathbf{x}^*(\mathbf{C}) = (x_i^*)$.

ods exist. For instance, one may compute symbolically once and for all, what is called a *comprehensive* Gröbner basis, i.e., a fixed basis that, for all instances of $\mathbf{y} \in \mathbf{Y}$, is a Gröbner basis of the ideal associated with the polynomials in the system of equations; see Weispfenning (1992) [131] and more recently Rostalski (2009) [105] for more details. Then, when needed, and for any specific value of the parameter $\mathbf{y}$, one may compute all complex solutions of the system of equations, e.g., by the eigenvalue method of Möller and Stetter (1995) [89]. However, one still needs to apply the latter method for *each* value of the parameter $\mathbf{y}$. A similar two-step approach is also proposed for homotopy (instead of Gröbner bases) methods in Rostalski (2009) [105].

Glossary

- $\mathbb{N}$, the set of natural numbers
- $\mathbb{Z}$, the set of integers
- $\mathbb{R}$, the set of real numbers
- $\mathbb{R}_+$, the set of nonnegative real numbers
- $\mathbb{C}$, the set of complex numbers
- $\leq$, less than or equal to
- $\leqq$, inequality "$\leq$" or equality "$=$"
- $\mathbf{A}$, matrix in $\mathbb{R}^{m\times n}$
- $\mathbf{A}_j$, column j of matrix A
- $\mathbf{A} \succeq 0$ ($\succ 0$), $\mathbf{A}$ is positive semidefinite (definite)
- x, scalar $x \in \mathbb{R}$
- $\mathbf{x}$, vector $\mathbf{x} = (x_1, \ldots, x_n) \in \mathbb{R}^n$
- $\boldsymbol{\alpha}$, vector $\boldsymbol{\alpha} = (\alpha_1, \ldots, \alpha_n) \in \mathbb{N}^n$
- $\mathbf{x}^{\boldsymbol{\alpha}}$, vector $\mathbf{x}^{\boldsymbol{\alpha}} = (x_1^{\alpha_1}, \ldots, x_n^{\alpha_n})$
- $\mathbb{R}[x]$, ring of real univariate polynomials
- $\mathbb{R}[\mathbf{x}] = \mathbb{R}[x_1, \ldots, x_n]$, ring of real multivariate polynomials
- $\mathbb{R}[\mathbf{x}]_t \subset \mathbb{R}[\mathbf{x}]$, real multivariate polynomials of degree at most t
- $(\mathbb{R}[\mathbf{x}])^*$, the vector space of linear forms on $\mathbb{R}[\mathbf{x}]$
- $(\mathbb{R}[\mathbf{x}]_t)^*$, the vector space of linear forms on $\mathbb{R}[\mathbf{x}]_t$
- $\text{co}(\mathbf{K})$, convex hull of $\mathbf{K} \subset \mathbb{R}^n$
- $P(g) \subset \mathbb{R}[\mathbf{x}]$, preordering generated by the polynomials $(g_j) \subset \mathbb{R}[\mathbf{x}]$
- $Q(g) \subset \mathbb{R}[\mathbf{x}]$, quadratic module generated by the polynomials $(g_j) \subset \mathbb{R}[\mathbf{x}]$
- $\mathcal{M}(\mathbf{X})$, vector space of finite signed Borel measures on $\mathbf{X}$
- $\mathcal{M}(\mathbf{X})_+ \subset \mathcal{M}(\mathbf{X})$, space of finite Borel measures on $\mathbf{X}$
- $\mathcal{P}(\mathbf{X}) \subset \mathcal{M}(\mathbf{X})_+$, space of Borel probability measures on $\mathbf{X}$
- $V_{\mathbb{C}}(I) \subset \mathbb{C}^n$, the algebraic variety associated with an ideal $I \subset \mathbb{R}[\mathbf{x}]$
- $V_{\mathbb{R}}(I) \subset \mathbb{R}^n$, the real variety associated with an ideal $I \subset \mathbb{R}[\mathbf{x}]$

- s.o.s., sum of squares
- LP, linear programming
- SDP, semidefinite programming
- $\hat{f}$, convex envelope of $f\colon \mathbb{R}^n \to \mathbb{R}$
- $(\mathbf{x}^{\boldsymbol{\alpha}})$, canonical monomial basis of $\mathbb{R}[\mathbf{x}]$
- $\mathbf{y} = (y_{\boldsymbol{\alpha}}) \subset \mathbb{R}$, moment sequence indexed by the canonical basis of $\mathbb{R}[\mathbf{x}]$
- $\mathbf{M}_i(\mathbf{y})$, moment matrix of order i associated with the sequence $\mathbf{y}$
- $\mathbf{M}_i(g\,\mathbf{y})$, localizing matrix of order i associated with the sequence $\mathbf{y}$ and $g \in \mathbb{R}[\mathbf{x}]$
- $L_1(\mathbf{X}, \mu)$, the Banach space of all measurable functions on $\mathbf{X}$ such that $\displaystyle\int_{\mathbf{X}} |f| d\mu < \infty$
- $L_\infty(\mathbf{X}, \mu)$, the Banach space of all measurable functions on $\mathbf{X}$ such that $\|f\|_\infty = \operatorname{ess\,sup} |f| < \infty$
- $B(\mathbf{X})$, the space of bounded measurable functions on $\mathbf{X}$
- $C(\mathbf{X})$, the space of bounded continuous functions on $\mathbf{X}$
- $\mu_n \Rightarrow \mu$, for $(\mu_n)_n \subset \mathcal{M}(\mathbf{X})_+$, weak convergence of probability measures

Bibliography

[1] W. Adams and P. Loustaunau. *An Introduction to Gröbner Bases.* American Mathematical Society, Providence, RI, 1994.

[2] N.I. Akhiezer. *The Classical Moment Problem.* Hafner Publishing Company, New York, 1965.

[3] E. Anderson and P. Nash. *Linear Programming in Infinite-Dimensional Spaces.* John Wiley and Sons, Chichester, 1987.

[4] V.G. Andronov, E.G. Belousov, and V.M. Shironin. On solvability of the problem of polynomial programming. *Izvestija Akadem. Nauk SSSR, Teckhnicheskaja Kibernetika,* 4:194–197, 1982.

[5] M. Anjos. *New Convex Relaxations for the Maximum Cut and VLSI Layout Problems.* PhD thesis, University of Waterloo, Ontario, Canada, 2001. orion.math.uwaterloo.ca/~hwolkowi.

[6] K. Arrow, S. Karlin, and H. Scarf. *Studies in the Mathematical Theory of Inventory and Production.* Stanford University Press, Stanford, CA, 1958.

[7] R.B. Ash. *Real Analysis and Probability.* Academic Press, Inc., San Diego, 1972.

[8] E. Balas, S. Ceria, and G. Cornuéjols. A lift-and-project cutting plane algorithm for mixed 0/1 programs. *Math. Prog.,* 58:295–324, 1993.

[9] S. Basu, R. Pollack, and M.-F. Roy. *Algorithms in Real Algebraic Geometry,* volume 10 of *Algorithms and Computations in Mathematics.* Springer-Verlag, 2003.

[10] E. Becker and N. Schwartz. Zum Darstellungssatz von Kadison–Dubois. *Arch. Math.,* 40:421–428, 1983.

[11] E.G. Belousov and D. Klatte. A Frank–Wolfe type theorem for convex polynomial programs. *Comp. Optim. Appl.,* 22:37–48, 2002.

[12] C. Berg. The multidimensional moment problem and semi-groups. In *Moments in Mathematics,* pages 110–124. American Mathematical Society, Providence, RI, 1987.

[13] S. Bernstein. Sur la représentation des polynômes positifs. *Math. Z.,* 9:74–109, 1921.

[14] D. Bertsimas, K. Natarajan, and C.-P. Teo. Persistence in discrete optimization under data uncertainty. *Math. Program.*, 108:251–274, 2005.

[15] G. Blekherman. There are significantly more nonnegative polynomials than sums of squares. *Israel J. Math.*, 153:355–380, 2006.

[16] J. Bochnak, M. Coste, and M.-F. Roy. *Real Algebraic Geometry.* Springer-Verlag, New York, 1998.

[17] J.F. Bonnans and A. Shapiro. *Perturbation Analysis of Optimization Problems.* Springer-Verlag, New York, 2000.

[18] J. Borwein and A. Lewis. Convergence of best entropy estimates. *SIAM J. Optim.*, 1:191–205, 1991a.

[19] J. Borwein and A.S. Lewis. On the convergence of moment problems. *Trans. Amer. Math. Soc.*, 325:249–271, 1991b.

[20] T. Carleman. *Les fonctions quasi-analytiques.* Gauthier-Villars, Paris, 1926.

[21] G. Cassier. Problème des moments sur un compact de $\mathbb{R}^n$ et représentation de polynômes à plusieurs variables. *J. Func. Anal.*, 58:254–266, 1984.

[22] E. Chlamtac. Approximation algorithms using hierarchies of semidefinite programming relaxations. In *48th Annual IEEE Symposium on Foundations of Computer Science (FOCS'07)*, pages 691–701, 2007.

[23] E. Chlamtac and G. Singh. Improved approximation guarantees through higher levels of SDP hierarchies. In *Approximation Randomization and Combinatorial Optimization Problems*, volume 5171 of *LNCS*, pages 49–62. Springer-Verlag, 2008.

[24] R. Curto and L. Fialkow. Recursiveness, positivity, and truncated moment problems. *Houston Math. J.*, 17:603–635, 1991.

[25] R. Curto and L. Fialkow. Solution of the truncated complex moment problem for flat data. *Mem. Amer. Math. Soc.*, 119, 1996.

[26] R. Curto and L. Fialkow. Flat extensions of positive moment matrices: recursively generated relations. *Mem. Amer. Math. Soc.*, 136, 1998.

[27] R. Curto and L. Fialkow. The truncated complex **K**-moment problem. *Trans. Amer. Math. Soc.*, 352:2825–2855, 2000.

[28] E. de Klerk, M. Laurent, and P. Parrilo. A PTAS for the minimization of polynomials of fixed degree over the simplex. *Theor. Comp. Sci.*, 361:210–225, 2006.

[29] E. de Klerk, D.V. Pasechnik, and A. Schrijver. Reduction of symmetric semidefinite programs using the regular $\star$-representation. *Math. Prog*, 109: 613–624, 2007.

[30] E. Delage and Y. Ye. Distributionally robust optimization under moment uncertainty with application to data-driven problems. *Oper. Res.*, 58:596–612, 2010.

[31] P. Diaconis and D. Freedman. The Markov moment problem and de Finetti's Theorem Part I. *Math. Z.*, 247:183–199, 2006.

[32] E.G. Dynkin and A.A. Yushkevich. *Controlled Markov Processes.* Springer-Verlag, New York, 1979.

[33] M. Fekete. Proof of three propositions of Paley. *Bull. Amer. Math. Soc.*, 41: 138–144, 1935.

[34] W. Feller. *An Introduction to Probability Theory and Its Applications, 2nd Edition.* John Wiley & Sons, New York, 1966.

[35] C.A. Floudas, P.M. Pardalos, C.S. Adjiman, W.R. Esposito, Z.H. Gümüs, S.T. Harding, J.L. Klepeis, C.A. Meyer, and C.A. Schweiger. *Handbook of Test Problems in Local and Global Optimization.* Kluwer, Boston, MA, 1999. titan.princeton.edu/TestProblems.

[36] K. Gaterman and P. Parrilo. Symmetry group, semidefinite programs and sums of squares. *J. Pure Appl. Alg.*, 192:95–128, 2004.

[37] G. Golub and C.V. Loan. *Matrix Computations.* The Johns Hopkins University Press, 3rd edition, New York, 1996.

[38] D. Handelman. Representing polynomials by positive linear functions on compact convex polyhedra. *Pacific J. Math.*, 132:35–62, 1988.

[39] B. Hanzon and D. Jibetean. Global minimization of a multivariate polynomial using matrix methods. *J. Global Optim.*, 27:1–23, 2003.

[40] F. Hausdorff. Summationsmethoden und Momentfolgen I. *Soobshch. Kharkov matem. ob-va ser.* 2, 14:227–228, 1915.

[41] E.K. Haviland. On the momentum problem for distributions in more than one dimension, I. *Amer. J. Math.*, 57:562–568, 1935.

[42] E.K. Haviland. On the momentum problem for distributions in more than one dimension, II. *Amer. J. Math.*, 58:164–168, 1936.

[43] J.W. Helton and J. Nie. Semidefinite representation of convex sets. *Math. Prog.*, 122:21–64, 2010.

[44] J.W. Helton and M. Putinar. Positive Polynomials in Scalar and Matrix Variables, the Spectral Theorem and Optimization. In M. Bakonyi, A. Gheondea, and M. Putinar, editors, *Operator Theory, Structured Matrices and Dilations*, pages 229–306. Theta, Bucharest, 2007.

[45] D. Henrion, J.B. Lasserre, and J. Löfberg. GloptiPoly 3: Moments, optimization and semidefinite programming. *Optim. Meth. Software*, 24:761–779, 2009.

[46] O. Hernández-Lerma and J.B. Lasserre. *Discrete-Time Markov Control Processes: Basic Optimality Criteria.* Springer-Verlag, New York, 1996.

[47] O. Hernández-Lerma and J.B. Lasserre. Approximation schemes for infinite linear programs. *SIAM J. Optim.*, 8:973–988, 1998.

[48] T. Jacobi and A. Prestel. Distinguished representations of strictly positive polynomials. *J. Reine. Angew. Math.*, 532:223–235, 2001.

[49] D. Jibetean and E. de Klerk. Global optimization of rational functions: a semidefinite programming approach. *Math. Prog.*, 106:93–109, 2006.

[50] D. Jibetean and M. Laurent. Semidefinite approximations for global unconstrained polynomial optimization. *SIAM J. Optim.*, 16:490–514, 2005.

[51] S. Kim, M. Kojima, and H. Waki. Exploiting sparsity in SDP relaxation for sensor network localization. *SIAM J. Optim.*, 20:192–215, 2009.

[52] M. Kojima and M. Maramatsu. An extension of sums of squares relaxations to polynomial optimization problems over symmetric cones. *Math. Prog.*, 110:315–336, 2007.

[53] M. Kojima and M. Maramatsu. A note on sparse SOS and SDP relaxations for polynomial optimization problems over symmetric cones. *Comp. Optim. Appl.*, 42:31–41, 2009.

[54] M. Kojima, S. Kim, and M. Maramatsu. Sparsity in sums of squares of squares of polynomials. *Math. Prog.*, 103:45–62, 2005.

[55] M.G. Krein and A. Nudel'man. *The Markov moment problem and extremal problems*, volume 50 of *Transl. Math. Monographs*. American Mathematical Society, Providence, RI, 1977.

[56] J.L. Krivine. Anneaux préordonnés. *J. Anal. Math.*, 12:307–326, 1964a.

[57] J.L. Krivine. Quelques propriétés des préordres dans les anneaux commutatifs unitaires. *C. R. Acad. Sci. Paris*, 258:3417–3418, 1964b.

[58] S. Kuhlmann and M. Putinar. Positive polynomials on fibre products. *C. R. Acad. Sci. Paris, Ser.* 1, 1344:681–684, 2007.

[59] S. Kuhlmann and M. Putinar. Positive polynomials on projective limits of real algebraic varieties. *Bull. Sci. Math.*, 2009.

[60] S. Kuhlmann, M. Marshall, and N. Schwartz. Positivity, sums of squares and the multi-dimensional moment problem ii. *Adv. Geom.*, 5:583–606, 2005.

[61] H. Landau. *Moments in Mathematics*, volume 37. Proc. Sympos. Appl. Math., 1987.

[62] J.B. Lasserre. Optimisation globale et théorie des moments. *C. R. Acad. Sci. Paris, Série I*, 331:929–934, 2000.

[63] J.B. Lasserre. Global optimization with polynomials and the problem of moments. *SIAM J. Optim.*, 11:796–817, 2001.

[64] J.B. Lasserre. Semidefinite programming vs. LP relaxations for polynomial programming. *Math. Oper. Res.*, 27:347–360, 2002a.

[65] J.B. Lasserre. An explicit equivalent positive semidefinite program for nonlinear 0-1 programs. *SIAM J. Optim.*, 12:756–769, 2002b.

[66] J.B. Lasserre. Polynomials nonnegative on a grid and discrete optimization. *Trans. Amer. Math. Soc.*, 354:631–649, 2002c.

[67] J.B. Lasserre. Polynomial programming: LP-relaxations also converge. *SIAM J. Optim.*, 15:383–393, 2004.

[68] J.B. Lasserre. SOS approximations of polynomials nonnegative on a real algebraic set. *SIAM J. Optim.*, 16:610–628, 2005.

[69] J.B. Lasserre. Robust global optimization with polynomials. *Math. Prog.*, 107:275–293, 2006a.

[70] J.B. Lasserre. Convergent SDP-relaxations in polynomial optimization with sparsity. *SIAM J. Optim.*, 17:822–843, 2006b.

[71] J.B. Lasserre. *Moment, Positive Polynomials and Their Applications.* Imperial College Press, London, 2009.

[72] J.B. Lasserre. A sum of squares approximation of nonnegative polynomials. *SIAM J. Optim*, 16:751–765, 2006d.

[73] J.B. Lasserre. Sufficient conditions for a real polynomial to a sum of squares. *Arch. Math.*, 89:390–398, 2007.

[74] J.B. Lasserre. Representation of nonnegative convex polynomials. *Arch. Math.*, 91:126–130, 2008.

[75] J.B. Lasserre. Convexity in semi-algebraic geometry and polynomial optimization. *SIAM J. Optim.*, 19:1995–2014, 2009.

[76] J.B. Lasserre. Certificates of convexity for basic semi-algebraic sets. *Appl. Math. Letters*, 23:912–916, 2010a.

[77] J.B. Lasserre. A "joint + marginal" approach to parametric polynomial optimization. *SIAM J. Optim.*, 20:1995–2022, 2010b.

[78] J.B. Lasserre and T. Netzer. SOS approximations of nonnegative polynomials via simple high degree perturbations. *Math. Z.*, 256:99–112, 2007.

[79] M. Laurent. A comparison of the Sherali–Adams, Lovász–Schrijver and Lasserre relaxations for 0-1 programming. *Math. Oper. Res.*, 28:470–496, 2003.

[80] M. Laurent. Revisiting two theorems of Curto and Fialkow on moment matrices. *Proc. Amer. Math. Soc.*, 133:2965–2976, 2005.

[81] M. Laurent. Semidefinite representations for finite varieties. *Math. Prog.*, 109:1–26, 2007a.

[82] M. Laurent. Strengthened semidefinite programming bounds for codes. *Math. Prog.*, 109:239–261, 2007b.

[83] M. Laurent. Sums of squares, moment matrices and optimization over polynomials. In M. Putinar and S. Sullivant, editors, *Emerging Applications of Algebraic Geometry*, volume 149, pages 157–270. Springer-Verlag, New York, 2008.

[84] L. Lovász and A. Schrijver. Cones of matrices and set-functions and 0-1 optimization. *SIAM J. Optim.*, 1:166–190, 1991.

[85] M. Marshall. *Positive polynomials and sums of squares*, volume 146 of *AMS Math. Surveys and Monographs*. American Mathematical Society, Providence, RI, 2008.

[86] M. Marshall. Representation of non-negative polynomials, degree bounds and applications to optimization. *Canad. J. Math.*, 61:205–221, 2009a.

[87] M. Marshall. Polynomials non-negative on a strip. *Proc. Amer. Math. Soc.*, 138:1559–1567, 2010.

[88] L. Mead and N. Papanicolaou. Maximum entropy in the problem of moments. *J. Math. Phys.*, 25:2404–2417, 1984.

[89] H.M. Möller and H.J. Stetter. Multivariate polynomial equations with multiple zeros solved by matrix eigenproblems. *Num. Math.*, 70:311–329, 1995.

[90] K. Natarajan, M. Song, and C.-P. Teo. Persistency and its applications in choice modelling. *Manag. Sci.*, 55:453–469, 2009a.

[91] K. Natarajan, C.-P. Teo, and Z. Zheng. Mixed zero-one linear programs under objective uncertainty: a completely positive representation. *Oper. Res.*, 59:713–728, 2011.

[92] Y. Nesterov. Squared functional systems and optimization problems. In H. Frenk, K. Roos, T. Terlaky, and S. Zhang, editors, *High Performance Optimization*. Kluwer, Dordrecht, 2000.

[93] J. Nie and M. Schweighofer. On the complexity of Putinars' Positivstellensatz. *J. Complexity*, 23:135–150, 2007.

[94] J. Nie, J. Demmel, and B. Sturmfels. Semidefinite approximations for global unconstrained polynomial optimization. *Math. Prog.*, 106:587–606, 2006.

[95] A.E. Nussbaum. Quasi-analytic vectors. *Ark. Mat.*, 5:179–191, 1966.

[96] P. Parrilo. An explicit construction of distinguished representations of polynomials nonnegative over finite sets. Technical report, IfA, ETH, Zürich, Switzerland, 2002. Technical report AUTO-02.

[97] P.A. Parrilo. *Structured Semidefinite Programs and Semialgebraic Geometry Methods in Robustness and Optimization*. PhD thesis, California Institute of Technology, Pasadena, CA, 2000.

[98] P.A. Parrilo. Semidefinite programming relaxations for semialgebraic problems. *Math. Prog.*, 96:293–320, 2003.

[99] G. Pólya. *Collected Papers. Vol. II*. MIT Press, Cambridge, Mass-London, 1974.

[100] G. Pólya and G. Szegö. *Problems and Theorems in Analysis II*. Springer-Verlag, 1976.

[101] V. Powers and B. Reznick. Polynomials that are positive on an interval. *Trans. Amer. Soc.*, 352:4677–4692, 2000.

[102] A. Prestel and C.N. Delzell. *Positive Polynomials*. Springer-Verlag, Berlin, 2001.

[103] M. Putinar. Positive polynomials on compact semi-algebraic sets. *Indiana Univ. Math. J.*, 42:969–984, 1993.

[104] B. Reznick. Some concrete aspects of Hilbert's 17th problem. In *Real algebraic geometry and ordered structures; Contemporary Mathematics,* 253, pages 251–272. American Mathematical Society, Providence, RI, 2000.

[105] P. Rostalski. *Algebraic Moments: Real Root Finding and Related Topics.* PhD thesis, Automatic Control Laboratory, ETH Zürich, Switzerland, 2009.

[106] M. Schäl. Conditions for optimality and the limit of n-stage optimal policies to be optimal. *Z. Wahrs. verw. Gerb*, 32:179–196, 1975.

[107] C. Scheiderer. Positivity and sums of squares: A guide to some recent results. In M. Putinar and S. Sullivant, editors, *Emerging application of Algebraic Geometry*, IMA Book Series, pages 271–324. Springer-Verlag, 2008.

[108] C.W. Scherer and C.W.J. Hol. Sum of qsquares relaxation for polynomial semidefinite programming. In *Proceedings of the* 16*th International Symposium on Mathematical Theory of Newtworks and Systems, Leuven*, pages 1–10, 2004.

[109] J. Schmid. *On the degree complexity of Hilbert's* 17*th problem and the real Nullstellensatz.* PhD thesis, University of Dortmund, 1998. Habilitationsschrift zur Erlangung der Lehrbefugnis für das Fach Mathematik an der Universität Dortmund.

[110] K. Schmüdgen. The K-moment problem for compact semi-algebraic sets. *Math. Ann.*, 289:203–206, 1991.

[111] R. Schneider. *Convex Bodies: The Brunn–Minkowski Theory.* Cambridge University Press, Cambridge, United Kingdom, 1994.

[112] G. Schoenebeck. Linear level Lasserre lower bounds for certain k-CSPs. In 49*th Annual IEEE Symposium on Foundations of Computer Science* (*FOCS*'08), pages 593–602, 2008.

[113] A. Schrijver. New codes upper bounds from the Terwilliger algebra and semidefinite pogramming. *IEEE Trans. Info. Theory*, 51:2859–2866, 2005.

[114] M. Schweighofer. On the complexity of Schmüdgen's positivstellensatz. *J. Complexity*, 20:529–543, 2005.

[115] M. Schweighofer. Global optimization of polynomials using gradient tentacles and sums of squares. *SIAM J. Optim.*, 17:920–942, 2006.

[116] H.D. Sherali and W.P. Adams. A hierarchy of relaxations between the continuous and convex hull representations for zero-one programming problems. *SIAM J. Discr. Math.*, 3:411–430, 1990.

[117] H.D. Sherali and W.P. Adams. *A Reformulation-Linearization Technique for Solving Discrete and Continuous Nonconvex Problems.* Kluwer, Dordrecht, MA, 1999.

[118] H.D. Sherali, W.P. Adams, and C.H. Tuncbilek. A global optimization algorithm for polynomial programming problems using a reformulation-linearization technique. *J. Global Optim.*, 2:101–112, 1992.

[119] N.Z. Shor. Quadratic optimization problems. *Tekhnicheskaya Kibernetika*, 1:128–139, 1987.

[120] N.Z. Shor. *Nondifferentiable Optimization and Polynomial Problems*. Kluwer, Dordrecht, 1998.

[121] B. Simon. The classical moment problem as a self-adjoint finite difference operator. *Adv. Math.*, 137:82–203, 1998.

[122] G. Stengle. A Nullstellensatz and a Positivstellensatz in semialgebraic geometry. *Math. Ann.*, 207:87–97, 1974.

[123] J.F. Sturm. Using SeDuMi 1.02, a matlab toolbox for optimizing over symmetric cones. *Opt. Meth. Softw.*, 11-12:625–653, 1999.

[124] V. Tchakaloff. Formules de cubature mécanique à coefficients non négatifs. *Bull. Sci. Math.*, 81:123–134, 1957.

[125] F. Vallentin. Symmetry in semidefinite programs. Technical report, CWI, Amsterdam, The Netherlands, 2007. arXiv:0706.4233.

[126] L. Vandenberghe and S. Boyd. Semidefinite programming. *SIAM Rev.*, 38: 49–95, 1996.

[127] F.-H. Vasilescu. Spectral measures and moment problems. In *Spectral Theory and Its Applications*. Theta Ser. Adv. Math., 2, Theta, Bucharest, 2003, pages 173–215.

[128] H.H. Vui and P.T. Son. Global optimization of polynomials using the truncated tangency variety and sums of squares. *SIAM J. Optim.*, 19:941–951, 2008.

[129] S. Waki, S. Kim, M. Kojima, and M. Maramatsu. Sums of squares and semidefinite programming relaxations for polynomial optimization problems witth structured sparsity. *SIAM J. Optim.*, 17:218–242, 2006.

[130] H. Waki, S. Kim, M. Kojima, M. Muramatsu, and H. Sugimoto. SparsePOP: a sparse semidefinite programming relaxation of polynomial optimization problems. *ACM Trans. Math. Softw.*, 35:90–104, 2009.

[131] V. Weispfenning. Comprehensive Gröbner bases. *J. Symb. Comp.*, 14:1–29, 1992.

Part II

Computation of Generalized Nash Equilibria: Recent Advancements

Francisco Facchinei

Acknowledgement

These notes are based on a series of talks given at the *Advanced Course on Optimization: Theory, Methods, and Applications* held at the Centre de Recerca Matemàtica (CRM), Bellaterra, Barcelona, Spain, in July 2009. In preparing them I relied heavily on [36, 37, 38, 41, 43, 91]. I am very grateful to Aris Daniilidis and Juan Enrique Martínez-Legaz for inviting me to the Advanced Course and for their extreme patience with my delays in preparing these chapters. I am also very thankful to Marco Rocco, who read with great care a preliminary version of this manuscript.

The work of the author has been partially supported by MIUR-PRIN 2005 Research Program no. 2005017083, "Innovative Problems and Methods in Nonlinear Optimization".

Introduction

Non-cooperative game theory is a branch of game theory for the resolution of conflicts among players (or economic agents), each behaving selfishly to optimize one's own well-being subject to resource limitations and other constraints that may depend on the rivals' actions. The Generalized Nash Equilibrium Problem (GNEP) is a central model in game theory that has been used actively in many fields in the past fifty years. But it is only since the mid-nineties that research on this topic gained momentum, especially in the Operations Research (OR) community. These notes are an introduction to this fascinating and challenging field and aim at presenting a selection of recent results that we believe are central to current research, with a special emphasis on algorithmic developments. We neither strive to full generality nor to completeness. Rather, after introducing the topic along with some basic facts, we will present some recent developments that appear to be promising. The aim of these notes, in keeping with their origins, is mainly didactic; no proofs are presented, some repetitions are allowed for maximum clarity, and many important issues are glossed over. The choice of the topics is clearly biased by the research interests of the author; nevertheless our hope is that the reading of the notes might be useful to young researchers with a background in optimization, but little knowledge of games, in order to understand what the main research topics and approaches are in the GNEP field and to provide organized bibliographic entry points for a more advanced study.

The GNEP lies at the intersection of many different disciplines (e.g., economics, engineering, mathematics, computer science, OR), and sometimes researchers in different fields worked independently and unaware of existing results. This explains why this problem has a number of different names in the literature, including *pseudo-game*, *social equilibrium problem*, *equilibrium programming*, *coupled constraint equilibrium problem*, and *abstract economy*. We will stick to the term "generalized Nash equilibrium problem", that seems the one favorite by OR researchers in recent years.

Formally, the GNEP consists of N players, each player ν controlling the variables $x^\nu \in \mathbb{R}^{n_\nu}$. We denote by $\boldsymbol{x}$ the vector formed by all these decision variables,

$$\boldsymbol{x} = \begin{pmatrix} x^1 \\ \vdots \\ x^N \end{pmatrix},$$

which has dimension $n = \sum_{\nu=1}^{N} n_\nu$, and by $\boldsymbol{x}^{-\nu}$ the vector formed by all the players' decision variables except those of player ν. To emphasize the νth player's variables within $\boldsymbol{x}$, we sometimes write $(x^\nu, \boldsymbol{x}^{-\nu})$ instead of $\boldsymbol{x}$. Note that this is still the vector $\boldsymbol{x} = (x^1, \dots, x^\nu, \dots, x^N)$ and that, in particular, the notation $(x^\nu, \boldsymbol{x}^{-\nu})$ does not mean that the block components of $\boldsymbol{x}$ are reordered in such a way that x^ν becomes the first block.

Each player has an objective function $\theta_\nu \colon \mathbb{R}^n \to \mathbb{R}$ that depends on both his own variables x^ν as well as on the variables $\boldsymbol{x}^{-\nu}$ of all other players. This mapping θ_ν is often called the *utility function* of player ν, sometimes also the *payoff function* or *loss function*, depending on the particular application in which the GNEP arises.

Furthermore, each player's strategy must belong to a set $X_\nu(\boldsymbol{x}^{-\nu}) \subseteq \mathbb{R}^{n_\nu}$ that depends on the rival players' strategies and that we call the *feasible set* or *strategy space* of player ν. The aim of player ν, given the other players' strategies $\boldsymbol{x}^{-\nu}$, is to choose a strategy x^ν that solves the minimization problem

$$\min_{x^\nu} \theta_\nu(x^\nu, \boldsymbol{x}^{-\nu}) \quad \text{s.t.} \quad x^\nu \in X_\nu(\boldsymbol{x}^{-\nu}). \tag{1}$$

Note that we interpret the objective function as a loss function that therefore is to be minimized. Alternatively, we could have considered the maximization of the utility of each player, this being obviously equivalently, after a change of sign, to a minimization problem. For any $\boldsymbol{x}^{-\nu}$, the solution set of problem (1) is denoted by $\mathcal{S}_\nu(\boldsymbol{x}^{-\nu})$. The GNEP is the problem of finding a vector $\boldsymbol{x}_*$ such that

$$x_*^\nu \in \mathcal{S}_\nu(\boldsymbol{x}_*^{-\nu}) \quad \text{for all } \nu = 1, \dots, N.$$

Such a point $\boldsymbol{x}_*$ is called a (*generalized Nash*) *equilibrium* or, more simply, a *solution* of the GNEP. A point $\boldsymbol{x}_*$ is therefore an equilibrium if no player can decrease his objective function by changing *unilaterally* x_*^ν to any other feasible point. If we denote by $\mathcal{S}(\boldsymbol{x})$ the set $\mathcal{S}(\boldsymbol{x}) = \Pi_{\nu=1}^N \mathcal{S}_\nu(\boldsymbol{x}^{-\nu})$, we see that we can say that $\bar{\boldsymbol{x}}$ is a solution if and only if $\bar{\boldsymbol{x}} \in \mathcal{S}(\bar{\boldsymbol{x}})$, i.e., if and only if $\bar{\boldsymbol{x}}$ is a fixed point of the point-to-set mapping $\mathcal{S}$. If the feasible sets $X_\nu(\boldsymbol{x}^{-\nu})$ do not depend on the rival players' strategies, so we have $X_\nu(\boldsymbol{x}^{-\nu}) = X_\nu$ for some set $X_\nu \subseteq \mathbb{R}^{n_\nu}$ and all $\nu = 1, \dots, N$, then the GNEP reduces to the standard Nash equilibrium problem (NEP for short).

It is useful to illustrate the above definitions with a simple example.

Example 0.1. Consider a game with two players, i.e., $N = 2$, with $n_1 = 1$ and $n_2 = 1$, so that each player controls one variable (for simplicity we therefore set $x_1^1 = x^1$ and $x_1^2 = x^2$). Assume that the players' problems are

$$\begin{array}{llcll} \min_{x^1} & (x^1 - 1)^2 & \qquad & \min_{x^2} & (x^2 - \frac{1}{2})^2 \\ \text{s.t.} & x^1 + x^2 \le 1, & & \text{s.t.} & x^1 + x^2 \le 1. \end{array}$$

The optimal solution sets are given by

$$\mathcal{S}_1(x^2) = \begin{cases} 1, & \text{if } x^2 \le 0, \\ 1 - x^2, & \text{if } x^2 \ge 0, \end{cases} \quad \text{and} \quad \mathcal{S}_2(x^1) = \begin{cases} \frac{1}{2}, & \text{if } x^1 \le \frac{1}{2}, \\ 1 - x^1, & \text{if } x^1 \ge \frac{1}{2}. \end{cases}$$

Then it is easy to check that the solutions of this problem are given by $(\alpha, 1-\alpha)$ for every $\alpha \in [1/2, 1]$. Note that the problem has infinitely many solutions.

In the example above, the sets $X_\nu(\boldsymbol{x}^{-\nu})$ are defined explicitly by inequality constraints. This is the most common case and we will often use such an explicit representation in the sequel. More precisely, in order to fix notation, we will assume that the sets $X_\nu(\boldsymbol{x}^{-\nu})$ are given by

$$X_\nu(\boldsymbol{x}^{-\nu}) = \{x^\nu \in \mathbb{R}^{n_\nu} : g^\nu(x^\nu, \boldsymbol{x}^{-\nu}) \leq 0\}, \tag{2}$$

where $g^\nu(\cdot, \boldsymbol{x}^{-\nu})\colon \mathbb{R}^{n_\nu} \to \mathbb{R}^{m_\nu}$. Equality constraints can easily be incorporated – we omit them for notational simplicity. Furthermore, we will usually not make any distinction between "private constraints" of player ν, i.e., constraints that depend on the player's variables x^ν only, and constraints that depend also on the other players' variables; in fact, the former can formally be included in the latter without loss of generality.

We say that a point $\boldsymbol{x}$ is *feasible* if it is feasible for all players, i.e., if $x^\nu \in X_\nu(\boldsymbol{x}^{-\nu})$ for all ν. We denote by $\mathbf{X}$ the *feasible set* of the game, i.e., the set of all feasible points, and by $\mathbf{X}_{-\nu}$ the projection of $\mathbf{X}$ on the space of the $\boldsymbol{x}^{-\nu}$variables. In other words, a point $\boldsymbol{x}^{-\nu}$ belongs to $\mathbf{X}_{-\nu}$ if and only if an x^ν exists such that $(x^\nu, \boldsymbol{x}^{-\nu})$ belongs to $\mathbf{X}$. In order to simplify the exposition, from now on we will assume that the following blanket assumptions hold, unless otherwise stated:

(A1) For all ν, θ_ν is continuosly differentiable on $\mathbf{X}$.
(A2) For all ν and all $\boldsymbol{x}^{-\nu} \in \mathbf{X}_{-\nu}$, the function $\theta_\nu(\cdot, \boldsymbol{x}^{-\nu})$ is convex on $X_\nu(\boldsymbol{x}^{-\nu})$.
(A3) For all ν, the sets $X_\nu(\boldsymbol{x}^{-\nu})$ are described by (2), where the functions g^ν are continuously differentiable on $\mathbf{X}$ and, for all $\boldsymbol{x}^{-\nu} \in \mathbf{X}_{-\nu}$, the functions $g_i^\nu(\cdot, \boldsymbol{x}^{-\nu})$ are convex on $X_\nu(\boldsymbol{x}^{-\nu})$.

These assumptions are rather standard, even if many efforts have been made in order to relax them. We do not deal with the issue of relaxing these assumptions here, and simply note that, at this stage of the development of solution algorithms, they seem more or less essential in order to develop efficient numerical methods.

The paper is organized as follows. In the first chapter we present a short historical account of GNEPs and briefly describe the important economic equilibrium model of Arrow and Debreu that "started it all". In the following chapter we give some modern applications of GNEPs in telecommunications. These applications have a rather different flavour from more traditional economic applications and represent well one of the modern trends in the study of GNEPs. Chapter 2 is divided in two parts. In the first we discuss NEPs and, in particular, their reduction to a *variational inequality* (VI for short). This reduction is central to our algorithmic approach also to more general GNEPs. In the second half of this chapter we discuss *jointly convex* generalized Nash equilibrium problems (JC GNEPs) that are an important and much studied class of GNEPs. In Chapter 4 we finally consider the challenging problem of GNEPs with no particular structure and describe some algorithms for their solution. We assume that the reader of these notes is familiar with classical optimization concepts and has some essential notions about variational inequalities. Since this is a relatively more advanced topic, we report in an Appendix some basic facts about VIs.

A few words regarding the notation: We denote by $\|\cdot\|$ the Euclidean norm. The Euclidean projection of a vector $\boldsymbol{x}$ onto a set $\mathbf{X}$ is denoted by $\Pi_{\mathbf{X}}(\boldsymbol{x})$. We say that a function f is C^0 if it is continuous, and C^k if it is k-times continuously differentiable. For a real-valued C^1-function f, we denote its gradient at a point $\boldsymbol{x}$ by $\nabla f(\boldsymbol{x})$. Similarly, for a vector-valued C^1-function F, we write $JF(\boldsymbol{x})$ for its Jacobian at a point $\boldsymbol{x}$ and $\nabla F(\boldsymbol{x})$ for its transpose.

History

The celebrated Nash equilibrium problem (NEP), where $X_\nu(\boldsymbol{x}^{-\nu}) = X_\nu$ for all $\nu = 1, \ldots, N$, was formally introduced by Nash in his 1950/1 papers [81, 82]. Nash's 1950 paper [81] in the Proceedings of the National Academy of Sciences contains the main results of his 28-page long PhD thesis and was submitted just fourteen months after he started his graduate studies! Obviously, as usually happens, ideas related to what we currently call game theory can be found in the work of much older researchers. Cournot [23], in the first half of the XIX century, proposed, in the context of an oligopolistic economy, a model that can now be interpreted as a two-players game and whose solution is a Nash equilibrium. But it was only at the beginning of the XX century that major developments occured, with the work of von Neumann [83] and von Neumann and Morgenstern, who published in 1944 the book *Theory of Games and Economic Behavior* [84], that essentially deals with zero-sum, two-player games, and that established game theory as an important field of study. The notion of Nash equilibrium introduced in [81] expanded enormously the scope of game theory, that had previously been essentially limited to zero-sum, two-player games, and proved to be the fundamental springboard for all successive developments. Nash's papers [81, 82] are a landmark in the scientific history of the twentieth century and the notion of Nash equilibrium has extensively proved to be powerful, flexible, and rich of consequences.

In spite of all its importance, the need of an extension of the NEP, where the players interact also at the level of the feasible sets, soon emerged as necessary. The GNEP was first formally introduced in 1952 by Debreu in [26], where the term *social equilibrium* was coined. This paper was actually intended to be just a mathematical preparation for the famous 1954 Arrow and Debreu paper [9] about economic equilibria. In this latter paper, Arrow and Debreu termed the GNEP "an abstract economy" and explicitly noted that "... In a game, the pay-off to each player depends upon the strategies chosen by all, but the domain from which strategies are to be chosen is given to each player independently of the strategies chosen by other players. An abstract economy, then, may be characterized as a generalization of a game in which the choice of an action by one agent affects both the pay-off and the domain of actions of other agents"; see [9, p. 273]. It is safe to say that [9] and the subsequent book [27] provided the rigorous foundation for the contemporary development of mathematical economics.

The mathematical-economic origin of the GNEP explains why the GNEP has long been (let us say up to the beginning of the nineties) the almost exclusive domain of economists and game-theory experts. In truth, it must also be noted that in this community some reserves have been advanced on GNEPs, on the grounds that a GNEP is not a game. For example, Ichiishi states, in his influential 1983 book [63, p. 60], "It should be emphasized, however, that an abstract economy is *not* a game, ... since player j must know the others' strategies in order to know his own feasible strategy set ... but the others cannot determine their feasible strategies without knowing j's strategy. Thus an abstract economy is a *pseudo-game* and it is useful only as a mathematical tool to establish existence theorems in various applied contexts".

The point here is that one cannot imagine a game where the players make their choices *simultaneously* and then, for some reason, it happens that the constraints are satisfied. But, indeed, this point of view appears to be rather limited, and severely undervalues:

(a) the descriptive and explanatory power of the GNEP model;

(b) its *normative* value, i.e., the possibility to use GNEPs to design rules and protocols, set taxes and so forth, in order to achieve certain goals, a point of view that has been central to recent applications of GNEPs outside the economic field (see below and Chapter 1);

(c) the fact that, in any case, different paradigms for games can be and have been adopted, where it is possible to imagine that, although in a noncooperative setting, there are mechanisms that make the satisfactions of the constraints possible.

Following the founding paper [9], researchers dedicated most of their energies to the study of the existence of equilibria under weaker and weaker assumptions and to the analysis of some structural properties of the solutions (for example, uniqueness or local uniqueness). However, and with few exceptions, it was not until the beginning of the 1990s that applications of the GNEP outside the economic field started to be considered along with algorithms for calculation of equilibria. In this respect, possibly one of the early contributions was given by Robinson in 1993 in [95, 96]. In these twin papers, Robinson considers the problem of measuring effectiveness in optimization-based combat models, and gives several formulations that are nothing else but, in our terminlogy, GNEPs. For some of these GNEPs, Robinson provides both existence results and computational procedures.

More or less at the same time, Scotti (see [100] and references therein) introduced GNEPs in the study and solution of complex structural design problems as an evolution of the more standard use of nonlinear programming techniques promoted by Schmit in the 1960s (see [99] for a review) and motivated by some early suggestions in the previous decade; see [93, 112].

After these pioneering contributions, in the last decade the GNEP became a relatively common paradigm, used to model problems from many different fields.

In fact, GNEPs arise quite naturally from standard NEPs if the players share some common resource (a communication link, an electrical transmission line, a transportation link, etc.) or limitations (for example, a common limit on the total pollution in a certain area). More in general, the ongoing process of liberalization of many markets (electricity, gas, telecommunications, transportation and others) naturally leads to GNEPs. But GNEPs have also been employed to model more technical problems that do not fit any of the categories listed above, and it just seems likely that now that the model is winning more and more popularity, many other applications will be uncovered in the near future. It is impossible to list here all relevant references for these applications; we limit ourselves to a few that, in our view, are either particularly interesting or good entry points to the literature [1, 2, 5, 9, 11, 12, 19, 21, 32, 33, 48, 49, 52, 53, 54, 59, 60, 64, 66, 67, 69, 87, 90, 91, 92, 106, 107, 109, 115, 117]. In very recent years GNEPs seem to have gained an especially important place in two fields: Computer Science and Telecommunications. The book *Algorithmic Game Theory* [85] and a special issue of the IEEE Magazine in Signal Processing [65] constitute fundamental and rich sources for these exciting developments. We finally mention that the very recent reference [41] presents a rather comprehensive and sophisticated treatment of the GNEP based on a variational inequality approach that allows to establish some interesting new results.

At this stage it may be useful to describe more in detail the economic equilibrium model of Arrow and Debreu, both because of the importance it played in the development of the field and because it serves well to illustrate the modelling power of GNEPs. The study of economic equilibrium models is a central theme to economics and deals with the problem of how commodities are produced and exchanged among individuals. Walras [114] was probably the first author to tackle this issue in a modern mathematical perspective. Arrow and Debreu [9] considered a general "economic system" along with a corresponding (natural) definition of equilibrium. They then showed that the equilibria of their model are those of a suitably defined GNEP (which they called an "abstract economy"); on this basis, they were able to prove important results on the existence of economic equilibria. Below we describe this economic model.

We suppose there are l distinct commodities (including all kinds of services). Each commodity can be bought or sold at a finite number of locations (in space and time). The commodities are produced in "production units" (companies), whose number is s. For each production unit j there is a set Y_j of possible production plans. An element $y^j \in Y_j$ is a vector in $\mathbb{R}^l$ whose hth component designates the output of commodity h according to that plan; a negative component indicates an input. If we denote by $p \in \mathbb{R}^l$ the prices of the commodities, the production units will naturally aim at maximizing the total revenue, $p^T y^j$, over the set Y_j.

We also assume the existence of "consumption units", typically families or individuals, whose number is t. Associated to each consumption unit i we have a vector $x^i \in \mathbb{R}^l$ whose hth component represents the quantity of the hth commodity consumed by the ith individual. For any commodity other than a labor service supplied by the individual, the consumption is nonnegative. More in general, x^i must belong to a certain set $X_i \subseteq \mathbb{R}^l$. The set X_i includes all consumption vectors among which the individual could choose if there were no budgetary constraints (the latter constraints will be explicitly formulated below). We also assume that the ith consumption unit is endowed with a vector $\xi^i \in \mathbb{R}^l$ of initial holdings of commodities and has a contractual claim to the share α_{ij} of the profit of the jth production unit. Under these conditions it is then clear that, given a vector of prices p, the choice of the ith unit is further restricted to those vectors $x^i \in X_i$ such that $p^T x^i \leq p^T \xi^i + \sum_{j=1}^s \alpha_{ij}(p^T y^j)$. As is standard in economic theory, the consumption units' aim is to maximize a utility function $u_i(x^i)$, which can be different for each unit.

Regarding the prices, obviously the vector p must be nonnegative; furthermore, after normalization, it is assumed that $\sum_{h=1}^l p_h = 1$. It is also expected that free commodities, i.e., commodities whose price is zero, are only possible if the supply exceeds the demand; on the other hand, it is reasonable to require that the demand is always satisfied. These two last requirements can be expressed in the form

$$\sum_{i=1}^t x^i - \sum_{j=1}^s y^j - \sum_{i=1}^t \xi^i \leq 0 \quad \text{and} \quad p^T \left(\sum_{i=1}^t x^i - \sum_{j=1}^s y^j - \sum_{i=1}^t \xi^i \right) = 0.$$

With the above setting in mind, Arrow and Debreu also make a series of further technical assumptions (which are immaterial to our discussion) on the properties of the sets Y_j, X_i, the functions u_i, etc., that correspond to rather natural economic features, and on this basis they define quite naturally a notion of an economic equilibrium. Essentially, an economic equilibrium is a set of vectors $(\bar{x}^1, \ldots, \bar{x}^t, \bar{y}^1, \ldots, \bar{y}^s, \bar{p})$ such that all the relations described above are satisfied. From our point of view, the interesting thing is that Arrow and Debreu show that the economic equilibria can also be described as the equilibria of a certain GNEP, and this reduction is actually the basis on which they can prove their key result: existence of equilibria. The GNEP they define has $s + t + 1$ players. The first s players correspond to the production units, the following t ones are the consumption units, and the final player is a fictious player who sets the prices and that is called "market participant". The jth production player controls the variables y^j, and his problem is

$$\max_{y^j} p^T y^j \quad \text{s.t.} \quad y^j \in Y_j. \tag{3}$$

The ith consumption player controls the variables x^i, and his problem is

$$\begin{aligned} &\max\nolimits_{x^i} \ u_i(x^i) \\ &\text{s.t.} \quad x^i \in X_i, \\ &\qquad p^T x^i \leq p^T \xi^i + \max\left\{0, \ \textstyle\sum_{j=1}^s \alpha_{ij}(p^T y^j)\right\}. \end{aligned} \tag{4}$$

Finally, the market participant's problem is

$$\begin{array}{ll}\max_p & p^T\left(\sum_{i=1}^t x^i - \sum_{j=1}^s y^j - \sum_{i=1}^t \xi^i\right) \\ \text{s.t.} & p \geq 0, \\ & \sum_{h=1}^l p_h = 1.\end{array} \tag{5}$$

Altogether, (3)–(5) represent a GNEP joint constraints coming from (4).

Chapter 1

Some Telecommunication Applications

As we have already mentioned, Computer Science and Telecommunications are probably the two overlapping fields where GNEPs are currently being used more extensively and with more effectiveness. In this chapter we describe some problems in telecommunications in order to give the reader a feel of current trends in the field. The description of these problems would benefit greatly from a knowledge of basic information theory. We tried to make our description as easy as possible to grasp, sometimes (over-)simplifying the description and we hope that even the reader with no background in telecommunications can grasp the basic ideas. In any case, this chapter is not a prerequisite for the remaining part of these notes and can be skipped at a first reading. For more information and background on these problems we refer the readers to [42, 90] and references therein.

1.1 Joint rate and power optimization in interference limited wireless communication networks

We consider the joint optimization of transmit powers and information rates in wireless (flat-fading) communication networks. Our description is mostly based on the setting of [86]; see also [20] for a broader picture.

We consider an N-user scalar Gaussian interference channel. In this model, there are N transmitter/receiver pairs, where each transmitter wants to communicate with its corresponding receiver over a scalar channel, which may represent time or frequency domains, affected by Gaussian noise. This mathematical model is sufficiently general to encompass many communication systems of practical interest (e.g., peer-to-peer networks, wireless flat-fading ad-hoc networks, and Code Division Multiple Access (CDMA) single/multicell cellular systems). Here, we assume without loss of generality that all pairs utilize a CDMA transmission scheme and share the same bandwidth (thus in principle they may interfere with each other). We denote by $G_{\nu\nu}(k) > 0$ the effective channel gain of link ν (including the multiplicative spreading gain, antenna gain and coding gain), while $G_{\nu\mu} \geq 0$

denotes the effective (cross-)channel gain between the transmitter μ and the receiver ν. The transmit power of each transmitter ν is denoted by p^ν. To avoid excessive signalling (i.e., exchange of information) and the need of coordination among users, we assume that encoding/decoding on each link is performed independently of the other links and no interference cancellation techniques are used. Hence, multiuser interference is treated as additive noise at each receiver. Within this setup, a common measure of quality is the so-called *transmission rate* over link ν. Very roughly speaking, this represents the number of bits that can be transmitted over the link and that can correctly be received with a positive probability of error that can be made as small as wanted. Under mild information theoretical assumptions (see, e.g., [103, 116]), the maximum transmission rate R^ν achievable over link ν for a given transmit power profile $\mathbf{p} = (p^\nu)_{\nu=1}^N$ of the users is

$$R^\nu(\mathbf{p}) = \log\left(\mathrm{SIR}^\nu(\mathbf{p})\right) \geq 0 \quad \text{with} \quad \mathrm{SIR}^\nu(\mathbf{p}) = \frac{G_{\nu\nu}p^\nu}{\sum_{\mu\neq\nu} G_{\nu\mu}p^\mu}, \tag{1.1}$$

where in the definition of the signal-to-noise-plus-interference ratio SIR^ν receiver noise is assumed to be dominated by the multiuser interference and is neglected in this model. This assumption is satisfied by most practical interference limited systems. Moreover, in these systems we have SIR $\gg 1$, since it represents the effective SIR after spreading gain, antenna gain, and coding gain. Note that, in view of these assumptions, the SIR is homogeneous in the transmitters' powers, and a scaling of the powers does not affect the SIR. Therefore, in the sequel we assume that the powers have been normalized so that $\sum_{\nu=1}^N p^\nu = 1$.

Within the setup above, the strategy of each player ν (user) is the transfer rate x^ν at which data are sent over the link and the transmit power p^ν that supports such a rate x^ν. A transfer rate profile $\boldsymbol{x} = (x^\nu)_{\nu=1}^N \in \mathbb{R}_+^N$ is feasible if it is possible for the system to simultaneously transfer data over the network at the specific rates $\mathbf{x}$ for some power vector $\mathbf{p}$. The rate-region of the system is the set of feasible transfer rates $\mathbf{x}$, which is formally defined as

$$\mathcal{R} = \left\{\boldsymbol{x} \in \mathbb{R}_+^N \mid \exists \mathbf{p} \geq 0 \text{ such that } x^\nu \leq R^\nu(\mathbf{p}),\ \forall \nu = 1,\dots,N\right\}, \tag{1.2}$$

where $R^\nu(\mathbf{p})$ is defined in (1.1). It is not difficult to show that the rate-region $\mathcal{R}$ is a convex set [86]. An equivalent characterization of the rate-region $\mathcal{R}$ useful for our purpose can be given through the so-called Perron–Frobenius eigenvalue of a nonnegative matrix [15], as detailed next. We recall that a square matrix M is *nonnegative*, $M \geq 0$, if all its elements are nonnegative and that if M is also primitive (meaning that all elements of M^k are positive for some k) then M has a unique strictly positive (real) eigenvalue $\lambda(M)$ which is larger in modulus than any other eigenvalue of M. Define two matrices D and $\tilde{G}$:

$$D(\boldsymbol{x}) = \operatorname{diag}\left(\frac{e^{x^\nu}}{G_{\nu\nu}}\right), \qquad \tilde{G}_{\nu\mu} = \begin{cases} G_{\nu\mu} & \text{if } \nu \neq \mu, \\ 0 & \text{otherwise.} \end{cases}$$

Using the above definition, the following equivalence can be shown to hold [86]:

$$x^\nu \leq R^\nu(\mathbf{p}),\ \forall \nu \quad \Longleftrightarrow \quad \mathbf{p} \geq D(\boldsymbol{x})\tilde{G}, \tag{1.3}$$

which, using the mentioned results on nonnegative matrices and recalling the normalization on $\mathbf{p}$, leads to the following equivalent expression for the rate-region $\mathcal{R}$:

$$\mathcal{R} = \left\{ \boldsymbol{x} \in \mathbb{R}^N_+ \mid \lambda\big(D(\boldsymbol{x})\tilde{G}\big) \leq 1 \right\}. \tag{1.4}$$

It is important to note that it can be proved that the function $\boldsymbol{x} \mapsto \lambda(D(\boldsymbol{x})\tilde{G})$ is continuously differentiable and convex [86].

The goal of each player is then to maximize his utility function $U_\nu(x^\nu)$, which is assumed to depend only on its own transmission rate x^ν, subject to the rate constraints in (1.4). The form of this utility function can range from very simple to complex expressions and may take into account several system level metrics and Quality of Service requirements (see, e.g., [20]). In general this function could even be nonconcave; we are not interested here in its precise expression[1]. Stated in mathematical terms, we have a game where each player's problem is

$$\begin{aligned} \max_{x^\nu} \ & U_\nu(x^\nu) \\ \text{s.t.} \ & \lambda\big(D(\boldsymbol{x})\tilde{G}\big) \leq 1, \\ & x^\nu \geq 0. \end{aligned} \tag{1.5}$$

Therefore, the problems (1.5) define a GNEP with (possibly) nonconcave objective potential function $P(\boldsymbol{x}) = \sum_{\nu=1}^N U_\nu(x^\nu)$ and a convex set $\mathbf{X} = \mathcal{R}$. The power profile $\mathbf{p}$ supporting the optimal rates $\mathbf{x}_\star$ that are solutions of this game can be computed solving the linear system of inequalities $\mathbf{p} \geq D(\mathbf{x}_\star)\tilde{G}$, $\sum_{\nu=1}^N p^\nu = 1$.

1.2 Rate maximization game over parallel Gaussian interference channels

The setting we consider next is somewhat similar to the previous one. The basic difference is that we now assume that each transmitter can use not only one channel, but N channels. Furthermore, we no longer assume that noise is negligible and represent the noise on the kth channel of player ν as a zero mean circularly symmetric complex Gaussian white noise with variance $\sigma_\nu^2(k)$. More formally, we consider a N-user Q-parallel Gaussian interference channel (IC). In this model, there are N transmitter-receiver pairs, where each transmitter wants to

[1]Note that, if the U is not concave, then we are not in the framework of these notes, since Assumption A1 is not met. But we stress that nonconvex games are emerging as a very important and challenging topic and we are sure that the next future will see huge efforts devoted to the study of nonconvex GNEPs.

communicate with its corresponding receiver over a set of Q parallel Gaussian subchannels, that may represent time or frequency bins (here we consider transmissions over a frequency-selective IC, without loss of generality). We denote by $H_{\nu\mu}(k)$ the (cross-)channel transfer function over the kth frequency bin between the transmitter μ and the receiver ν, while the channel transfer function of link ν is $H_{\nu\nu}(k)$. The transmission strategy of each user (pair) ν is the power allocation vector $p^\nu = \{p^\nu(k)\}_{k=1}^Q$ over the Q subcarriers, subject to the transmit power constraint

$$p^\nu \in \mathcal{P}^\nu = \left\{ p^\nu \in \mathbb{R}_+^Q \, : \, \sum_{k=1}^{Q} p^\nu(k) \le P_\nu \right\}. \tag{1.6}$$

Spectral mask constraints $p_{\max}^\nu = (p_{\max}^\nu(k))_{k=1}^Q$ in the form $\mathbf{0} \le p^\nu \le p_{\max}^\nu$ can also be included in the set $\mathcal{P}^\nu$ (see [74, 102, 103, 104] for more general results). Under basic information theoretical assumptions (see, e.g., [103, 116]), the maximum achievable rate on link ν for a specific power allocation profile $p^1, \ldots, p^N$ is

$$r^\nu(p^\nu, \mathbf{p}^{-\nu}) = \sum_{k=1}^{Q} \log \left(1 + \frac{|H_{\nu\nu}(k)|^2 p^\nu(k)}{\sigma_\nu^2(k) + \sum_{\mu \ne \nu} |H_{\nu\mu}(k)|^2 p^\mu(k)} \right), \tag{1.7}$$

where, as usual, $\mathbf{p}^{-\nu} = \left(p^1, \ldots, p^{\nu-1}, p^{\nu+1}, \ldots, p^N\right)$ is the set of all the users power allocation vectors, except the νth one, and $\sigma_\nu^2(k) + \sum_{\mu \ne \nu} |H_{\nu\mu}(k)|^2 p^\mu(k)$ is the overall power spectral density (PSD) of the noise plus multiuser interference (MUI) at each subcarrier measured by the receiver ν.

Given the above setup, we consider the following NEP [74, 90, 102, 103, 104] (see, e.g., [73, 75, 103] for a discussion on the relevance of this game-theoretical model in practical multiuser systems, such as digital subscriber lines, wireless adhoc networks, peer-to-peer systems, multicell OFDM/TDMA cellular systems):

$$\begin{aligned} &\max_{p^\nu} \, r^\nu(p^\nu, \mathbf{p}^{-\nu}) \\ &\text{s.t.} \quad p^\nu \in \mathcal{P}^\nu, \end{aligned} \tag{1.8}$$

for all $\nu = 1, \ldots, N$, where $\mathcal{P}^\nu$ and $r^\nu(p^\nu, \mathbf{p}^{-\nu})$ are defined in (1.6) and (1.8), respectively. Note that, since the feasible sets of the players' problems do not depend on the other players' variables, (1.8) actually describes a NEP.

1.3 GNEP: Power minimization game with Quality of Service constraints over interference channels

We consider now a game that can be viewed as a dual of game (1.8) just considered above. Under the same system model and assumptions, each player competes against the others by choosing the power allocation over the parallel channels that attains a desired information rate, with the minimum transmit power [90]. In the previous game, the transmission rate was maximized subject to a budget constraint. In the game we consider now, the power employed is minimized subject to a transmission rate constraint. This game-theoretical formulation is motivated by practical applications, where a prescribed Quality of Service in terms of achievable rate $R_\star^\nu$ for each user needs to be guaranteed. Stated in mathematical terms, we have the following optimization problem for each player ν [90]:

$$\begin{array}{ll} \min\limits_{\mathbf{p}^\nu} & \sum_{k=1}^N p^\nu(k) \\ \text{s.t.} & r^\nu\left(p^\nu, \mathbf{p}^{-\nu}\right) \geq R_\star^\nu, \end{array} \tag{1.9}$$

where the information rate $r^\nu(p^\nu, \mathbf{p}^{-\nu})$ is defined in (1.8).

The game in (1.9) is an example of GNEP in the general form. In spite of the apparent similarity with the NEP (1.8), the analysis of the GNEP (1.9) is extremely hard. This is principally due to the nonlinear coupling among the players' strategies and the unboundedness of the users' feasible regions. Nevertheless, in [90] a satisfactory answer to the characterization of the GNEP is provided. The analysis in [90] is mainly based on a proper nonlinear transformation that turns the GNEP in the power variables into a standard NEP in some new rate variables, thanks to which one can borrow from the more developed framework of standard VIs for a fruitful study of the game. This is just one of the many examples that show the importance of the VI reduction of games that is fundamental to our approach and that will be surveyed more in detail in the rest of these notes.

Chapter 2

NEPs and Jointly Convex GNEPs

The GNEP is a very general model, encompassing very many different kinds of problems. It may be useful then to preliminarily consider some special cases, which have a somewhat simpler structure. This will be useful also to the end of devising algorithms for the solution of general GNEPs, since in many cases the strategy adopted consists in some way of transforming the general GNEP to a simpler instance. In each of the cases we consider, we will briefly look at existence problems and at possible algorithms for calculating a solution.

We recall that we always assume that assumptions A1, A2 and A3 hold.

2.1 The Nash equilibrium problem

The Nash equilibrium problem is the best known and best studied class of GNEPs. It arises when the feasible regions of each player are independent of the choices of the other players. In other words, we have $X_\nu(x^{-\nu}) = X_\nu$ and $g^\nu(x^\nu, \boldsymbol{x}^{-\nu}) = g^\nu(x^\nu)$ for each player, so that the players' problem reduces to

$$\min_{x^\nu} \theta_\nu\big(x^\nu, \boldsymbol{x}^{-\nu}\big) \quad \text{s.t.} \quad x^\nu \in X_\nu = \big\{x^\nu : g^\nu(x^\nu) \leq 0\big\}. \tag{2.1}$$

To address existence issues, a useful point of view is to see a Nash equilibrium as a fixed-point of the best-response mapping for each player. We already briefly discussed this in the Introduction; let us present the approach again here. Let $\mathcal{S}_\nu(\mathbf{x}^{-\nu})$ be the (possibly empty) set of optimal solutions of the νth optimization problem (2.1) and set $\mathcal{S}(\mathbf{x}) = \mathcal{S}_1(\boldsymbol{x}^{-1}) \times \mathcal{S}_2(\boldsymbol{x}^{-2}) \times \cdots \times \mathcal{S}_N(\boldsymbol{x}^{-N})$. It is clear that a point $\boldsymbol{x}_*$ is a Nash equilibrium if and only if it is a fixed-point of $\mathcal{S}(\mathbf{x})$, i.e., if and only if $\boldsymbol{x}_* \in \mathcal{S}(\boldsymbol{x}_*)$. This observation is the key to the standard approach to the study of NEPs: the so-called fixed-point approach, which is based on the use of the well-developed machinery of fixed-point theory. However useful in establishing existence results and other theoretical properties, the fixed-point based analysis has severe limitations when it comes to the development of algorithms, unless it is possible to compute analytically the best-response mapping $\mathcal{S}(\mathbf{x})$.

There are at least two other ways to study GNEPs. The first is based on a reduction of the GNEP to a VI. This approach is pursued in detail, for example, in [41] and, resting on the well-developed theory of GNEPs and VIs, it has the advantage of permitting an easy derivation of many results about existence, uniqueness and stability of the solutions. But its main benefit is probably that of leading quite naturally to the derivation of implementable solution algorithms along with their convergence properties. It is this approach that will be at the basis of our exposition and will be exemplified next. In order to make these notes as self-contained as possible, we have collected a few basic definitions and results about variational inequalities in the Appendix. We stress once again that in our presentation we do not pursue maximum generality, but prefer to make somewhat strong assumptions in order to free the presentation from many technical questions so as to be able to concentrate on those issues that appear to be more relevant in practice. The second alternative approach is based on an *ad hoc* study of classes of games having a particular structure that can be exploited to facilitate their analysis. For example, this is the case of the so-called potential games [42, 77] and supermodular games [110]. These classes of games have recently received great attention in the signal processing and communication communities as a useful tool to model and solve various power control problems in wireless communications and networking [3, 4, 5, 62, 76].

VI reformulation of the NEP. At the basis of the VI approach to NEPs there is an easy equivalence between a NEP and a suitably defined VI. This equivalence hinges on the *minimum principle* for a convex optimization problem. When applied to (2.1), the minimum principle states that a point $\bar{x}^\nu$ is an optimal solution of (2.1) for a given $\bar{\boldsymbol{x}}^{-\nu}$ if and only if

$$\nabla_{x^\nu}\theta_\nu\big(\bar{x}^\nu, \bar{\boldsymbol{x}}^{-\nu}\big)^T\big(y^\nu - \bar{x}^\nu\big) \geq 0, \quad \forall y^\nu \in X_\nu. \tag{2.2}$$

Concatenating these conditions for all players and taking into account the Cartesian product structure of $\mathbf{X}$, we easily get (see [39]):

Theorem 2.1. *Consider the game defined by* (2.1), *and assume that every* X_ν *is closed and convex and every* θ_ν *is* C^1 *on* $\mathbf{X}$ *and such that* $\theta_\nu(\,\cdot\,, x^{-\nu})$ *is convex on* X_ν *for every fixed* $\boldsymbol{x}^{-\nu} \in \mathbf{X}_{-\nu}$. *Then a point* $\boldsymbol{x}_*$ *is a solution of the game* (2.1) *if and only if it is a solution of the variational inequality* $(\mathbf{X}, \mathbf{F})$, *where*

$$\mathbf{X} = \Pi_{\nu=1}^N X_\nu,$$
$$\mathbf{F}(\boldsymbol{x}) = \big(\nabla_{x^\nu}\theta_\nu(x^\nu, \boldsymbol{x}^{-\nu})\big)_{\nu=1}^N.$$

The function $\mathbf{F}$ introduced in the theorem is often called the *pseudo-gradient* of the game. The following example illustrates the theorem.

Example 2.2. Consider a NEP with two players, i.e., $N = 2$, with $n_1 = 1$ and $n_2 = 1$, so that each player controls one variable (for simplicity we therefore set $x_1^1 = x^1$ and $x_1^2 = x^2$). Assume that the players' problems are

$$\begin{array}{llcll} \min_{x^1} & x^2\left(x^1 - 1\right)^2 & \qquad\qquad & \min_{x^2} & \left(2x^2 + x^1\right)^2 \\ \text{s.t.} & 0 \le x^1 \le 1, & & \text{s.t.} & \frac{3}{4} \le x^2 \le 1. \end{array}$$

The first player's problem solution is always $x^1 = 1$, independent of the value of $x^2 \in [3/4, 1]$. Substituting this value in the second player's problem we easily see that the unique solution of the NEP is $(1, 3/4)$. It is readily seen that this NEP satisfies all the assumptions of Theorem 2.1 and so we may rewrite the problem as the variational equality $(\mathbf{X}, \mathbf{F})$, where

$$\mathbf{X} = \left\{(x^1, x^2) : 0 \le x^1 \le 1, 3/4 \le x^2 \le 1\right\}, \qquad \mathbf{F} = \begin{pmatrix} 2x^2(x^1 - 1) \\ 4(2x^2 + x^1) \end{pmatrix}.$$

It is now immediate to check that $(1, 3/4)$ is a solution of this VI. In fact, we have $\mathbf{F}(1, 3/4) = (0, 10/2)^T$ and

$$(0, 10/2)\begin{pmatrix} x^1 - 1 \\ x^2 - 3/4 \end{pmatrix} = \frac{10}{2}\left(x^2 - 3/4\right) \ge 0, \quad \forall \boldsymbol{x} \in \mathbf{X}, \quad \text{i.e.,} \quad \forall x^2 \in [3/4, 1].$$

We will verify shortly that this is actually the only solution of the VI. □

Existence and uniqueness of the NE based on VI. Given the equivalence between the NEP and the VI problem, conditions guaranteeing the existence of a NE follow readily from the existence of a solution of the VI. We refer the reader who is not familiar with basic definitions and results on VI to the Appendix for a brief review. On the basis of this results and of Theorem 2.1 the following result needs no further comment.

Theorem 2.3. *Consider the game defined by* (2.1), *and assume that every* X_ν *is closed and convex and every* θ_ν *is* C^1 *on* $\mathbf{X}$ *and such that* $\theta_\nu(\,\cdot\,, x^{-\nu})$ *is convex on* X_ν *for every fixed* $x^{-\nu} \in \mathbf{X}_{-\nu}$.

1. *If all* X_ν *are bounded, then* (2.1) *has a nonempty compact set of solutions.*
2. *If* $\mathbf{F}$ *is monotone on* $\mathbf{X}$, *then the solution set is convex.*
3. *If* $\mathbf{F}$ *is a strictly monotone function on* $\mathbf{X}$, *then* (2.1) *has at most one solution.*
4. *If* $\mathbf{F}$ *is a strongly monotone function on* $\mathbf{X}$, *then* (2.1) *has a unique solution.*

Example 2.4. Consider again the game of Example 2.2 and the equivalent VI defined there. First of all note that $\mathbf{X}$ is compact and $\mathbf{F}$ is continuous, so that we are sure that a solution exists (this we already knew from Example 2.2). What is more interesting is that it is easy to check that $\mathbf{F}$ is actually strictly monotone on $\mathbf{X}$. In fact we have

$$J\mathbf{F}(\boldsymbol{x}) = \begin{pmatrix} 2x^2 & 2(x^1-1) \\ 4 & 8 \end{pmatrix},$$

while its symmetric part is

$$J\mathbf{F}_{\rm sym}(\boldsymbol{x}) = \begin{pmatrix} 2x^2 & x^1+1 \\ x^1+1 & 8 \end{pmatrix}.$$

This matrix is clearly positive definite on $\mathbf{X}$ and therefore we conclude that $\mathbf{F}$ is strictly monotone on $\mathbf{X}$. Therefore, by Theorem 2.3, the VI $(\mathbf{X}, \mathbf{F})$ and the original game have one and only one solution. □

More refined existence results can certainly be derived by using more sophisticated existence results for VI. We refer the reader to [39] for more details. These results are nothing else than the application of general results of the VI theory to games, and as such can easily be worked out. More advanced developments, based on the use of degree theory, can also be found in [41]. Instead of devoting our efforts to a listing of such results, we are happy to have illustrated how the reduction of a NEP to a VI can be used in order to get a basic existence result and prefer to focus our attention on two issues that are specific to the game environment we are dealing with. Further discussion of existence issues for general GNEPs will be offered in Section 4.1.

The first such issue is strictly related to the results in Theorem 2.3. The natural question posed by this theorem is: When is the mapping $\mathbf{F}$ strongly monotone? In fact, it is not at all clear what kind of condition on the original game one should impose in order for $\mathbf{F}$ to be strongly monotone. It can be shown that a sufficient condition for such an $\mathbf{F}$ to be strongly monotone on $\mathbf{X}$ is that the following matrix be positive definite [41, 91]:

$$\Upsilon = \begin{bmatrix} \alpha_1^{\min} & -\beta_{1\,2}^{\max} & -\beta_{1\,3}^{\max} & \cdots & -\beta_{1\,N}^{\max} \\ -\beta_{2\,1}^{\max} & \alpha_2^{\min} & -\beta_{2\,3}^{\max} & \cdots & -\beta_{2\,N}^{\max} \\ \vdots & \vdots & \ddots & \vdots & \vdots \\ -\beta_{N-1\,1}^{\max} & -\beta_{N-1\,2}^{\max} & \cdots & \alpha_{N-1}^{\min} & -\beta_{N-1\,N}^{\max} \\ -\beta_{N\,1}^{\max} & -\beta_{N\,2}^{\max} & \cdots & -\beta_{N\,N-1}^{\max} & \alpha_N^{\min} \end{bmatrix} \succ \mathbf{0}, \qquad \text{(C1)}$$

with

$$\alpha_\nu^{\min} = \inf_{\boldsymbol{z}\in\mathbf{X}} \left\{ \lambda_{\min}\left(\nabla^2_{x^\nu x^\nu}\theta_\nu(\boldsymbol{z})\right) \right\} \quad \text{and} \quad \beta_{\nu\mu}^{\max} = \sup_{\boldsymbol{z}\in\mathbf{X}} \left\{ \left\| \nabla^2_{x^\nu x^\mu}\theta_\nu(\boldsymbol{z}) \right\| \right\}, \tag{2.3}$$

where $\nabla^2_{x^\nu,x^\nu}\theta_\nu(\boldsymbol{x})$ is the Hessian of $\theta_\nu(\boldsymbol{x})$, $\nabla^2_{x^\nu x^\mu}\theta_\nu(\boldsymbol{x})$ is the Jacobian of $\nabla_{x^\nu}\theta_\nu(\boldsymbol{x})$ with respect to $\boldsymbol{x}^\mu$, $\lambda_{\min}(\mathbf{A})$ denotes the minimum eigenvalue of $\mathbf{A}$, and $\|\mathbf{A}\| = \sqrt{\lambda_{\max}(\mathbf{A}^T\mathbf{A})}$ is the spectral norm of $\mathbf{A}$.

To the best of our knowledge, condition (C1) is the first condition in the literature easy to be checked and valid for NEPs with *arbitrary* strategy sets and payoff functions that guarantee the strong monotonicity of $\mathbf{F}$. Interestingly, the same condition also has some important algorithmic consequences to be discussed later on.

We conclude by observing that, even if we do not go into the technical proof of these results, we believe that the rationale behind it can easily be grasped. The diagonal terms in the matrix Υ, the $\alpha_\nu^{\min}$, tell us "how much convex" the objective functions are, while the off-diagonal terms in the same matrix, $\beta_{\nu\mu}^{\max}$, say "how much large" the block-diagonal terms in the Jacobian of $\mathbf{F}$ are. We then see that, very roughly speaking, condition (C1), i.e., the positive definiteness of Υ, simply states that the objective functions of the players must be sufficiently strongly convex with respect to the "influence" of the other players on their own objective functions. Note that, if the game is such that each player's objective function does not depend on the other players' variables – that is, if the game is simply a collection of N independent optimization problems – then the βs in matrix Υ are all 0 and condition (C1) simply says that the N optimization problems have one and only one solution if all the objective functions are strongly convex. Condition (C1) becomes nontrivial when there is an interaction between the various optimization problems, i.e., when some βs are nonzero. Condition (C1) gives a quantitative measure of how large the interaction can become, before the existence and uniqueness of the solution is lost.

Solution methods. In order to find an equilibrium of a game, the most effective methods are based on the VI reduction described above. Once the original game has been transformed into a VI $(\mathbf{X}, \mathbf{F})$ (under the assumptions of Theorem 2.1), we can apply any of the array of methods described in [39]. We do not go into too many details and refer the interested reader to [39] for a complete description and analysis of these methods. However, for our purposes here, it may be useful to recall a few more facts. The main theoretical requirement needed in establishing convergence of algorithms for the solution of a VI is that $\mathbf{F}$ enjoy some kind of monotonicity (details will be given below). In order to describe these algorithms, we will make reference to a VI (K, F). Note the change of notation, which is due to the fact that the results below hold for a generic VI (K, F), where K is any convex closed set and $F\colon K \to \mathbb{R}^n$ is continuous, and not only to the VI $(\mathbf{X}, \mathbf{F})$ arising form the reformulation of a NEP.

We describe below two algorithms that belong to the well-studied class of "projection algorithms". These algorithms are relatively simple and yet enjoy some strong theoretical convergence properties, even if practical efficiency might be lacking, in practice. The first algorithm we describe is probably the one that, if properly implemented, is often the more practically efficient among projection algorithms. We denote by $\Pi_K(y)$ the projection of y on the set K.

Algorithm 1: Projection Algorithm with Variable Steps

(S.0) : Choose $x_0 \in K$ and set $k = 0$.
(S.1) : If $x_k = \Pi_K(x_k - F(x_k))$ stop.
(S.2) : Choose $\tau_k > 0$. Set $x_{k+1} = \Pi_K(x_k - \tau_k F(x_k))$ and $k \leftarrow k+1$; go to Step 1.

As can be seen, the algorithm is really simple and essentially only requires, at each iteration, one evaluation of the function F and one projection on the set K. So, unless the evaluation of F is extremely complex, the main computational burden per iteration is the calculation of the projection of a point $y = x_k - \tau_k F(x_k)$, which is equivalent to the solution of the strongly convex optimization problem

$$\begin{aligned} &\min_z \ \tfrac{1}{2}\|z - y\|^2 \\ &\text{s.t.} \quad z \in K. \end{aligned}$$

If K is a polyhedron, this is a strongly convex quadratic optimization problem and can be solved in a finite number of steps. Otherwise, the computation of the projection requires the solution of a convex optimization problem by some iterative method. The algorithm is illustrated in Figure 2.1, that clearly shows the rather simple structure of the iterations of the method.

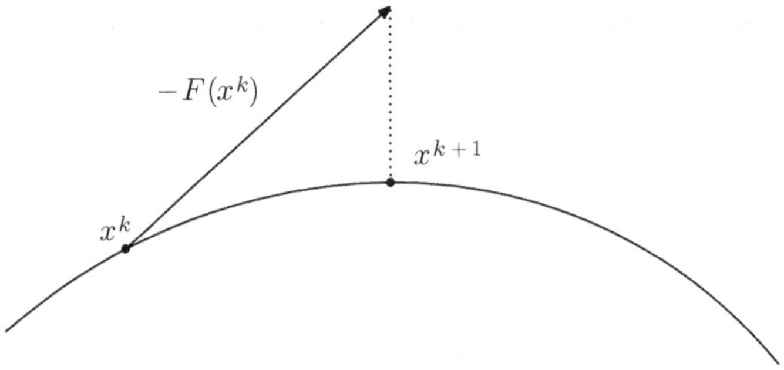

Figure 2.1: Projection Algorithm with Variable Steps ($\tau_k = 1$).

It is worth mentioning that the practical efficiency of Algorithm 1 depends on a proper choice of the sequence τ_k that must be made on the basis of heuristic considerations. We need not go into these implementative details here, and consider instead the theoretical convergence properties of the algorithm. In order to show convergence, we need F to be co-coercive with constant $c > 0$, meaning that a constant c exists such that

$$(F(x) - F(y))^T (x - y) \geq c\|F(x) - F(y)\|^2, \quad \forall x, y \in K.$$

Co-coercivity is equivalent to the strong monotonicity of the inverse of F (which is, in general, a point-to-set mapping, so the definition of strong monotonicity needs to be adapted, but this is out of the scope of these notes), and therefore the term "inverse strong monotonicity" has also been used for this property, along with those of "strong-F-monotonicity" and "Dunn property", among others. It can be shown that F is co-coercive if F is strongly monotone and Lipschitz on K, but the converse is not true. Furthermore, co-coercivity implies monotonicity, so that, loosely speaking, co-coercivity can be viewed as an intermediate condition between monotonicity and strong monotonicity (plus some Lipschitzianity, which is automatically satisfied if K is compact). Note, for example, that, if F is constant, then F is monotone and co-coercive, but not even strictly monotone.

Theorem 2.5. *Let K be a closed convex subset of $\mathbb{R}^n$ and F be a mapping from K into $\mathbb{R}^n$ that is co-coercive on K with constant c. Suppose that the VI (K, F) has at least a solution. If*

$$0 < \inf_k \tau_k \leq \sup_k \tau_k < 2c,$$

Algorithm 1 produces a sequence $\{x^k\}$ converging to a solution of the VI (K, F).

The second projection method we consider is the one with the best theoretical convergence properties among projection-type methods (and indeed, among all kinds of solution methods for VIs), even if this theoretical superiority is often not backed up by corresponding numerical results. But it is nevertheless worth exposing in order to establish theoretical limits to what is possible to achieve.

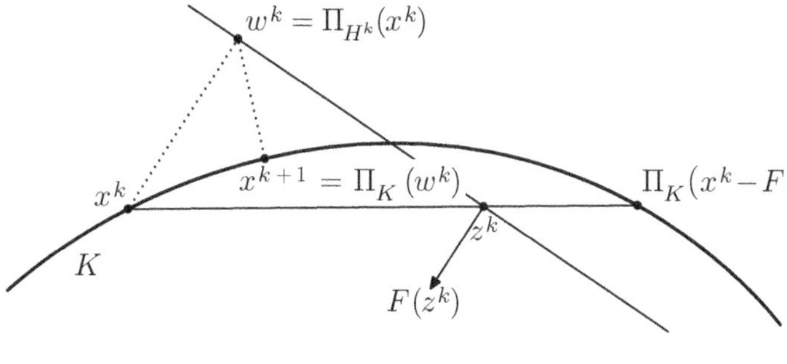

Figure 2.2: One iteration of the Hyperplane Projection Algorithm.

The Hyperplane Projection Method is rather more complex than Algorithm 1 and it requires three projections per iteration – two onto K and one onto H^k (the latter projection is easily performed and is given by an explicit formula), in addition to a line-search that might require several evaluations of F.

Algorithm 2: Hyperplane Projection Algorithm

(S.0) : Choose $x_0 \in K$, $\tau > 0$, and $\sigma \in (0,1)$.
(S.1) : If x_k is a solution, stop.
(S.2) : Compute

$$y_k = \Pi_K(x_k - \tau F(x_k)),$$

and find the smallest nonnegative integer i_k such that, with $i = i_k$,

$$F(2^{-i} y_k + (1 - 2^{-i}) x_k)^T (x_k - y_k) \geq \frac{\sigma}{\tau} \|x_k - y_k\|^2. \tag{2.4}$$

(S.3) : Set

$$z_k = 2^{-i_k} y_k + (1 - 2^{-i_k}) x_k,$$

and

$$w_k = \Pi_{H_k}(x_k) = x_k - \frac{F(z_k)^T (x_k - z_k)}{\|F(z_k)\|^2} F(z_k),$$

(S.4) : Set $x_{k+1} = \Pi_K(w_k)$ and $k \leftarrow k+1$; go to S.1.

Some comments are in order. In S.2 a point y_k is calculated which is exactly the new iteration x_{k+1} in Algorithm 1. So the two algorithms start exactly in the same way. But in Algorithm 2 this point y_k is taken only as a starting point for a more complex procedure. In S.3 a line-search is performed in order to identify on the segment joining x_k and y_k a point z_k where the angle between $F(z_k)$ and $x_k - y_k$ is "sufficiently less than 90 degrees". Then we project x_k on the hyperplane H_k passing through z_k and having $F(z_k)$ as normal. This projection can be computed analytically and the resulting formula is the one used in the second half of S.3, where we denoted by w_k the projection of x_k on H_k. In S.4, we finally obtain the new iteration x_{k+1} by projecting w_k on the set K. The iterations of the Hyperplane Projection Algorithm are illustrated in Figure 2.2. It might also be useful to note that, having set $z^k = \tau_k y^k + (1 - \tau_k) x^k$, with $\tau_k = 2^{-i_k}$, we have

$$w^k = x^k - \tilde{\tau}_k F(z^k)$$

where

$$\tilde{\tau}_k = \tau_k \frac{F(z^k)^T (x^k - y^k)}{\|F(z^k)\|^2}.$$

Hence the new iteration can be written as

$$x^{k+1} = \Pi_K(x^k - \tilde{\tau}_k F(\tau_k y^k + (1 - \tau_k) x^k)).$$

Note that, at difference with Algorithm 1, it is not even clear that Algorithm 2 is well defined, since it is not clear that a point z_k satisfying (2.4) can always be found along the segment joining x_k and y_k. The following theorem shows that indeed, under some adequate assumptions, Algorithm 2 is well defined and converges to a solution.

Theorem 2.6. *Let K be a closed convex set in $\mathbb{R}^n$ and let F be a continuous mapping from K into $\mathbb{R}^n$ that is monotone on K and assume the VI (K, F) has a solution. Algorithm 2 is well defined and, if $\{x^k\}$ is a sequence produced by Algorithm 2, then $\{x^k\}$ converges to a solution of the VI (K, F).*

For those versed in function classes, it might be useful to know that actually Theorem 2.6 holds even under a mere *pseudo monotonicity* assumption on F with respect to the solution set (meaning that, if $\bar{x}$ is a solution, then $F(x)^T(x - \bar{x}) \geq 0$ for all $x \in K$; note that this condition is rather weaker than the assumed monotonicity on K). These are somewhat technical issues though, and we believe that the two theorems above, with their convergence properties, give a good picture of what one can get by employing VI methods.

Instead of insisting on more details on standard algorithms, in the remaining part of this section we focus our attention on distributed algorithms. Since in a Nash game every player is trying to minimize his own objective function, a natural approach is to consider an iterative algorithm based, e.g., on the Jacobi (simultaneous) or Gauss–Seidel (sequential) schemes, where at each iteration every player, given the strategies of the others, updates his own strategy by solving his optimization problem (2.1). These kind of methods arise quite naturally when analyzing real situations (for example the telecommunication applications described in Chapter 1) and have great practical significance.

The Gauss–Seidel implementation of the best-response based algorithm for the solution of a NEP is formally described in Algorithm 3.

Algorithm 3: Gauss–Seidel Best-response Based Algorithm

(S.0) : Choose any feasible starting point $\mathbf{x}_0 = (\mathbf{x}_0^\nu)_{\nu=1}^N$, and set $k = 0$.
(S.1) : If $\boldsymbol{x}_k$ satisfies a suitable termination criterion, stop.
(S.2) : for $\nu = 1, \ldots, N$, compute a solution $\mathbf{x}_{k+1}^\nu$ of

$$\begin{aligned} &\min_{\mathbf{x}^\nu} \; \theta_i\left(\mathbf{x}_{k+1}^1, \ldots, \mathbf{x}_{k+1}^{\nu-1}, \mathbf{x}^\nu, \mathbf{x}_k^{\nu+1}, \ldots, \mathbf{x}_k^N\right) \\ &\mathbf{x}^\nu \in X_\nu, \end{aligned} \tag{2.5}$$

end
(S.3) : Set $\boldsymbol{x}_{k+1} = (\mathbf{x}_{k+1}^\nu)_{\nu=1}^N$ and $k \leftarrow k+1$; go to S.1.

Building on the VI-framework, one can prove that Algorithm 3, as well as its Jacobi version, globally converge to the NE of the game, under condition (C1) guaranteeing the uniqueness of the equilibrium [41, 91].

In many practical multiuser communication systems, such as wireless ad-hoc networks or CR systems, the synchronization requirements imposed by the sequential and simultaneous algorithms described above might be not always acceptable. In fact, in Algorithm 3 it is supposed that players coordinate in order to solve sequentially their problems. It is rather intuitive that in many real applications it may be expected that the players behave in a rather chaotic way. It is possible to show that under conditions similar to those above, a *totally asynchronous* implementation (in the sense of [18]) converges to the unique NE of the game (see, e.g., [102, 105, 106] for details).

Nice as the theory above may look, it has a severe drawback. In fact, condition (C1) requires that the objective functions of all players are strongly convex and "uniformly so with respect to the other players' decisions". This can be seen by observing that the positive definiteness of the matrix Υ entails the positivity of its diagonal terms and by remembering the definition of these diagonal terms. While this is a perfectly fine mathematical requirement, it turns out that in many applications, while we can expect the objective functions to be convex, the same functions cannot be expected to be strongly convex for some structural reasons. So the need arises for a decomposition method that can relax assumption (C1). To this end we will require that $\mathbf{F}(\boldsymbol{x})$ be monotone on $\mathbf{X}$ or, equivalently under our differentiability assumptions, that $J\mathbf{F}(\boldsymbol{x})$ be positive semidefinite for all $\boldsymbol{x}$ in $\mathbf{X}$. Note that this requirement is much weaker than (C1) and exactly the same that was needed to prove convergence of the centralized Algorithm 2; see Theorem 2.6.

The tool we use to design a decentralized algorithm for the solution of a NEP is classical: regularization. We know that the NEP is equivalent to the VI $(\mathbf{X}, \mathbf{F})$. We can regularize this VI and, given an iteration $\boldsymbol{x}_k$, compute $\boldsymbol{x}_{k+1}$ as the solution of the regularized VI $(\mathbf{X}, \mathbf{F}(\cdot) + \tau(\cdot - \boldsymbol{x}_k))$. Note that, since we assumed that $\mathbf{F}$ is monotone, $\mathbf{F}(\cdot) + \tau(\cdot - \boldsymbol{x}_k)$ is strongly monotone and therefore the VI $(\mathbf{X}, \mathbf{F}(\cdot) + \tau(\cdot - \boldsymbol{x}_k))$ has one and only one solution, so that $\boldsymbol{x}_{k+1}$ is well defined. Those familiar with regularization methods know that such a procedure will lead (under some assumptions) to convergence of the sequence $\{\boldsymbol{x}_k\}$ to a solution of the VI $(\mathbf{X}, \mathbf{F})$ and hence to a solution of the NEP. We will come back to this in a moment. What we want to stress now is the implications of such a procedure. First of all, note that finding a solution of the VI $(\mathbf{X}, \mathbf{F}(\cdot)+\tau(\cdot - \boldsymbol{x}_k))$ is equivalent to finding a solution of a "regularized" NEP where each player's problem has been regularized and is given by

$$\begin{aligned} \min_{x^\nu} \quad & \theta_\nu(x^\nu, \boldsymbol{x}^{-\nu}) + \tfrac{\tau}{2}\left\|x^\nu - x_k^\nu\right\|^2 \\ \text{s.t.} \quad & x^\nu \in X_\nu. \end{aligned} \tag{2.6}$$

To see that this regularized game is equivalent to the regularized VI, just use Theorem 2.1.

Based on this, the procedure we are considering can be described more in detail in the following way, where we indicate by $S_\tau(\boldsymbol{x}_k)$ the (unique) solution of the VI $(\mathbf{X}, \mathbf{F}(\cdot) + \tau(\cdot - \boldsymbol{x}_k))$ or, equivalently, of the game (2.6).

Algorithm 4: Exact Proximal Decomposition Algorithm

(S.0) : Choose a starting point $\boldsymbol{x}_0$ and a positive τ. Set $k = 0$.
(S.1) : If $\boldsymbol{x}_k$ satisfies a suitable termination criterion, stop.
(S.2) : Set $\boldsymbol{x}_{k+1} = S_\tau(\boldsymbol{x}_k)$.
(S.3) : Set $k \leftarrow k + 1$, and go to S.1.

Note that, in order to solve the original game, we now ask to solve a (an infinite) sequence of problems with a similar structure. This might seem nonsense, but the crucial point is that, by choosing τ "large enough", one can ensure (under mild conditions) that the regularized VI game $(\mathbf{X}, \mathbf{F}(\cdot) + \tau(\cdot\ - \boldsymbol{x}_k))$ does satisfy condition (C1), even if the original game does not satisfy that condition. Therefore we can compute $S_\tau(\boldsymbol{x}_{k+1})$ by using the distributed Algorithm 3. This is the big advantage we get by regularization.

As it is stated, Algorithm 4 is essentially conceptual, since in S.2 we require the exact solution of a VI, a task that usually will entail an infinite number of operations. As usual, it is possible to truncate the computation of $S_\tau(\boldsymbol{x}_k)$ and use an *approximate* solution $\boldsymbol{z}_k$ of the regularized VI. This is described in the following, more practical, version of Algorithm 4, where we also added another extra feature in S.3, where the new iteration $\boldsymbol{x}_{k+1}$ is obtained as a combination of $\boldsymbol{z}_k$ and $\boldsymbol{x}_k$. This added flexibility can usually be exploited to enhance the numerical performance of the algorithm. We do not go into a discussion of these issues here, though.

Algorithm 5: Inexact Proximal Decomposition Algorithm

(S.0) : Choose a starting point $\boldsymbol{x}_0$, a positive τ, a positive ε_0, and a $\rho_0 \in (0, 2)$. Set $k = 0$.
(S.1) : If $\boldsymbol{x}_k$ satisfies a suitable termination criterion, stop.
(S.2) : Find a point $\boldsymbol{z}_k$ such that $\|\boldsymbol{z}_k - S_\tau(\boldsymbol{x}_k)\| \leq \varepsilon_k$.
(S.3) : Set $\boldsymbol{x}_{k+1} = (1 - \rho_k)\boldsymbol{x}_k + \rho_k \boldsymbol{z}_k$, select ε_{k+1}, ρ_{k+1}, set $k \leftarrow k + 1$, and go to S.1.

Theorem 2.7. *In the general setting of this section, assume that* $\mathbf{F}$ *is monotone on* $\mathbf{X}$ *and that the game* (2.1) *has at least one solution. Let* $\{\varepsilon_k\} \subset [0, \infty)$ *be a sequence such that* $\sum_{k=1}^{\infty} \varepsilon_k < \infty$, *and let* ρ_k *be such that* $\{\rho_k\} \subset [R_m, R_M]$ *with* $0 < R_m \leq R_M < 2$. *Then the sequence* $\boldsymbol{x}^k$ *generated by Algorithm* 5 *converges to a solution of* (2.1).

Note once again that, in S.2, $\boldsymbol{z}_k$ can be computed in a finite number of iterations in principle. The summability condition on the allowed errors ε_k given in Theorem 2.7 essentially says that the regularized games have to be solved more and more accurately as the iterations proceed.

There are many numerical and theoretical questions that should be discussed, but those are beyond the scope of these notes. One important theoretical issue is the value of τ, that should be large enough so that (C1) is satisfied; but, how large exactly? Another important question is: How can we know when the error in calculating $S_\tau(\boldsymbol{x}_k)$ is below a desired value? We leave these issues to a more advanced study and refer the interested reader to [39] and [101].

Chapter 3

Jointly Convex Nash Equilibrium Problems

Jointly convex (JC) GNEPs are by far the most studied class of GNEPs where there actually is an interaction among players at the feasible set level. In spite of this (relative) popularity, we will see that there are several open problmes yet. We begin with a formal definition.

Definition 3.1. We say that a GNEP is *jointly convex* if, for some closed convex set $\mathbf{X} \subseteq \mathbb{R}^n$ and all $\nu = 1, \dots, N$, we have

$$X_\nu(\boldsymbol{x}^{-\nu}) = \{x^\nu \in \mathbb{R}^{n_\nu} : (x^\nu, \boldsymbol{x}^{-\nu}) \in \mathbf{X}\}. \tag{3.1}$$

We remark that, in order not to introduce further symbols, we use the symbol $\mathbf{X}$ with a meaning that is different from the one in the previous section. Since it will always be clear from the context whether we are speaking of JC GNEPs or NEPs, hopefully this will not give rise to any confusion.

When the sets $X_\nu(\boldsymbol{x}^{-\nu})$ are defined explicitly by a system of inequalities as in (2), then it is easy to check that (3.1) is equivalent to the requirement that $g^1 = g^2 = \cdots = g^N = g$ and that $g(\boldsymbol{x})$ be (componentwise) convex with respect to all variables $\boldsymbol{x}$; furthermore, in this case, it obviously holds that $\mathbf{X} = \{\boldsymbol{x} \in \mathbb{R}^n : g(\boldsymbol{x}) \leq 0\}$.

Note that the Example 0.1 and the model described in Section 1.1 are instances of jointly convex GNEPs. This class of problems has been first studied in detail in a seminal paper by Rosen [98] and has often been identified with the whole class of GNEPs. Jointly convex GNEPs are also often termed as GNEPs with "coupled constraints"; however, we prefer the more descriptive definition of jointly convex.

The single more important fact about JC games is probably the one described in the following theorem, which shows that we can still try to recover a solution of a JC game by solving a VI.

Theorem 3.2. *Let a* JC GNEP *be given and assume that* A1, A2 *and* A3 *hold. Then every solution of the VI* $(\mathbf{X}, \mathbf{F})$, *where* $\mathbf{X}$ *is the set in the definition of joint convexity and, as usual,* $\mathbf{F}(\boldsymbol{x}) = (\nabla_{x^\nu}\theta_\nu(\boldsymbol{x}))_{\nu=1}^N$, *is also a solution of the* GNEP.

Note that, in general, there is no VI through which we can compute *all* the solutions of a JC GNEP. Indeed, the above theorem does not say that any solution of a jointly convex GNEP is also a solution of the VI $(\mathbf{X}, \mathbf{F})$, and actually in the passage from the GNEP to the VI it is not difficult to see that "most" solutions are lost. We illustrate this with a simple example.

Example 3.3. Consider again the JC game with two players in Example 0.1. In the Introduction we have shown that this game has infinitely many solutions given by $(\alpha, 1-\alpha)$ for every $\alpha \in [1/2, 1]$. Consider now the VI $(\mathbf{X}, \mathbf{F})$ where

$$\mathbf{X} = \{(x^1, x^2) \in \mathbb{R}^2 \,:\, x^1 + x^2 \leq 1\}, \qquad \mathbf{F}(\boldsymbol{x}) = \begin{pmatrix} 2x^1 - 2 \\ 2x^2 - 1 \end{pmatrix}.$$

Here $\mathbf{F}$ is clearly strictly monotone and therefore this VI has a unique solution, which is given by $(3/4, 1/4)$ as can be checked by using the definition of VI. Note that, as expected, this is a solution of the original GNEP. However, all the other infinite solutions of the GNEP are not solutions of the VI, since we already remarked this VI has only one solution.

Definition 3.4. Let a jointly convex GNEP be given with C^1-functions θ_ν. We call a solution of the GNEP that is also a solution of the VI $(\mathbf{X}, \mathbf{F})$ a *variational equilibrium.*

The alternative name *normalized equilibrium* is also frequently used in the literature instead of *variational equilibrium.* In view of its close relation to a certain variational inequality problem, however, we prefer to use the term "variational equilibrium" here.

With this new terminology, the point $(3/4, 1/4)$ is the (unique) variational equilibrium of the problem in Example 0.1. Note that, by Theorem 2.1, in the case of NEPs the set of solutions and of variational solutions coincide. For GNEPs, however, every variational equilibrium is a generalized Nash equilibrium, but Example 0.1 shows that the converse is not true in general.

The results above indicate that the calculation of a variational equilibrium could be a valuable target for an algorithm. Furthermore, in some applicative contexts, variational equilibria can also have further practical interest; see, for example, the comments in [52].

The question naturally arises of what characterizes variational equilibria. The answer is rather simple and neat. Suppose, as usual, that the set $\mathbf{X}$ is defined by the inequalities $g(\boldsymbol{x}) \leq 0$, where $g\colon \mathbb{R}^n \to \mathbb{R}^m$, with g convex in $\boldsymbol{x}$ (and continuously differentiable). The player's problem is then given by

$$\begin{aligned} &\min_{x^\nu} \ \theta_\nu(x^\nu, \boldsymbol{x}^{-\nu}) \\ &\text{s.t.} \quad g(x^\nu, \boldsymbol{x}^{-\nu}) \leq 0. \end{aligned}$$

Assume that at each solution of a player's problem, the KKT conditions are satisfied. For simplicity, actually assume that the KKT conditions are satisfied for only one set of multipliers (for example assume that linear independence of the active constraints holds). Let $\boldsymbol{x}_*$ be a solution of the game. Then, for each player ν, a vector λ^ν of multipliers exists such that

$$\begin{aligned} \nabla_{x^\nu}\theta_\nu(\boldsymbol{x}_*) + \nabla_{x^\nu} g(\boldsymbol{x}_*)\lambda^\nu &= 0, \\ 0 \le \lambda^\nu \perp g(\boldsymbol{x}_*) &\le 0. \end{aligned}$$

Note that, in general, λ^ν need not be equal to the multipliers λ^μ of another player. It turns out that variational equilibria are precisely those equilibria for which the multipliers are the same for all players.

Theorem 3.5. *In the setting just described above, an equilibrium of the* JC GNEP *is a variational equilibrium if and only if* $\lambda^1 = \lambda^2 = \cdots = \lambda^N$.

This characterization can both be improved on – see [72] – and put to algorithmic use – see [34, 101].

We conclude by observing that the jointly convex GNEP has been the subject of much analysis and certainly covers some very interesting applications. However, it should be noted that the jointly convex assumption on the constraints is strong, and practically is likely to be satisfied only when the joint constraints $g^\nu = g$, $\nu = 1, \dots, N$ are linear, i.e., of the form $A\boldsymbol{x} \le b$ for some suitable matrix A and vector b.

Solution algorithms. The calculation of a variational solution by a centralized algorithm does not need any further comment and can be carried out exactly as the calculation of an equilibrium of a standard Nash equilibrium problem since, in both cases, we have reduced the problem to a VI. Algorithms 1 and 2 can be used, for example. The situation is more complicated if we want to use a decentralized algorithm. In this case, in fact, we cannot apply directly Algorithm 5, since the feasible set $\mathbf{X}$ does not have the cartesian product structure underlying the use of Algorithm 5. Luckily this problem can be fixed relatively easily. We consider below, for simplicity, just the case of decomposable constraints:

$$g(\boldsymbol{x}) = \sum_{\nu=1}^{N} g^\nu(x^\nu)$$

where each g^ν is convex and defined on the whole space. The most common case in which this assumption holds is when the constraints g are linear: $g(\boldsymbol{x}) = A\boldsymbol{x} - b$. In turn, as discussed above, this is the most important case in which the joint convexity assumption may be expected to hold. So suppose that the players's problems are

$$\begin{aligned} &\min_{x^\nu} \; \theta_\nu\big(x^\nu, \boldsymbol{x}^{-\nu}\big) \\ &\text{s.t.} \quad g(\boldsymbol{x}) \le 0 \end{aligned}$$

and that we want to calculate a variational solution. The "trick" that allows us to use Algorithm 5 is the observation that variational solutions are also the solutions of the following NEP, where we have introduced an additional player, controlling the variables $\lambda \in \mathbb{R}^m$:

$$\begin{aligned} &\min_{x^\nu} \ \theta_\nu(x^\nu, \boldsymbol{x}^{-\nu}) + \lambda^T g^\nu(x^\nu), \qquad \nu = 1, \ldots N, \\ &\min_\lambda \ -\lambda^T g(\boldsymbol{x}) \\ &\text{s.t.} \quad \lambda \geq 0. \end{aligned} \tag{3.2}$$

The verification of the fact that a solution of this game is a variational solution of the original JC NEP is very simple and amounts to an application of the definition of solution to both game (3.2) and the original JC GNEP; we leave this verification to the reader. The important thing to note is that the game (3.2) is a NEP and so, in principle, we can apply Algorithm 5, provided we can ensure the required monotonicity conditions are satisfied. If we convert the NEP (3.2), the corresponding pseudo-gradient function, that we denote by $\widetilde{\mathbf{F}}$, is given by

$$\widetilde{\mathbf{F}}(\mathbf{x}, \lambda) = \begin{pmatrix} \vdots \\ \nabla_{x^\nu}\theta_\nu(\mathbf{x}) + \nabla_{x^\nu} g^\nu(x^\nu)\lambda \\ \vdots \\ -g(\boldsymbol{x}) \end{pmatrix} = \begin{pmatrix} \mathbf{F}(\mathbf{x}) + \nabla_x g(\mathbf{x})\lambda \\ -g(\mathbf{x}) \end{pmatrix}. \tag{3.3}$$

The interesting point is that if we assume that $\mathbf{F}$ is monotone then also $\widetilde{\mathbf{F}}$ is monotone. This is stated in the following proposition, whose proof is once again just a mere and simple verification of the definition.

Proposition 3.6. *If $\mathbf{F}(\mathbf{x}) = (\nabla_{x^\nu}\theta_\nu(\mathbf{x}))_{\nu=1}^N$ is monotone on $\{\boldsymbol{x} : g(\boldsymbol{x}) \leq 0\}$, with g convex, then also $\widetilde{\mathbf{F}}(\mathbf{x}, \lambda)$ is monotone on the same set.*

We then see that, with a little overburden, we can solve game (3.2) by using the distributed Algorithm 5 and find in this way a variational equilibrium.

Having seen that in a more or less direct way the tools developed for the computation of equilibria of NEPs can be adapted to the computation of variational equilibria of JC GNEPs, we devote the rest of this section to a more difficult issue: How can we calculate solutions that are not variational solutions? It is well understood that GNEPs (including JC GNEP) have, as a rule, a continuum of solutions and it has been long recognized that in many cases it is of great interest to develop algorithms that potentially are able to calculate multiple if not all equilibria. Unfortunately, there are only a handful of proposals that attempt to calculate non-variational solutions of a jointly convex GNEP; see, e.g., [80]. Here we briefly describe some recent results from [43] that appear to be rather effective, from the practical point of view. However, we must stress that theoretical interesting results can be obtained only for restricted classes of JC GNEPs.

We begin by considering the case in which the players share some common linear *equality* constraints, i.e., we assume that each player's problem has the form

$$\begin{aligned} \min\nolimits_{x^\nu} \quad & \theta_\nu(\boldsymbol{x}) \\ \text{s.t.} \quad & A\boldsymbol{x} = b, \\ & g^\nu(x^\nu) \le 0, \end{aligned} \tag{3.4}$$

where $g^\nu \colon \mathbb{R}^{n_\nu} \to \mathbb{R}^{m_\nu}$, $b \in \mathbb{R}^l$, $A = (A_1 \cdots A_N)$, with $A_\nu \in \mathbb{R}^{l \times n_\nu}$. We denote the corresponding game as GNEP$_=$ and, as usual, we recall that assumptions A1, A2 and A3 are assumed to hold. We recall that in a general JC GNEP the shared constraints can also be defined by convex inequalities. We will consider the case of JC GNEPs with shared *inequality constraints* later on.

We indicate by $\mathbf{X}_= \subseteq \mathbb{R}^n$ the common strategy space of all players, i.e., $\mathbf{X}_= = \{\boldsymbol{x} \in \mathbb{R}^n : A\boldsymbol{x} = b,\, g^\nu(x^\nu) \le 0,\, \nu = 1, \dots, N\}$ (the $=$ in $\mathbf{X}_=$ serves to remember that the shared constraints are equalities). A point $\boldsymbol{x}_* \in \mathbf{X}_=$ is a solution of the GNEP$_=$ if, for all ν, x_*^ν is an optimal solution of problem (3.4) when one fixes $\boldsymbol{x}^{-\nu}$ to $\boldsymbol{x}_*^{-\nu}$.

We assume that if $\bar{x}^\nu$ is a solution of problem (3.4) for a given $\bar{\boldsymbol{x}}^{-\nu}$, then the KKT conditions hold at $\bar{x}^\nu$.

Assumption 3.7. Let $\bar{\boldsymbol{x}}^{-\nu}$ be given, and suppose that $\bar{x}^\nu$ is a solution of problem (3.4). Then multipliers $\mu^\nu \in \mathbb{R}^l$ and $\lambda^\nu \in \mathbb{R}^{m_\nu}$ exist such that

$$\begin{aligned} & \nabla_{x^\nu} \theta_\nu(\bar{\boldsymbol{x}}) + A_\nu^T \mu^\nu + \nabla_{x^\nu} g^\nu(\bar{x}^\nu) \lambda^\nu = 0, \\ & A\bar{\boldsymbol{x}} = b, \\ & 0 \le \lambda^\nu \perp g^\nu(\bar{x}^\nu) \le 0. \end{aligned}$$

It is well known that the KKT conditions will hold under any of a host of constraint qualifications; among them the *Mangasarian–Fromovitz constraint qualification* (MFCQ). Assumption 3.7 implies that if $\boldsymbol{x}_*$ is a solution of the GNEP$_=$, then there exist multipliers $\boldsymbol{\mu}$ and $\boldsymbol{\lambda}$ such that

$$\begin{aligned} & \begin{pmatrix} \nabla_{x^1} \theta_1(\boldsymbol{x}_*) \\ \vdots \\ \nabla_{x^N} \theta_N(\boldsymbol{x}_*) \end{pmatrix} + \begin{pmatrix} A_1^T & & 0 \\ & \ddots & \\ 0 & & A_N^T \end{pmatrix} \boldsymbol{\mu} \\ & \qquad + \begin{pmatrix} \nabla_{x^1} g^1(x_*^1) & & 0 \\ & \ddots & \\ 0 & & \nabla_{x^N} g^N(x_*^N) \end{pmatrix} \boldsymbol{\lambda} = 0, \end{aligned} \tag{3.5}$$

$$A\boldsymbol{x}_* = b,$$

$$0 \le \lambda^\nu \perp g^\nu(x_*^\nu) \le 0, \quad \nu = 1, \dots, N,$$

where $\boldsymbol{\mu} = \begin{pmatrix} \mu^1 \\ \vdots \\ \mu^N \end{pmatrix}$ and $\boldsymbol{\lambda} = \begin{pmatrix} \lambda^1 \\ \vdots \\ \lambda^N \end{pmatrix} \in \mathbb{R}^m$, with $m = m_1 + \cdots + m_N$ and where, we recall, ∇ denotes the transposed Jacobian of a function.

Note that, under Assumption 3.7, a point $\boldsymbol{x}_*$ is a solution of the GNEP$_=$ if and only if it satisfies the KKT conditions (3.5).

We now proceed to show that we can define a (an infinite) family of standard NEPs whose solutions will give the whole set of solutions of the original JC GNEP. Once again, this will allow us to calculate non-variational equilibria by solving standard VIs.

For any fixed $\bar{\boldsymbol{x}} \in \mathbb{R}^n$, we define NEP$_=(\bar{\boldsymbol{x}})$ as the game in which every player's subproblem is

$$\begin{aligned} &\min_{x^\nu} \ \theta_\nu(\boldsymbol{x}) \\ &\text{s.t.} \quad A_\nu x^\nu = A_\nu \bar{x}^\nu, \\ &\qquad\ \ g^\nu(x^\nu) \leq 0. \end{aligned} \tag{3.6}$$

We denote by $\mathbf{X}_=(\bar{\boldsymbol{x}})$ the feasible set of this game: $\mathbf{X}_=(\bar{\boldsymbol{x}}) = \{\boldsymbol{x} \in \mathbb{R}^n \mid A_\nu x^\nu = A_\nu \bar{x}^\nu,\ g^\nu(x^\nu) \leq 0,\ \nu = 1, \ldots, N\}$. Note that, if $\bar{\boldsymbol{x}} \in \mathbf{X}_=$, then $\mathbf{X}_=(\bar{\boldsymbol{x}})$ is nonempty since it is trivial to check that $\bar{\boldsymbol{x}} \in \mathbf{X}_=(\bar{\boldsymbol{x}})$. Note also that the NEP$_=(\bar{\boldsymbol{x}})$ is a standard Nash game, since there is no coupling in the constraints. Note also the difference between the VI $(\mathbf{X}_=, \mathbf{F})$ (where $\mathbf{F}$ is the usual function given by the partial gradients of the players' objective functions) and the VI $(\mathbf{X}_=(\bar{\boldsymbol{x}}), \mathbf{F})$. The following theorem shows that, as mentioned above, we can in principle recover all the solutions of a JC GNEP by solving a family of standard NEPs that, in turn, can be solved using the techniques described in the previous section. We denote by SOL$_=$ the solution set of the GNEP$_=$ and by SOL$_=(\bar{\boldsymbol{x}})$ the solution set of the NEP$_=(\bar{\boldsymbol{x}})$.

Theorem 3.8. *It holds that*

$$\bigcup_{\boldsymbol{x} \in \mathbf{X}_=} \text{SOL}_=(\boldsymbol{x}) = \text{SOL}_=.$$

The theorem therefore shows that by picking any feasible point belonging to $\mathbf{X}_=$ we can find a solution of the JC GNEP by solving a NEP, and that all and only the solutions of the JC GNEP can be obtained this way. In view of the discussion above, the following procedure will produce a solution of GNEP$_=$ which is not necessarily variational.

Algorithm 6: Two Stages Method for the Computation of Non-variational Equilibria

---

(S.1) : Find any point $\bar{\boldsymbol{x}} \in \mathbf{X}_=$.
(S.2) : Find any solution of the VI $(\mathbf{X}_=(\bar{\boldsymbol{x}}), (\nabla_{x^\nu}\theta_\nu(\boldsymbol{x}))_{\nu=1}^N)$.

The results above essentially cover all JC GNEPs where the shared constraints are convex equalities, since in this case essentially only linear shared equalities are possible. If we consider instead the case of shared inequalities, then we could have both linear and nonlinear inequalities. The case of shared inequalities is much more difficult; nevertheless we partially address the case of separable shared inequalities below.

We assume that the optimization problem faced by each player is given by

$$\begin{aligned} \min_{x^\nu}\ & \theta_\nu(\boldsymbol{x}), \\ \text{s.t.}\quad & A\boldsymbol{x} = b, \\ & h(\boldsymbol{x}) \le 0, \\ & g^\nu(x^\nu) \le 0, \end{aligned} \tag{3.7}$$

where $h(\boldsymbol{x}) = \sum_{\nu=1}^{N} h^\nu(x^\nu) \in \mathbb{R}^p$, with each h^ν continuously differentiable and convex. Obviously, the most common case of separable h is when h is linear. We call this new game GNEP$_\le$ and denote its common feasible set by $\mathbf{X}_\le = \{\boldsymbol{x} \in \mathbb{R}^n : A\boldsymbol{x} = b,\ h(\boldsymbol{x}) \le 0,\ g^\nu(x^\nu) \le 0,\ \nu = 1,\dots,N\}$. We still assume that the KKT conditions hold at the solution of each player subproblem.

Assumption 3.9. Let $\bar{\boldsymbol{x}}^{-\nu}$ be given, and suppose that $\bar{x}^\nu$ is a solution of problem (3.7). Then multipliers $\mu^\nu \in \mathbb{R}^l, \gamma^\nu \in \mathbb{R}^p$ and $\lambda^\nu \in \mathbb{R}^{m_\nu}$ exist such that

$$\begin{aligned} & \nabla_{x^\nu}\theta_\nu(\bar{\boldsymbol{x}}) + A_\nu^T\mu^\nu + \nabla_{x^\nu}h^\nu(\bar{x}^\nu)\gamma^\nu + \nabla_{x^\nu}g^\nu(\bar{x}^\nu)\lambda^\nu = 0, \\ & A\bar{\boldsymbol{x}} = b, \\ & 0 \le \gamma^\nu \perp h(\bar{\boldsymbol{x}}) \le 0, \\ & 0 \le \lambda^\nu \perp g^\nu(\bar{x}^\nu) \le 0. \end{aligned}$$

Again, we have that, under Assumption 3.9, a point $\boldsymbol{x}_*$ is a solution of the GNEP$_\le$ if and only if it satisfies the KKT conditions (3.8):

$$\begin{aligned} & \begin{pmatrix} \nabla_{x^1}\theta_1(\boldsymbol{x}_*) \\ \vdots \\ \nabla_{x^N}\theta_N(\boldsymbol{x}_*) \end{pmatrix} + \begin{pmatrix} A_1^T & & 0 \\ & \ddots & \\ 0 & & A_N^T \end{pmatrix} \boldsymbol{\mu} \\ & \qquad + \begin{pmatrix} \nabla_{x^1}h^1(x_*^1) & & 0 \\ & \ddots & \\ 0 & & \nabla_{x^N}h^N(x_*^N) \end{pmatrix} \boldsymbol{\gamma} \\ & \qquad + \begin{pmatrix} \nabla_{x^1}g^1(x_*^1) & & 0 \\ & \ddots & \\ 0 & & \nabla_{x^N}g^N(x_*^N) \end{pmatrix} \boldsymbol{\lambda} = 0, \\ & A\boldsymbol{x}_* = b, \\ & 0 \le \gamma^\nu \perp h(\boldsymbol{x}_*) \le 0, \quad \nu = 1,\dots,N, \\ & 0 \le \lambda^\nu \perp g^\nu(x_*^\nu) \le 0, \quad \nu = 1,\dots,N. \end{aligned} \tag{3.8}$$

In order to find many (possibly all) solutions of the game $\text{GNEP}_{\leq}$, we follow lines similar to those adopted for the equalities-only problem. However, the presence of shared inequality constraints complicates things considerably.

For any $\bar{\boldsymbol{x}} \in \mathbb{R}^n$, we define a new, standard Nash game, that we denote by $\text{NEP}_{\leq}(\bar{\boldsymbol{x}})$, in which each player ν has to solve the following minimization problem:

$$\begin{aligned} \min_{x^\nu} \ & \theta_\nu(\boldsymbol{x}), \\ \text{s.t.} \quad & A_\nu x^\nu = A_\nu \bar{x}^\nu, \\ & h^\nu(x^\nu) \leq h^\nu(\bar{x}^\nu) - \frac{h(\bar{\boldsymbol{x}})}{N}, \\ & g^\nu(x^\nu) \leq 0. \end{aligned} \tag{3.9}$$

Once again the game $\text{NEP}_{\leq}(\bar{\boldsymbol{x}})$ has decoupled constraints and can therefore be solved by converting it to a VI. If we denote by $\mathbf{X}_{\leq}(\bar{\boldsymbol{x}})$ the feasible set of the $\text{NEP}_{\leq}(\bar{\boldsymbol{x}})$,

$$\mathbf{X}_{\leq}(\bar{\boldsymbol{x}}) = \Big\{ \boldsymbol{x} \in \mathbb{R}^n \mid A_\nu x^\nu = A_\nu \bar{x}^\nu, \ h^\nu(x^\nu) \leq h^\nu(\bar{x}^\nu) - \frac{h(\bar{\boldsymbol{x}})}{N}, \\ g^\nu(x^\nu) \leq 0, \ \nu = 1, \ldots, N \Big\},$$

then the solutions of the $\text{NEP}_{\leq}(\bar{\boldsymbol{x}})$ are the solutions of the variational inequality $(\mathbf{X}_{\leq}(\bar{\boldsymbol{x}}), (\nabla_{x^\nu}\theta_\nu(\boldsymbol{x}))_{\nu=1}^N)$. It may be useful to remark that, if $\bar{\boldsymbol{x}} \in \mathbf{X}_{\leq}$, then $\mathbf{X}_{\leq}(\bar{\boldsymbol{x}})$ is nonempty since it is easily seen that $\bar{\boldsymbol{x}} \in \mathbf{X}_{\leq}(\bar{\boldsymbol{x}})$. We can give an extension of Theorem 3.8. To this end we denote by $\text{SOL}_{\leq}$ the solution set of $\text{GNEP}_{\leq}$ and by $\text{SOL}_{\leq}(\boldsymbol{x})$ the solution set of $\text{NEP}_{\leq}(\boldsymbol{x})$.

Theorem 3.10. *It holds that*

$$\text{SOL}_{\leq} \subseteq \bigcup_{\boldsymbol{x} \in \mathbf{X}_{\leq}} \text{SOL}_{\leq}(\boldsymbol{x}).$$

We see from this theorem that the situation in the case of shared inequality constraints is more complex than that of shared equality constraints. In fact, when $\bar{\boldsymbol{x}}$ varies in $\mathbf{X}_{\leq}$ we can obtain "spurious solutions" from the $\text{NEP}_{\leq}(\bar{\boldsymbol{x}})$. This is shown by the following example.

Example 3.11. There are two players each controlling one variable that we call, for simplicity, x and y. The players' problems are

$$\begin{aligned} \min_x \ & x^2 + \tfrac{8}{3}xy - 34x \\ \text{s.t.} \ & x + y \leq 15, \\ & 0 \leq x \leq 10, \end{aligned} \qquad \begin{aligned} \min_y \ & y^2 + \tfrac{5}{4}xy - 24,25y \\ \text{s.t.} \ & x + y \leq 15, \\ & 0 \leq y \leq 10. \end{aligned}$$

This jointly convex GNEP has the following equilibria:

- $x = 5$ and $y = 9$, which is a variational equilibrium;
- a continuum of non-variational equilibria given by

$$\begin{pmatrix} t \\ 15 - t \end{pmatrix}, \qquad t \in [9, 10].$$

Set $h^1(x) = x - 15$ and $h^2(y) = y$. If we take $\bar{\boldsymbol{x}} = (0,0)$, the feasible set $\mathbf{X}_{\leq}(\bar{\boldsymbol{x}})$ is given by those points (x,y) such that $0 \leq x \leq 7.5$, $0 \leq y \leq 7.5$. It is clear that any solution of the VI $(\mathbf{X}_{\leq}(\bar{\boldsymbol{x}}), \mathbf{F})$ cannot be a solution of the original game, since in any solution of the original game there is at least one component that is greater than or equal to 9. Therefore we have

$$\text{SOL}_{\leq} \subsetneq \bigcup_{\boldsymbol{x} \in \mathbf{X}_{\leq}} \text{SOL}_{\leq}(\boldsymbol{x}).$$

We conclude that the inclusion in Theorem (3.10) is strict. □

The problem of understanding when a solution of $\text{NEP}_{\leq}(\bar{\boldsymbol{x}})$ is also a solution of the original $\text{GNEP}_{\leq}$ immediately arises. To deal with this issue we need to look at the KKT conditions of $\text{NEP}_{\leq}(\bar{\boldsymbol{x}})$, that are

$$\begin{aligned}
&\begin{pmatrix} \nabla_{x^1}\theta_1(\boldsymbol{x}_*) \\ \vdots \\ \nabla_{x^N}\theta_N(\boldsymbol{x}_*) \end{pmatrix} + \begin{pmatrix} A_1^T & & 0 \\ & \ddots & \\ 0 & & A_N^T \end{pmatrix} \boldsymbol{\mu}_* \\
&\qquad + \begin{pmatrix} \nabla_{x^1}h^1(x_*^1) & & 0 \\ & \ddots & \\ 0 & & \nabla_{x^N}h^N(x_*^N) \end{pmatrix} \gamma_* \\
&\qquad + \begin{pmatrix} \nabla_{x^1}g^1(x_*^1) & & 0 \\ & \ddots & \\ 0 & & \nabla_{x^N}g^N(x_*^N) \end{pmatrix} \boldsymbol{\lambda}_* = 0, \\
&A_\nu x_*^\nu = A_\nu \bar{x}^\nu, \quad \nu = 1, \dots, N, \\
&0 \leq \gamma_*^\nu \perp h^\nu(x_*^\nu) - \left(h^\nu(\bar{x}^\nu) - \frac{h(\bar{\boldsymbol{x}})}{N} \right) \leq 0, \quad \nu = 1, \dots, N, \\
&0 \leq \lambda_*^\nu \perp g^\nu(x_*^\nu) \leq 0, \quad \nu = 1, \dots, N.
\end{aligned} \tag{3.10}$$

The following theorem clarifies when a solution of $\text{NEP}_{\leq}(\bar{\boldsymbol{x}})$ is a solution of the original GNEP.

Theorem 3.12. *Let $\boldsymbol{x}_*$ be a solution of the* GNEP$_\leq(\bar{\boldsymbol{x}})$, *with* $\bar{\boldsymbol{x}} \in \mathbf{X}_\leq$. *Then the following holds:*

$$\left\{ \begin{array}{c} \textit{Multipliers } (\boldsymbol{\mu}_*, \boldsymbol{\gamma}_*, \boldsymbol{\lambda}_*) \textit{ that satisfy with } \boldsymbol{x}_* \\ \textit{the KKT conditions (3.10) exist such that} \\ \left\{ \begin{array}{c} \textit{either} \\ h_i^\nu(x_*^\nu) - \left(h_i^\nu(\bar{x}^\nu) - \frac{h_i(\bar{\boldsymbol{x}})}{N} \right) = 0, \ \forall \nu, \\ \textit{or} \\ \gamma_{*i}^\nu = 0, \ \forall \nu \end{array} \right\} \forall i = 1, \dots, p \end{array} \right\} \Longleftrightarrow \boldsymbol{x}_* \in \mathrm{SOL}_\leq .$$

What this theorem says is, loosely speaking, that a solution of the NEP$_\leq(\bar{\boldsymbol{x}})$ is a solution of the original JC GNEP if and only if each corresponding ith constraint is either active for all players or nonactive for all players. The previous theorem simplifies considerably when the KKT conditions of the NEP$_\leq(\bar{\boldsymbol{x}})$ at $\boldsymbol{x}_*$ are satisfied for only one triple of multipliers.

Corollary 3.13. *Let $\boldsymbol{x}_*$ be a solution of the* NEP$_\leq(\bar{\boldsymbol{x}})$, *with* $\bar{\boldsymbol{x}} \in \mathbf{X}_\leq$, *and let* $(\boldsymbol{\mu}_*, \boldsymbol{\gamma}_*, \boldsymbol{\lambda}_*)$ *be the unique triple of multipliers that together with $\boldsymbol{x}_*$ satisfies the KKT conditions* (3.10). *Then $\boldsymbol{x}_*$ is a solution of the* GNEP$_\leq$ *if and only if, for all* $i = 1, \dots, p$, *either* $h_i^\nu(x_*^\nu) - \left(h_i^\nu(\bar{x}^\nu) - \frac{h_i(\bar{\boldsymbol{x}})}{N} \right) = 0$ *for all* ν, *or* $\gamma_{*i}^\nu = 0$ *for all* ν.

The above corollary holds in particular when the *linear independence constraint qualification* (LICQ) holds for the NEP$_\leq(\bar{\boldsymbol{x}})$. In this case the corollary shows that we can decide whether $\boldsymbol{x}_*$ is a solution of the GNEP$_\leq$ simply by looking at the multipliers $\boldsymbol{\gamma}_*$ and the constraints $h^\nu(x_*^\nu) - \left(h^\nu(\bar{x}^\nu) - \frac{h(\bar{\boldsymbol{x}})}{N} \right) \leq 0$.

The computational difficulty with shared inequality constraints is that one does not know a priori which points $\bar{\boldsymbol{x}} \in \mathbf{X}_\leq$ will give rise to solutions of the original GNEP, i.e., to solutions satisfying the conditions in Theorem 3.12. If one samples the set $\mathbf{X}_\leq$, many points can lead to useless solutions. It would clearly be desirable to have a simple test predicting which points give rise to solutions of the VI $(\mathbf{X}_\leq(\bar{\boldsymbol{x}}), \mathbf{F})$ that are also solutions of the GNEP$_\leq$. Such a test should, for example, lead to discard the point $(0, 0)$ in Example 3.11 without the need of actually solving the corresponding VI. While we do not have such a test, the following heuristic procedure has been found to work well in practice.

Algorithm 7: Iterative Two-stage Method in the Case of Inequality Constraints

---

(S.1) : Find any point $\boldsymbol{x}_0 \in \mathbf{X}_\leq$, and set $k = 0$.
(S.2) : Find any solution $\boldsymbol{x}_{k+1}$ of the VI $(\mathbf{X}_\leq(\boldsymbol{x}_k), (\nabla_{x^\nu}\theta_\nu(\boldsymbol{x}))_{\nu=1}^N)$.
(S.3) : If $\boldsymbol{x}_{k+1}$ is a solution of the GNEP$_\leq$ then return $\boldsymbol{x}_{k+1}$ and stop; else set $k = k + 1$ and go to S.2.

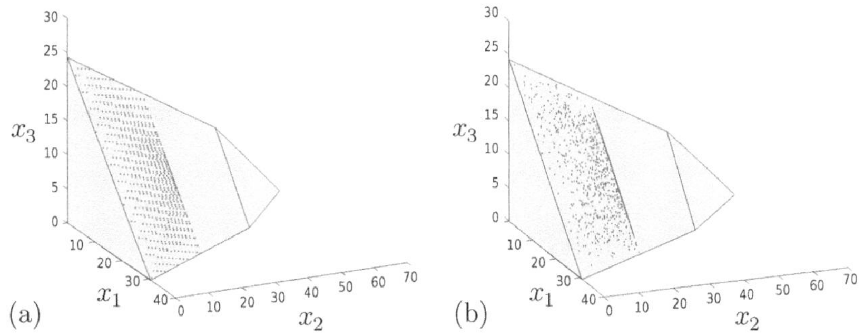

Figure 3.3: Equilibria found for Example 3.14 with sampling on a grid (a) and randomly (b).

The following example illustrates the procedure.

Example 3.14. We consider the river basin pollution game described in detail in [80]. In this problem there are 3 players, $N = 3$, each controlling 1 variable, x^1, x^2, x^3 respectively. The problem of player $\nu \in \{1,2,3\}$ is

$$\begin{aligned} \min_{x^\nu} \ & (c_{1,\nu} + c_{2,\nu} x^\nu) x^\nu - \big(3 - 0.01(x^1 + x^2 + x^3)\big) x^\nu \\ \text{s.t.} \ & \textstyle\sum_{\mu=1}^3 u_{\mu l} e_\mu x^\mu \le 100, \quad l = 1,2, \\ & x^\nu \ge 0, \end{aligned}$$

where

$$c_1 = \begin{pmatrix} 0.10 \\ 0.12 \\ 0.15 \end{pmatrix}, \ c_2 = \begin{pmatrix} 0.01 \\ 0.05 \\ 0.01 \end{pmatrix}, \ e = \begin{pmatrix} 0.50 \\ 0.25 \\ 0.75 \end{pmatrix}, \ u = \begin{pmatrix} 6.5 & 4.583 \\ 5.0 & 6.250 \\ 5.5 & 3.750 \end{pmatrix}.$$

From the constraints of the problem it is clear that the variables must belong to $[0,31] \times [0,64] \times [0,25]$. Therefore we generated points in this rectangle as starting points for Algorithm 7. We tested two strategies for generating these points. In the first strategy, points where generated on a grid with step 2; in the second strategy they were generated randomly. Below we report the results obtained. Moreover, in Figure 3.3 we can see that the equilibria found are spatially rather uniformly distributed both in case of points generated on a grid or randomly.

	Starting points	VIs solved	Solutions	Different solutions
Grid	1518	9432	1518	1494
Random	1367	8215	1367	1367

The computation of all solutions or, more realistically, of a representative sampling is computationally very intensive and there are still many open problems that are current object of research.

Chapter 4

GNEPs with No Special Structure

As we saw in the previous chapter, there are some sound proposals for the analysis and solution of standard Nash problems and of jointly convex problems. However, when it comes to the analysis and solution of a GNEP in its full generality, the situation is much bleaker. A GNEP cannot be reduced to a VI, but it can be reduced to a quasi-variational inequality, as first noted in [14]. A quasi-variational inequality is a problem formally similar to a VI where, however, the feasible set K is not fixed but depends on x. Since the theory of quasi-variational inequalities is not nearly as developed as that of VIs, this reduction is not very useful, especially from the computational point of view. So we do not pursue this issue further. In this chapter, instead, after discussing very briefly some existence results, we consider in some detail two recent algorithmic developments that seem promising.

4.1 Existence of solutions

Existence of solutions has been the main focus of early research in GNEPs. The 1952 Debreu paper [26], where the GNEP was formally introduced, also gives the first existence theorem. This existence result was based on fixed-point arguments, and this turned out to be the main proof tool used in the literature. Essentially this approach is based on the very definition of equilibrium that states that a point $\boldsymbol{x}$ is an equilibrium if $\boldsymbol{x} \in \mathcal{S}(\boldsymbol{x})$, where $\mathcal{S} = \Pi_{\nu=1}^{N} \mathcal{S}_\nu$ with the solution mappings $\mathcal{S}_\nu$ of problem (1) as stated in the Introduction. This shows clearly that $\boldsymbol{x}$ is a fixed point of $\mathcal{S}$, thus paving the way to the application of the fixed-point machinery to establish existence of an equilibrium. There also exist some other approaches, an interesting one being the one presented in [50], where a continuation approach is used. The main existence result is probably the one established in [9]. We report below a slightly simplified version given by Ichiishi [63]. We state this theorem with very basic assumptions in place; in particular we do not assume that A1, A2, and A3 hold.

Theorem 4.1. *Let a* GNEP *be given and suppose the following:*

(a) *There exist N nonempty, convex and compact sets $K_\nu \subset \mathbb{R}^{n_\nu}$ such that, for every $\boldsymbol{x} \in \mathbb{R}^n$ with $x^\nu \in K_\nu$ for all ν, $X_\nu(\boldsymbol{x}^{-\nu})$ is nonempty, closed and convex, $X_\nu(\boldsymbol{x}^{-\nu}) \subseteq K_\nu$, and X_ν, as a point-to-set map, is both upper and lower semicontinuous*[1]*.*

(b) *For every player ν, the function $\theta_\nu(\,\cdot\,, \boldsymbol{x}^{-\nu})$ is quasi-convex*[2] *on $X_\nu(\boldsymbol{x}^{-\nu})$.*

Then a generalized Nash equilibrium exists.

Remark 4.2. When the sets X_ν are defined by inequality constraints as in (2), the lower and upper semicontinuity requirements translate into reasonably mild conditions on the functions g^ν. See for example [10, 97].

The relaxation of the assumptions in the previous theorem has been the subject of a fairly intense study. Relaxations of the (a) continuity assumption; (b) compactness assumption and (c) quasi-convexity assumption have all been considered in the literature. The relaxation of the continuity assumption is the most interesting one, since it is peculiar to GNEPs. In fact, a number of classical problems in economics can be formulated as games with discontinuous objective functions. The best known of these are probably Bertrand's model of duopolistic price competition [17] and Hotelling's model of duopolistic spatial competition [61]. In the Bertrand model, firms choose prices, and the firm that charges the lower price supplies the whole market. In the Hotelling model, instead, firms choose locations and each firm monopolizes the part of the market closer to that firm than to the others. In each case, discontinuities arise when firms charge the same price or locate at the same point. There are, however, a host of other problems that give rise to games with discontinuous objective functions; good entry points to the literature on the subject are [13, 24, 25]. There are several papers where the relaxation of continuity is pursued; the seminal one is the 1986 paper by Dasgupta and Maskin [24]. Further developments and applications are discussed in [13, 25, 94, 108, 113] and references therein. However, with the (partial) exception of [13], where jointly convex GNEPs are discussed, all these papers deal only with pure NEPs. A rather general result for GNEPs seems to be the one in [79]. Other recent, valuable results can also be found in [41]. The interest of this latter reference is that, in contrast to all previous works, existence results are established by using some VI reduction. We do not go into these details here and pass instead to the discussion of algorithms.

---

[1] A point-to-set mapping $G\colon \mathbb{R}^n \rightrightarrows \mathbb{R}^s$ is *upper semicontinuous* at $y \in \mathbb{R}^n$ if for every sequence $\{y^k\}$ in $\mathbb{R}^n$ converging to y, and for every neighborhood U of $G(y)$ in $\mathbb{R}^s$, there exists $\bar{k}$ such that $G(y^k) \subseteq U$ for all $k \geq \bar{k}$. The mapping G is *lower semicontinuous* at $y \in \mathbb{R}^n$ if for every sequence $\{y^k\}$ in $\mathbb{R}^n$ converging to y, and for every open subset U of $\mathbb{R}^s$ with $G(y) \cap U \neq \emptyset$, there exists $\bar{k}$ such that $G(y^k) \cap U \neq \emptyset$ for all $k \geq \bar{k}$. We say that G is upper (lower) semicontinuous on a set Y if it is upper (lower) semicontinuous at each point of Y.

[2] A function $f\colon \mathbb{R}^t \to \mathbb{R}$ is *quasi-convex* if the level sets $\mathcal{L}(\alpha) = \{x \in \mathbb{R}^t : f(x) \leq \alpha\}$ are convex for every $\alpha \in \mathbb{R}$.

4.2 Penalty approach

The GNEPs in their full generality are extremely difficult and the computation of a solution is, in general, a formidable task to accomplish, and very few proposals, all very recent, are available in the literature. In these notes we discuss two approaches: a penalty approach and the use of the KKT conditions. As far as we are aware of, and if we exclude a handful of methods tailored to very specific and structured applications – see [90] for an example – these are essentially the only two approaches for which it is possible to establish theoretical convergence, albeit under extremely strong conditions.

Penalty approaches are probably the most prominent class of methods that have been proposed in the literature for the solution of general GNEPs; see [37, 40, 46, 71, 89]. However, as far as we are aware of, in none of these papers classes of GNEPs were exhibited for which the penalty approach is guaranteed to converge to a solution. What we discuss below is the method discussed in [38], a variant of [37] that, in turn, is inspired by [28, 29].

The aim of penalty methods is that of reducing a general GNEP to a conceptually simpler NEP, We assume that the feasible sets of the players are defined by

$$K_\nu \cap \{x^\nu \,:\, g^\nu(x^\nu, \boldsymbol{x}^{-\nu}) \le 0\},$$

thus distinguishing between the "private" constraints represented by K_ν and those constraints g^ν that depend on both the player's variables and his rivals' ones. Note that usually the set K_ν will in turn be defined by further constraints, $K_\nu = \{x^\nu : h^\nu(x^\nu) \le 0\}$, but this representation is, for our purposes, immaterial. With this definition, the players' problems become

$$\begin{aligned} \min_{x^\nu} \;& \theta_\nu(x^\nu, \boldsymbol{x}^{-\nu}) \\ \text{s.t.} \;& g^\nu(x^\nu, \boldsymbol{x}^{-\nu}) \le 0, \\ & x^\nu \in K_\nu. \end{aligned} \tag{4.1}$$

We assume that all the sets K_ν are convex and compact.

We now consider a partial penalization scheme where only the coupling, "difficult" constraints g^ν are penalized so that the penalized problem becomes:

$$\begin{aligned} \min_{x^\nu} \;& \theta_\nu(x^\nu, \boldsymbol{x}^{-\nu}) + \rho_\nu \left\| g^\nu_+(x^\nu, \boldsymbol{x}^{-\nu}) \right\|_3 \\ \text{s.t.} \;& x^\nu \in K_\nu. \end{aligned} \tag{4.2}$$

We recall that, if $v \in \mathbb{R}^p$, then $\|v\|_3$ denotes the vector norm defined by $\|v\|_3 = (|v_1|^3 + \cdots + |v_p|^3)^{\frac{1}{3}}$ and that g_+ denotes the vector $\max\{0, g\}$, where the max is taken componentwise.

By setting

$$P_\nu(\boldsymbol{x}, \rho_\nu) = \theta_\nu(x^\nu, \boldsymbol{x}^{-\nu}) + \rho_\nu \left\| g^\nu_+(x^\nu, \boldsymbol{x}^{-\nu}) \right\|_3,$$

problem (4.2) can be rewritten as a (standard) Nash equilibrium problem, where each player's problem is given by

$$\begin{array}{ll} \min_{x^\nu} & P_\nu(\boldsymbol{x}, \rho_\nu) \\ \text{s.t.} & x^\nu \in K_\nu. \end{array} \tag{4.3}$$

We refer to problem (4.2) or, equivalently, (4.3) as the Penalized Nash Equilibrium Problem with parameters ρ (PNEP_ρ). The basic idea of penalization techniques is that we can recover a solution of the original GNEP by finding a solution of the PNEP_ρ for suitable values of the penalty parameters ρ_ν. Since the PNEP_ρ is a GNEP, we also hope that we are actually able to solve the PNEP_ρ. However, we point our from the outset that the PNEP_ρ has nondifferentiable objective functions, a fact that, as we will see, causes some troubles. In a sense we traded the difficulty of the variable sets $X_\nu(\boldsymbol{x}^{-\nu})$ with the difficulty given by a nondifferentiable objective function. We also want to point out, for those who have familiarity with penalty methods, that the penalty term used the $\|\cdot\|_3$ norm, and not the more usual $\|\cdot\|_1$ or $\|\cdot\|_\infty$ norms; this plays a crucial role in our developments.

In order to show that indeed the GNEP and the PNEP_ρ have strong relations, we need to introduce a constraint qualification.

Definition 4.3. We say that the GNEP (4.1) satisfies the *extended Mangasarian–Fromovitz constraint qualification* (EMFCQ) at a point $\bar{\boldsymbol{x}} \in \mathbf{K}$ if, for every player $\nu = 1, \dots, N$, there exists a vector $d^\nu \in T_{K_\nu}(\bar{\boldsymbol{x}}^\nu)$ such that

$$\nabla_{x^\nu} g_i^\nu(\bar{x}^\nu, \bar{\boldsymbol{x}}^{-\nu})^T d^\nu < 0, \quad \forall i \in I_+^\nu(\bar{\boldsymbol{x}}), \tag{4.4}$$

where $I_+^\nu(\bar{\boldsymbol{x}}) = \{i \in \{1, \dots, m_\nu\} \mid g_i^\nu(\bar{x}^\nu, \bar{\boldsymbol{x}}^{-\nu}) \geq 0\}$ is the index set of all active and violated constraints at $\bar{\boldsymbol{x}}$ and $T_{K_\nu}(\bar{\boldsymbol{x}}^\nu)$ is the tangent cone of K_ν at $\bar{\boldsymbol{x}}^\nu$.

By standard reasonings, it is easily seen that condition (4.4) is equivalently satisfied if, for every player ν, there is no nonzero vector $v^\nu \in \mathbb{R}_+^{m_\nu}$, with $v_i^\nu = 0$ if $g_i^\nu(\bar{x}^\nu, \bar{\boldsymbol{x}}^{-\nu}) < 0$, such that

$$-\left[v_1^\nu \nabla_{x^\nu} g_1^\nu(\bar{\boldsymbol{x}}) + \cdots + v_{m_\nu}^\nu \nabla_{x^\nu} g_{m_\nu}^\nu(\bar{\boldsymbol{x}})\right] \in N_{K_\nu}(\bar{x}^\nu), \tag{4.5}$$

where $N_{K_\nu}(\bar{x}^\nu)$ denotes the normal cone to K_ν at $\bar{x}^\nu$.

We observe that it is classical to show that if the EMFCQ holds at a solution of the game, then the KKT conditions are satisfied for each player at this point.

The EMFCQ for optimization problems has been used often in analyzing solution algorithms. The definition given here is the natural extension of the EMFCQ for optimization problems to games. We mention that some of the results below can be proved under weaker assumptions, but this would lead us to very technical developments that are not appropriate for the level of our discussion.

The following theorem shows precisely the correspondence between the solutions of the GNEP and of its penalization.

Theorem 4.4. *In the setting of this section, suppose that the* EMFCQ *holds at every point of* $\mathbf{K} = \prod_{\nu=1}^{N} K_\nu$*; then there exists a positive* $\bar{\rho}$ *such that, for every* ρ *with* $\rho_\nu \geq \bar{\rho}$ *for all* ν*, the solutions of the* PNEP$_\rho$ (4.2) *coincide with the solution of the* GNEP (4.1).

If we knew the value of the parameter $\bar{\rho}$ in Theorem 4.4, our only task would be that of studying an algorithm for the solution of the PNEP$_\rho$ (4.2), which has the peculiarity of having nondifferentiable objective functions. We stress that the nondifferentiability of the objective functions makes the direct application of the methods discussed in the previous section impossible, since they require differentiability of the objective functions as a prerequisite. However, this difficulty apart, in general, the value of $\bar{\rho}$ is not known in advance and so we see that our task is twofold: on the one hand we have to (iteratively) find the "right" values for the penalty parameters and, on the other hand, we must also be able to solve the nondifferentiable penalized Nash game. Therefore we first propose a way to algorithmically update the penalty parameters to an appropriate value, while subsequently we deal with the nondifferentiability issue. For the time being, in order to "decouple" the two issues, we assume that we have an algorithm for the solution of the penalized Nash PNEP$_\rho$ (4.2) *for given fixed values of the penalty parameters* ρ_ν. More precisely, we suppose that an iterative algorithm $\mathcal{A}$ is available, which, given a point $\boldsymbol{x}_k$, generates a new point $\boldsymbol{x}_{k+1} = \mathcal{A}[\boldsymbol{x}_k]$ such that $\boldsymbol{x}_k \in \mathbf{K}$ for all k. We suppose that $\mathcal{A}$ enjoys the following property.

Property 4.5. For every $\boldsymbol{x}_0 \in \mathbf{K}$, the sequence $\{\boldsymbol{x}_k\}$ obtained by setting $\boldsymbol{x}_{k+1} = \mathcal{A}[\boldsymbol{x}_k]$ is such that $x_k \in \mathbf{K}$ for all k and every limit point is a solution of (4.2).

The meaning of the property is plain: Algorithm $\mathcal{A}$ solves the PNEP$_\rho$ when the ρ_ν are fixed. As we already said, a suitable algorithm $\mathcal{A}$ will be described in the next section. Having this algorithm $\mathcal{A}$ at hand, the scheme we propose for the updating of the penalty parameters is described in Figure 4.4.

At each iteration we perform two steps: First we test whether we should update the penalty parameters and then we perform *a single step* of algorithm $\mathcal{A}$. The hope is that eventually the outcome of the test on the updating of the penalty parameters is always not to update, so that the scheme reduces to the application of algorithm $\mathcal{A}$ to the PNEP$_\rho$ with fixed values of penalty parameters. What we must then be sure of is that the value of the penalty parameters is the "right" one. In order to achieve this and also to understand the rationale behind the updating test, we note that if, for any value of the penalty parameters, we find a solution of PNEP$_\rho$ (4.2) that is feasible for the original problem (4.1), then it is easily seen that this solution solves problem (4.1) itself. So the updating rule for the penalty parameters should aim at avoiding convergence of algorithm $\mathcal{A}$ to solutions that are unfeasible for the original GNEP by increasing the penalty parameters if this "dangerous" situation seems to be occurring.

To this end, let us consider a solution $\bar{\boldsymbol{x}}$ of (4.2) infeasible for the GNEP. This means that there exists a player $\nu \in \{1, \ldots, N\}$ such that $\|g_+^\nu(\bar{x}^\nu, \bar{\boldsymbol{x}}^{-\nu})\|_3 > 0$. By the closedness and the convexity of the feasible set K_ν and by the continuous differentiability of $P_\nu(x^\nu, x^{-\nu})$ in $\bar{\boldsymbol{x}}$, we have

$$\bar{x}^\nu = \Pi_{K_\nu}[\bar{x}^\nu - \nabla_{x^\nu} P_\nu(\bar{\boldsymbol{x}}, \rho_\nu)], \tag{4.6}$$

where $\Pi_{K_\nu}(\cdot)$ is the Euclidean projector on the closed convex set K_ν. Our updating rule tries to detect when condition (4.6) is nearly satisfied (see test (4.7) below), and in such case, tries to force feasibility by increasing the penalty parameters.

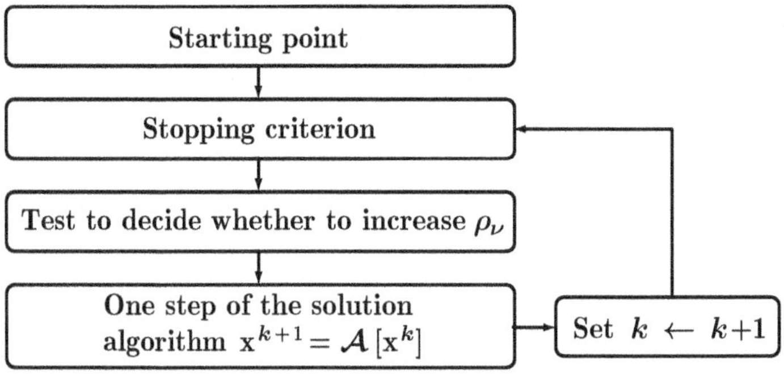

Figure 4.4: The algorithmic scheme.

Algorithm 8: Penalty Updating Scheme for GNEPs

(S.0) : $\boldsymbol{x}_0 \in \mathbf{K}$, $\rho_0^\nu > 0$ and $c_\nu \in (0, 1)$ for all $\nu = 1, \ldots, N$. Set $k = 0$.
(S.1) : If $\boldsymbol{x}_k$ is a solution of the GNEP (4.1), stop.
(S.2) : Let $I_k = \{\nu \mid x_k^\nu \notin X_\nu(\boldsymbol{x}_k^{-\nu})\}$. For every $\nu \in I_k$, if

$$\left\| x_k^\nu - \Pi_{K_\nu}[x_k^\nu - \nabla_{x^\nu} P_\nu(\boldsymbol{x}_k, \rho_k^\nu)] \right\| \le \frac{c_\nu}{\rho_k^\nu}, \tag{4.7}$$

then double the penalty parameters ρ_k^ν.
(S.3) : Compute $\boldsymbol{x}_{k+1} = \mathcal{A}[\boldsymbol{x}_k]$, set $k \leftarrow k + 1$, and go to S.1.

Based on Theorem 4.4, we know that the EMFCQ is sufficient to guarantee the existence of suitable values of the penalty parameters that allow to recover a solution of the GNEP from a solution of the PNEP$_\rho$. We would therefore expect that the EMFCQ should be all we need to have Algorithm 8 work properly. Luckily, this is the case, as shown by the following theorem.

Theorem 4.6. *Let a* GNEP *be given satisfying the* EMFCQ *Let* $\{\boldsymbol{x}_k\}$ *be the sequence generated by Algorithm* 8. *The following two assertions hold:*

1. *The penalty parameters are updated a finite number of times only.*
2. *Every limit point* $\bar{\boldsymbol{x}}$ *of the sequence* $\{\boldsymbol{x}_k\}$ *is a solution of the* GNEP.

If the EMFCQ is not satisfied everywhere, it could happen that the penalty parameters are updated an infinite number of times. In this case, we could still prove that the limit points of the sequence $\{\boldsymbol{x}_k\}$ are still meaningful to the GNEP. However, this analysis is somewhat involved and, in the end, of little *practical* interest, since, if some penalty parameters go to infinity, the algorithm is anyway becoming numerically unstable. We therefore omit this kind of analysis here and refer the interested reader to [37] for details.

Theorem 4.6 says that, under appropriate assumptions, the updating scheme will increase the penalty parameter only a finite number of times, so that eventually Algorithm 8 reduces, as we had hoped, to the application of Algorithm $\mathcal{A}$ to the PNEP$_\rho$ for a fixed value of the penalty parameters. The overall convergence properties of Algorithm 8 are therefore based on those of Algorithm $\mathcal{A}$ and in particular on Property 4.5. Next we show how, under reasonable assumptions, we can develop an algorithm $\mathcal{A}$ that enjoys Property 4.5. We stress once more that we are only interested in algorithms for the solution of PNEP$_\rho$ for fixed values of the penalty parameters.

Problem (4.3) is a standard Nash equilibrium problem whose solution is problematic because of the nondifferentiability of the objective functions $P_\nu(\boldsymbol{x}, \rho_\nu)$. We deal with this difficulty by using smoothing techniques. Certainly this is not the only option, but it is probably the simplest and one that seems to work well in practice.

We recall that the objective function of player ν in (4.3) is given by

$$\begin{aligned} P_\nu(\boldsymbol{x}, \rho_\nu) &= P_\nu(x^\nu, \boldsymbol{x}^{-\nu}, \rho_\nu) \\ &= \theta_\nu(x^\nu, \boldsymbol{x}^{-\nu}) + \rho_\nu \| g_+^\nu(x^\nu, \boldsymbol{x}^{-\nu}) \|_3 \\ &= \theta_\nu(x^\nu, \boldsymbol{x}^{-\nu}) + \rho_\nu \left(\sum_{i=1}^{m_\nu} \max\{0, g_i^\nu(x^\nu, \boldsymbol{x}^{-\nu})\}^3 \right)^{1/3}. \end{aligned}$$

We can approximate these functions by the smooth mappings

$$\begin{aligned} \tilde{P}_\nu(\boldsymbol{x}, \rho_\nu, \varepsilon) &= \tilde{P}_\nu(x^\nu, \boldsymbol{x}^{-\nu}, \rho_\nu, \varepsilon) \\ &= \theta_\nu(x^\nu, \boldsymbol{x}^{-\nu}) + \rho_\nu \left(\sum_{i=1}^{m_\nu} \max\{0, g_i^\nu(x^\nu, \boldsymbol{x}^{-\nu})\}^3 + \varepsilon \right)^{1/3} + \frac{\varepsilon}{2} \|x^\nu\|^2. \end{aligned}$$

From now on we assume that θ_ν and g^ν are twice continuously differentiable on $\mathbf{K}$. Note that the assumption above makes the max-term twice continuously differentiable, so that $\tilde{P}_\nu$ itself is twice continuously differentiable. We are therefore naturally led to define a smooth "approximation" of the PNEP_ρ (4.3), namely the $\text{PNEP}_\rho(\varepsilon)$ where the problem of player ν is minimizing the function $\tilde{P}_\nu$ on the set K_ν:

$$\begin{aligned} &\min_{x^\nu} \ \tilde{P}_\nu(\boldsymbol{x}, \rho_\nu, \varepsilon) \\ &\text{s.t.} \quad x^\nu \in K_\nu. \end{aligned}$$

The presence of the regularization term $(\varepsilon/2)\|x^\nu\|^2$ guarantees that $\tilde{P}_\nu(\boldsymbol{x}, \rho_\nu, \varepsilon)$ is strongly convex as a function of x^ν.

We observe that the $\text{PNEP}_\rho(\varepsilon)$ is a game where each player's problem is a continuously differentiable, constrained (the feasible sets K_ν are compact for all ν), convex optimization problem. We are now able to resort to our usual VI reduction technique: the game $\text{PNEP}_\rho(\varepsilon)$ can equivalently be cast as a variational inequality $(\mathbf{K}, \mathbf{F}_\varepsilon(\boldsymbol{x}))$, with pseudo-gradient given by

$$\mathbf{F}_\varepsilon(\boldsymbol{x}) = \begin{pmatrix} \nabla_{x^1} \tilde{P}_1(x^1, \boldsymbol{x}^{-1}, \rho_1, \varepsilon) \\ \vdots \\ \nabla_{x^N} \tilde{P}_N(x^N, \boldsymbol{x}^{-N}, \rho_N, \varepsilon) \end{pmatrix}. \tag{4.8}$$

We remark that, thanks to the assumptions made, $F_\varepsilon(\boldsymbol{x})$ is continuously differentiable. The simple idea, common to all smoothing methods, is now to solve PNEP_ρ by solving, inaccurately, but with increasing accuracy, a sequence of variational inequalities $(\mathbf{K}, \mathbf{F}_{\varepsilon_i})$ for values of the smoothing parameter going to zero (we use the parameter i to denote iterations here in order to have no confusion with the iteration counter k in the previous section). The following result substantiates this approach.

Proposition 4.7. *Let $\{\varepsilon_i\}$ and $\{\eta_i\}$ be two sequences of positive numbers converging to 0 and, for every i, let $\boldsymbol{x}(\varepsilon_i) \in \mathbf{K}$ be a point such that*

$$\left\|\boldsymbol{x}(\varepsilon_i) - \Pi_{\mathbf{K}}[\boldsymbol{x}(\varepsilon_i) - \mathbf{F}_{\varepsilon_i}(\boldsymbol{x}(\varepsilon_i))]\right\| \leq \eta_i. \tag{4.9}$$

Then every limit point of the sequence $\boldsymbol{x}(\varepsilon_i)$ is a solution of the PNEP_ρ (4.3).

The developments so far have shown that it is possible to iteratively find "correct" values for the penalty parameters and that we are able to solve the penalized problem if we are able to solve the smoothed problems or, from a different point of view, the VI $(\mathbf{K}, \mathbf{F}_\varepsilon)$. Note however that, at this stage, it is not clear whether we are capable to solve this VI. We know that, theoretically, in order to be able to solve this VI we need some kind of monotonicity property for the operator F_ε; see Chapter 2. This is not a trivial point to analyze, but some valuable results can be obtained. Although very technical, these results in the end all reduce to the verification that the assumed properties of the problem imply that F_ε is monotone.

The first such result shows that the penalty approach allows us to solve those problems for which we have already developed convergent methods in the previous section.

Theorem 4.8. *Assume that* $\mathbf{F}(\boldsymbol{x})$ *is monotone. If the original* GNEP (4.1) *reduces to either a* NEP *or to a jointly convex* GNEP, *then* $\mathbf{F}_\varepsilon(\boldsymbol{x})$ *is strongly monotone.*

It is comforting that, by adopting our nontrivial penalty approach, we can at least solve those problems that we can solve by using other methods. But it would obviously be disappointing to discover that we cannot go beyond that. Fortunately this is not so. The following theorem goes in this direction.

Let us suppose that constraints in (4.1) have the following form:

$$g^\nu(\boldsymbol{x}) = h^\nu(x^\nu) + l^\nu(\boldsymbol{x}^{-\nu}), \quad \forall \nu, \tag{4.10}$$

where $h^\nu : \mathbb{R}^{n_\nu} \to \mathbb{R}^{m_\nu}$ is a convex function, $l^\nu : \mathbb{R}^{n-n_\nu} \to \mathbb{R}^{m_\nu}$, and both are twice continuously differentiable. Linear constraints, i.e.,

$$g^\nu(\boldsymbol{x}) = A^{\nu 1}x^1 + \cdots + A^{\nu N}x^N, \quad \forall \nu,$$

with $A^{\nu i} \in M_{m_\nu, n_i}$, are a particular kind of this class of constraints. In the next theorem we use the following two matrices:

$$\mathbf{diag}\, J\mathbf{g}(\boldsymbol{x}) = \begin{pmatrix} J_{x^1} g^1(\boldsymbol{x}) & 0_{m_1 n_2} & \cdots & 0_{m_1 n_N} \\ 0_{m_2 n_1} & J_{x^2} g^2(\boldsymbol{x}) & \cdots & 0_{m_2 n_N} \\ \vdots & \vdots & & \vdots \\ 0_{m_N n_1} & 0_{m_N n_2} & \cdots & J_{x^N} g^N(\boldsymbol{x}) \end{pmatrix},$$

$$\mathbf{off}\, J\mathbf{g}(\boldsymbol{x}) = \begin{pmatrix} 0_{m_1 n_1} & J_{x^2} g^1(\boldsymbol{x}) & \cdots & J_{x^N} g^1(\boldsymbol{x}) \\ J_{x^1} g^2(\boldsymbol{x}) & 0_{m_2 n_2} & \cdots & J_{x^N} g^2(\boldsymbol{x}) \\ \vdots & \vdots & & \vdots \\ J_{x^1} g^N(\boldsymbol{x}) & J_{x^2} g^N(\boldsymbol{x}) & \cdots & 0_{m_N n_N} \end{pmatrix},$$

that represent the diagonal and off-diagonal parts of $J_{\boldsymbol{x}}\mathbf{g}(\boldsymbol{x})$ respectively, where we set $\mathbf{g}(\boldsymbol{x}) = (g^\nu(\boldsymbol{x}))_{\nu=1}^N$. We actually use also a third matrix, denoted by $\bar{M}(\boldsymbol{x}, \rho_1, \ldots, \rho_N)$. This is a positive semidefinite matrix whose exact definition can be found in [38]. For our purposes it is enough to know that $\|\bar{M}\|$ is continuous in $(\boldsymbol{x}, \rho_1, \ldots, \rho_N)$ and an increasing function in $(\rho_1, \ldots, \rho_N)$.

Theorem 4.9. *In the setting of this section, for a* GNEP *with constraints defined in* (4.10), *if there exists a constant* $\alpha > 0$ *such that*

$$d^T J\mathbf{F}(\boldsymbol{x}) d \geq \begin{cases} 0 & \forall d \in \left\{\ker(\mathbf{diag}\, J\mathbf{g}(\boldsymbol{x})) \cup \ker(\mathbf{off}\, J\mathbf{g}(\boldsymbol{x}))\right\} \\ \alpha \|d\|^2 & \textit{otherwise} \end{cases} \tag{4.11}$$

for all $\boldsymbol{x} \in \mathbf{K}$, *with*

$$\alpha \geq 2\big\|\mathbf{diag}\, J\mathbf{g}(\boldsymbol{x})\big\|\big\|\bar{M}(\boldsymbol{x}, \rho_1, \ldots, \rho_N)\big\|\big\|\mathbf{off}\, J\mathbf{g}(\boldsymbol{x})\big\|, \tag{4.12}$$

then $\mathbf{F}_\varepsilon(\boldsymbol{x})$ *is strongly monotone.*

Note that, supposing that $\mathbf{F}$ is monotone, the assumptions of the previous theorem are automatically satisfied if the game reduces to a pure NEP (or equivalently if, for all players, $l^\nu(\boldsymbol{x}^{-\nu})$ in (4.10) is a constant), since in this case $\|\mathbf{off}\, J\mathbf{g}(\boldsymbol{x})\| = 0$. What Theorem 4.9 tells us then is that, if the coupling in the constraints is "compensated by sufficient curvature of $\mathbf{F}$ in the right subspaces", then we may still be able to solve the resulting GNEP. Although these conditions are strong, they are quite natural. The widening of the class of solvable GNEPs given by the theorem above may seem not so impressive, but one should take into account the difficulty of the problem and also consider that not much more has been established so far, except for the very recent results to be discussed in the next section. Note also that there is a conceptual difference between the requirements made in Theorem 4.8 and those in Theorem 4.9 in that the latter theorem requires a condition that depends on the penalty parameters ρ_ν, while the former theorem has no such dependence. Conceptually, this means that while under the conditions given in Theorem 4.8 we can ensure that the overall penalty scheme of Algorithm 8 will converge to a solution, we can draw the same conclusions under the condition of Theorem 4.9 only provided that (4.12) holds for all penalty parameters generated by the algorithm. On the positive side, we can add that, under the EMFCQ, we know in advance that the algorithm will generate penalty parameters that have a common upper bound. This means that, conceptually, condition (4.12) must only be verified for this upper bound. Although the upper bound is not known in general, and so condition (4.12) is not checkable *a priori*, the qualitative value of Theorem 4.9 is still not to be underestimated.

4.3 Using the KKT condition

We conclude our algorithmic survey by considering what is probably the most obvious approach to the solution for the GNEP: the calculation of a point satisfying the KKT conditions for all players. Let $\boldsymbol{x}_*$ be a solution of the GNEP (1) with feasible set given by (2). Assuming that any standard constraint qualification holds, it is well known that the following KKT conditions will be satisfied for every player ν:

$$\nabla_{x^\nu}\theta_\nu(x_*^\nu, \boldsymbol{x}_*^{-\nu}) + \sum_{i=1}^{m_\nu} \lambda_i^\nu \nabla_{x^\nu} g_i^\nu(x_*^\nu, \boldsymbol{x}_*^{-\nu}) = 0, \tag{4.13}$$

$$\lambda_i^\nu \geq 0, \quad g_i^\nu(x_*^\nu, \boldsymbol{x}_*^{-\nu}) \leq 0, \quad \lambda_i^\nu g_i^\nu(x_*^\nu, \boldsymbol{x}_*^{-\nu}) = 0, \quad \forall i = 1, \ldots, m_\nu,$$

where λ^ν is the vector of Lagrange multipliers of player ν. Vice versa, recalling that the player's problems are convex (see A1), we have that if a point $\boldsymbol{x}_*$ satisfies, together with a suitable vector of multipliers $\boldsymbol{\lambda} = (\lambda^1, \lambda^2, \ldots, \lambda^N)$, the KKT conditions (4.13) for every ν, then $\boldsymbol{x}_*$ is a solution of the GNEP. It then seems rather natural to try to solve the GNEP by solving the system obtained by concatenating the N systems (4.13). In spite of its seeming simplicity, until very recently there were no theoretical results relative to this approach. The results we present here are from [30]. In order to use a more compact notation, we introduce some further definitions.

We denote by L^ν the Lagrangian of player ν:

$$L^\nu(\boldsymbol{x}, \lambda^\nu) = L^\nu(x^\nu, \boldsymbol{x}^{-\nu}, \lambda^\nu) = \theta_\nu(x^\nu, \boldsymbol{x}^{-\nu}) + \sum_{i=1}^{m_\nu} \lambda_i^\nu g_i^\nu(x^\nu, \boldsymbol{x}^{-\nu}).$$

If we set $\mathbf{F}(\boldsymbol{x}, \boldsymbol{\lambda}) = (\nabla_{x^\nu} L^\nu(\boldsymbol{x}, \lambda^\nu))_{\nu=1}^N$ and $\mathbf{g}(\boldsymbol{x}) = (g^\nu(\boldsymbol{x}))_{\nu=1}^N$, the concatenated KKT system can be written as

$$\mathbf{F}(\boldsymbol{x}, \boldsymbol{\lambda}) = 0, \quad \boldsymbol{\lambda} \geq 0, \quad \mathbf{g}(\boldsymbol{x}) \leq 0, \quad \boldsymbol{\lambda}^T \mathbf{g}(\boldsymbol{x}) = 0. \tag{4.14}$$

We formulate this system as a constrained nonlinear system of equations (CE) of the form

$$H(z) = 0, \quad z \in Z, \tag{4.15}$$

for a given function $H\colon \mathbb{R}^l \to \mathbb{R}^l$ and a given set $Z \subseteq \mathbb{R}^l$ that we define below.

We introduce slack variables

$$\boldsymbol{w} = \begin{pmatrix} w^1 \\ \vdots \\ w^N \end{pmatrix},$$

where $w^\nu \in \mathbb{R}^{m_\nu}$, and set

$$\boldsymbol{\lambda} \circ \boldsymbol{w} = \left(\lambda_1^1 w_1^1, \ldots, \lambda_{m_N}^N w_{m_N}^N\right)^T.$$

Then we define

$$H(z) = H(\boldsymbol{x}, \boldsymbol{\lambda}, \boldsymbol{w}) = \begin{pmatrix} \mathbf{F}(\boldsymbol{x}, \boldsymbol{\lambda}) \\ \mathbf{g}(\boldsymbol{x}) + \boldsymbol{w} \\ \boldsymbol{\lambda} \circ \boldsymbol{w} \end{pmatrix} \tag{4.16}$$

and

$$Z = \{z = (\boldsymbol{x}, \boldsymbol{\lambda}, \boldsymbol{w}) \mid \boldsymbol{x} \in \mathbb{R}^n,\ \boldsymbol{\lambda} \in \mathbb{R}_+^m,\ \boldsymbol{w} \in \mathbb{R}_+^m\}. \tag{4.17}$$

It is immediate then to verify that a point $(\boldsymbol{x}, \boldsymbol{\lambda})$ solves the KKT system (4.14) if and only if this point, together with a suitable $\boldsymbol{w}$, solves the constrained equation (4.15) with H and Z given by (4.16) and (4.17).

In order to solve this constrained equation problem, we use an interior point approach that generates points in the interior of Z. In other words, our method will generate a sequence $(\boldsymbol{x}_k, \boldsymbol{\lambda}_k, \boldsymbol{w}_k)$ with $\boldsymbol{\lambda}_k > 0$ and $\boldsymbol{w}_k > 0$ for every k.

The particular method that we base our analysis on is the potential reduction method from [78], also discussed in detail in [39]. We generalize this potential reduction method allowing inexact solutions of the subproblems and study in detail its implication in the case of our specific system (4.16) and (4.17).

To this end, we define the following subset of the range of H on Z:

$$T = \mathbb{R}^n \times \mathbb{R}^{2m}_+,$$

as well as a potential function on $\operatorname{int} T$,

$$p(u,v) = \zeta \log\left(\|u\|^2 + \|v\|^2\right) - \sum_{i=1}^{2m} \log(v_i), \quad (u,v) \in \mathbb{R}^n \times \mathbb{R}^{2m}_+, \quad \zeta > m.$$

The properties of this function are well known from the literature on interior point methods. Basically, the function p is defined in the interior of T and penalizes points that are near the boundary of T, but are far from the origin.

Based on p, we obtain a potential function for the CE which is defined on the nonempty set $Z_I = H^{-1}(\operatorname{int} T) \cap \operatorname{int} Z$ by setting

$$\psi(z) = p(H(z)) \quad \text{for } z \in Z_I.$$

Throughout this section, p and ψ always denote these two potential functions.

We are now in a position to formulate the inexact version of the potential reduction method from [39, 78] for the solution of GNEPs.

Algorithm 9: Inexact Potential Reduction Method for GNEPs

(S.0) : Choose $z_0 \in Z_I$ and $\beta, \gamma \in (0,1)$, and set $k = 0$, $\bar{\sigma} = 1$, and $a^T = (0_n^T, 1_{2m}^T)$.

(S.1) : If $H(z_k) = 0$, stop.

(S.2) : Choose $\sigma_k \in [0, \bar{\sigma})$, $\eta_k \geq 0$ and compute a vector $d_k \in \mathbb{R}^{n+2m}$ such that

$$\left\| H(z_k) + JH(z_k)d_k - \sigma_k \frac{a^T H(z_k)}{\|a\|^2} a \right\| \leq \eta_k \left\| H(z_k) \right\| \tag{4.18}$$

and

$$\nabla \psi(z_k)^T d_k < 0. \tag{4.19}$$

(S.3) : Compute a stepsize $t_k = \max\left\{\beta^\ell \mid \ell = 0, 1, 2, \dots\right\}$ such that

$$z_k + t_k d_k \in Z_I \tag{4.20}$$

and

$$\psi(z_k + t_k d_k) \leq \psi(z_k) + \gamma t_k \nabla \psi(z_k)^T d_k. \tag{4.21}$$

(S.4) : Set $x_{k+1} = x_k + t_k d_k$, $k \leftarrow k+1$, and go to S.1.

Remark 4.10.

(a) By construction, all iterates z_k generated by Algorithm 9 belong to the set Z_I; hence we have $z_k \in \operatorname{int} Z$ and $H(z_k) \in \operatorname{int} T$ for all $k \in \mathbb{N}$.

(b) If $JH(z_k)$ is a nonsingular $(n+2m) \times (n+2m)$ matrix for all k, it follows that the linear system of equations

$$H(z_k) + JH(z_k)d_k = \sigma_k \frac{a^T H(z_k)}{\|a\|^2} a$$

always has an exact solution $\hat{d}_k$. In particular, this exact solution satisfies the inexactness requirement from (4.18) for an arbitrary number $\eta_k \geq 0$. Furthermore, this exact solution also satisfies the descent property $\nabla\psi(z_k)^T \hat{d}_k < 0$; see [39]. It therefore follows that one can always find a vector d_k satisfying the two requirements (4.18) and (4.19), i.e., S.2 is well defined.

(c) Since, by induction, we have $z_k \in Z_I$ for an arbitrary fixed iteration $k \in \mathbb{N}$ and since Z_I is an open set, we see that the test (4.20) holds for all sufficiently small stepsizes t_k. Furthermore, the Armijo line search from (4.21) is eventually satisfied since d_k is a descent direction of the potential function ψ in view of the construction in S.2; cf. (4.19). In particular, this means that S.3 is also well defined.

The following is the main convergence result for Algorithm 9, where, implicitly, we assume that Algorithm 9 does not terminate within a finite number of iterations with a solution of the constrained nonlinear system (H, Z).

Theorem 4.11. *Assume that $JH(z)$ is nonsingular for all $z \in Z_I$, and that the two sequences $\{\sigma_k\}$ and $\{\eta_k\}$ from S.2 of Algorithm 9 satisfy the conditions*

$$\limsup_{k\to\infty} \sigma_k < \bar{\sigma} \quad \textit{and} \quad \lim_{k\to\infty} \eta_k = 0. \tag{4.22}$$

Let $\{z_k\}$ be any sequence generated by Algorithm 9. Then the following statements hold:

(a) *The sequence $\{H(z_k)\}$ is bounded.*

(b) *Any accumulation point of $\{z_k\}$ is a solution of the constrained nonlinear system of equations (4.15).*

The critical issue in applying Theorem 4.11 is establishing the nonsingularity of JH. The next theorem is devoted to this point. We recall that the definition of the matrix $\mathbf{diag}\, J\mathbf{g}(\boldsymbol{x})$ is given before Theorem 4.9.

Theorem 4.12. *Let $z = (\boldsymbol{x}, \boldsymbol{\lambda}, \boldsymbol{w}) \in \mathbb{R}^n \times \mathbb{R}^m_+ \times \mathbb{R}^m_+$ be given such that $J_{\boldsymbol{x}}\mathbf{F}(\boldsymbol{x}, \boldsymbol{\lambda})$ is nonsingular and*

$$M(\boldsymbol{x}, \boldsymbol{\lambda}) = J_{\boldsymbol{x}}\mathbf{g}(\boldsymbol{x})\, J_{\boldsymbol{x}}\mathbf{F}(\boldsymbol{x}, \boldsymbol{\lambda})^{-1}\, \mathbf{diag}\, J\mathbf{g}(\boldsymbol{x}) \tag{4.23}$$

is a P_0 matrix. Then the Jacobian $JH(z)$ is nonsingular.

We recall that a matrix M is a P_0 matrix if all its principal minors are nonnegative. If M is symmetric, being P_0 or being positive semidefinite is exactly the same thing. However, if M is nonsymmetric, then the class of P_0 matrices strictly contains the class of positive semidefinite matrices. The interested reader can refer [22] for more information. The following theorem gives another sufficient condition for the nonsingularity of JH. This condition is stronger than that in Theorem 4.12; nevertheless it is interesting because it gives a quantitative insight into what is necessary to guarantee the nonsingularity of JH. We use the notation $\text{eig}_{\min}(A)$ for the smallest eigenvalue of a symmetric matrix A.

Theorem 4.13. *Let $z = (\boldsymbol{x}, \boldsymbol{\lambda}, \boldsymbol{w}) \in \mathbb{R}^n \times \mathbb{R}^m_+ \times \mathbb{R}^m_+$ be given such that $J_{\boldsymbol{x}}\mathbf{F}(\boldsymbol{x}, \boldsymbol{\lambda})$ is nonsingular and*

$$\text{eig}_{\min}\left(\frac{1}{2}\textbf{diag}\, J\mathbf{g}(\boldsymbol{x})^T \left(J_{\boldsymbol{x}}\mathbf{F}(\boldsymbol{x},\boldsymbol{\lambda})^{-1} + J_{\boldsymbol{x}}\mathbf{F}(\boldsymbol{x},\boldsymbol{\lambda})^{-T}\right) \textbf{diag}\, J\mathbf{g}(\boldsymbol{x})\right) \geq \left\|J_{\boldsymbol{x}}\mathbf{g}(\boldsymbol{x}) - \textbf{diag}\, J\mathbf{g}(\boldsymbol{x})^T\right\| \left\| J_{\boldsymbol{x}}\mathbf{F}(\boldsymbol{x},\boldsymbol{\lambda})^{-1}\right\| \left\|\textbf{diag}\, J\mathbf{g}(\boldsymbol{x})\right\|.$$

Then the Jacobian $JH(z)$ is nonsingular.

The condition in this theorem may look technical, but in reality, roughly speaking, it only prescribes that a certain matrix, $J_{\boldsymbol{x}}\mathbf{F}(\boldsymbol{x},\boldsymbol{\lambda})^{-1} + J_{\boldsymbol{x}}\mathbf{F}(\boldsymbol{x},\boldsymbol{\lambda})^{-T}$, be "sufficiently positive definite" on a certain subspace. We remark that in the case of NEPs the term $\|J_{\boldsymbol{x}}\mathbf{g}(\boldsymbol{x}) - \textbf{diag}\, J\mathbf{g}(\boldsymbol{x})^T\|$ is zero. Note that Theorem 4.11 does not guarantee the existence of an accumulation point of the sequence generated by Algorithm 9. The following result therefore considers precisely this question and provides conditions under which the entire sequence generated by our algorithm remains bounded.

Theorem 4.14. *Assume that*

(a) *$JH(z)$ is nonsingular for all $z \in Z_I$;*

(b) *$\lim_{\|\boldsymbol{x}\| \to \infty} \left\|\mathbf{g}_+(\boldsymbol{x})\right\| = +\infty$;*

(c) *The extended Mangasarian–Fromovitz constraint qualification holds for each player, i.e., for all $\nu = 1, \ldots, N$ and for all $\boldsymbol{x} \in \mathbb{R}^n$,*

$$\exists d^\nu \in \mathbb{R}^{n_\nu} : \nabla_{\boldsymbol{x}^\nu} g_i^\nu(\boldsymbol{x})^T d^\nu < 0 \;\; \forall i \in I_+^\nu(\boldsymbol{x}), \tag{4.24}$$

where $I_+^\nu(\boldsymbol{x}) = \left\{i \in \{1, \ldots, m_\nu\} \mid g_i^\nu(\boldsymbol{x}) \geq 0\right\}$ denotes the set of active or violated constraints for player ν.

Then any sequence $\{(\boldsymbol{x}^k, \boldsymbol{\lambda}^k, \boldsymbol{w}^k)\}$ generated by Algorithm 9 remains bounded.

Note that the EMFCQ considered here is exactly the same condition used in the previous subsection, the only difference being that here there are no abstract constraints of the type $x^\nu \in K_\nu$. The condition in point (b), on the other hand, is a mild coercivity condition that will automatically be satisfied if each player has, for example, bound constraints on its own variables among the private constraints.

Conclusions

We have presented an overview of some recent results in the computation of solutions of Generalized Nash Equilibrium Problems. Game Theory is a venerable field with a long history and a wealth of results that is not possible to summarize in a few pages. What we have tried to do is to give a glimpse of some recent algorithmic developments from the point of view of mathematical programming. There are very many aspects and approaches that we did not even mention in these notes. Restricting attention to algorithmic issues, we want to at least mention [6, 7, 8, 31, 35, 44, 45, 51, 55, 56, 57, 58, 68, 70, 111] as an interesting sample of the available literature.

The interest of the mathematical programming community in game theory is possibly one of the big novelties in the field, and it promises to change many points of view, with its emphasis on numerical tractability and the emergence of new priorities. The attention of optimizers to Game Theory has also been fuelled by numerous engineering applications that required the effective computation of equilibria. The development of new models, different from traditional ones, has brought new challenges that need to be addressed. To attack these new problems a strong background in optimization techniques is needed. These notes were addressed precisely to optimizers that want to expand their interests in the hope to provide good, albeit somewhat informal and incomplete, entry point into this fascinating world.

Appendix

Elements of variational inequality theory

In this appendix we give a very concise introduction to variational inequalities, which is an essential tool used in our approach to the analysis and solution of games. We do not go much beyond some basic definitions and properties and refer the reader to [39] for a broader and deeper analysis of variational inequalities and also for proofs of all statements in this appendix.

Let a convex closed subset K of $\mathbb{R}^n$ be given, together with a function $F: K \to \mathbb{R}^n$. The *variational inequality* (VI) (K, F) is the problem of finding a point x belonging to K such that

$$F(x)^T(y - x) \geq 0, \quad \forall y \in K. \tag{A.1}$$

The set of all solutions of a VI (K, F), i.e., the set of all points x satisfying (A.1) is denoted by $\mathrm{SOL}(K, F)$. Note that, if we assume, as we shall always assume from now on, that F is continuous on K, it is readily seen that $\mathrm{SOL}(K, F)$ is a (possibly empty) closed set. The simplest geometric interpretation of a VI or, more precisely, of (A.1), is that a point $x \in K$ is a solution if an only if $F(x)$ forms a non-obtuse angle with all vectors of the form $y - x$, for all y in K.

VIs provide a unified mathematical model for a host of applied equilibrium problems and include many special cases that are important in their own right. Although some of these special cases can and should be dealt with directly and using more specific tools, it is important to list them here, in order to understand the nature of VIs.

- If $K = \mathbb{R}^n$, it is easy to see that solving the VI (K, F) is equivalent to finding the solution of the system of equations $F(x) = 0$. In fact, loosely speaking, we are looking for a point x such that (A.1) is satisfied for all $y \in \mathbb{R}^n$. But since in this case $y - x$ can be any vector, it is readily seen that the only possibility is that $F(x)$ be zero.
- If $F = \nabla f$ for some continuously differentiable convex function $f: K \to \mathbb{R}$, it can be shown that a point x is a solution of the VI (K, F) if and only if it

is a minimum point of the constrained, convex optimization problem

$$\begin{aligned} &\min f(x) \\ &\text{s.t. } x \in K. \end{aligned}$$

In fact, it is known that the *minimum principle* states that a point $x \in K$ is a minimum point of this optimization problem if and only if $\nabla f(x)^T(y-x) \geq 0$ for all $y \in K$, which shows our equivalence.

- If K is the nonnegative orthant, i.e., if $K = \mathbb{R}^n_+$, then a point x is a solution of the VI $(\mathbb{R}^n_+, F)$ if and only if

$$0 \leq x \perp F(x) \geq 0,$$

where by $x \perp F(x)$ we mean that the two vectors are perpendicular, that is, $x^T F(x) = 0$. This problem is known as the *nonlinear complementarity problem* and is often denoted by NCP(F). The *linear* complementarity problem (i.e., the case in which $F(x) = Ax - b$) was introduced by Cottle as a tool to study in a unified way linear programs, quadratic problems, and bimatrix games. We refer the reader to [22] and [39] for a detailed history and analysis of this problem.

- If K is a cartesian product of the form $\mathbb{R}^{n_1} \times \mathbb{R}^{n_2}_+$, then we have a problem that can be viewed as a mixture of a system of equations and a nonlinear complementarity problem. Assume that $x = (u, v)$ with $u \in \mathbb{R}^{n_1}$, $v \in \mathbb{R}^{n_2}$, and F is partitioned accordingly as $F = (H, G)$, with $H\colon \mathbb{R}^{n_1} \times \mathbb{R}^{n_2} \to \mathbb{R}^{n_1}$ and $G\colon \mathbb{R}^{n_1} \times \mathbb{R}^{n_2} \to \mathbb{R}^{n_2}$. It can be shown that the VI (K, F) is in this case equivalent to the problem of finding a (u, v) such that

$$\begin{aligned} H(u,v) &= 0, \\ 0 \leq v \perp G(u,v) &\geq 0. \end{aligned}$$

This problem is know as the *mixed complementarity problem* and, as we will see shortly, it is strictly related to KKT systems.

A solution of a VI can be characterized, under some standard assumptions, by its KKT conditions, that are formally very similar to the KKT conditions of an optimization problem. In order to describe these KKT conditions, we assume that the set K is defined through a system of equalities and inequalities

$$K = \{x \in \mathbb{R}^n \,:\, g(x) \leq 0,\, Ax = b\},$$

where $g\colon \mathbb{R}^n \to \mathbb{R}^m$ is a vector of continuously differentiable convex functions and $A \in \mathbb{R}^{p \times n}$, $b \in \mathbb{R}^p$. Note that, fine points apart, if K must be convex, equality constraints involved in its definitions must be linear. In this setting, the following theorem holds.

Theorem A.1. *Let the set K be defined as above. The following statements hold:*

(a) *Let $x \in \mathrm{SOL}(K, F)$. If any standard constraint qualification holds at x (for example, linear independence of active constraints, Mangasarain–Fromovitz constraint qualifications, or Abadie condition) then there exist multipliers $\lambda \in \mathbb{R}^m$ and $\mu \in \mathbb{R}^p$ such that*

$$\begin{aligned} &F(x) + \nabla g(x)\lambda + A^T\mu = 0, \\ &Ax = b, \\ &0 \leq \lambda \perp g(x) \leq 0. \end{aligned} \tag{A.2}$$

(b) *Conversely, if a triple (x, λ, ν) satisfies system* (A.2)*, then x solves the VI (K, F).*

Note that the KKT system (A.2) is very similar to the KKT system of an optimization problem, but it is important to observe that, assuming that some constraint qualification holds and assuming that the set K is defined as above, the KKT system is totally equivalent to the VI, in the sense that a point x is a solution of the KKT system if and only if $x \in \mathrm{SOL}(K, F)$. This is in contrast to optimization problems where, in order to have a similar equivalence, one must assume that the problem is convex. As a further remark, it may be interesting to note that the KKT system (A.2) is an instance of a mixed complementarity problem; just take

$$u = \begin{pmatrix} x \\ \mu \end{pmatrix}, \quad v = \lambda,$$

$$H(u, v) = \begin{pmatrix} F(x) + \nabla g(x)\lambda + A^T\mu \\ Ax - b \end{pmatrix}, \quad G(u, v) = -g(x).$$

The first important issue one has to analyze when studying VIs is, obviously, existence of solutions. The basic result one can establish parallels the famous Weierstrass theorem for optimization problems, although it should be noted that while the Weierstrass theorem can be proved using elementary tools, the proof of the following theorem, providing the basic existence result for VIs, is rather sophisticated and by no means trivial.

Theorem A.2. *Let a VI (K, F) be given and assume that K is convex and compact and that F is continuous on K. Then $\mathrm{SOL}(K, F)$ is nonempty and compact.*

This theorem plays a central role in establishing existence of a solution to a VI, but it is not (directly) applicable to all those cases in which the set K is unbounded; for example, it is of no use when dealing with (mixed) nonlinear complementarity problems. There are obviously many ways to deal with the difficulty of an unbounded set K – here we only mention the possibility to introduce special classes of functions that, among other things, have some bearing to existence issues.

Let a function $F\colon K \to \mathbb{R}^n$ be given, with K a closed convex subset of $\mathbb{R}^n$. We say that F is

- *monotone* on K if

$$(F(x) - F(y))^T(y - x) \geq 0, \quad \forall x, y \in K;$$

- *strictly monotone* on K if

$$(F(x) - F(y))^T(y - x) > 0, \quad \forall x, y \in K,\ x \neq y;$$

- *strongly monotone* on K if a positive constant α exists such that

$$(F(x) - F(y))^T(y - x) \geq \alpha \|x - y\|^2, \quad \forall x, y \in K.$$

Two simple considerations can illustrate better the nature of these definitions. In the unidimensional case, $n = 1$, a function is monotone if and only if it is nondecreasing, and strictly monotone if and only if it is increasing. In this setting, strongly monotone functions can be seen, roughly speaking, as increasing functions whose slope is bounded away from zero. In the general case of $n \geq 1$, there is an important relation to convex functions that also illustrates the various definitions of monotonicity. Let $f\colon K \to \mathbb{R}$ be a continuously differentiable function on K. Then the gradient ∇f is (strictly, strongly) monotone on K if and only if f is (strictly, strongly) convex on K. In a sense, monotonicity plays a role in the VI theory that is very similar to that of convexity in optimization theory. Having this in mind, the following theorem is rather "natural".

Theorem A.3. *Let a VI (K, F) be given, with K closed and convex and F continuous on K. Then:*

(a) *If F is monotone on K then* SOL(K, F) *is a (possibly empty) closed, convex set.*

(b) *If F is strictly monotone on K then the VI (K, F) has at most one solution.*

(c) *If F is strongly monotone on K then the VI (K, F) has one and only one solution.*

Note that point (c) in the above theorem guarantees existence (and uniqueness) of a solution even in the case of unbounded sets K. In view of the importance of monotonicity, it is important to have some easy way to check whether a function is monotone. The following proposition gives an easy test for C^1 functions.

Proposition A.4. *Let K be a closed convex subset of $\mathbb{R}^n$ and let $F\colon K \to \mathbb{R}^n$ be a C^1 function. The:*

(a) *F is monotone on K if and only if $JF(x)$ is positive semidefinite for all $x \in K$.*

(b) *If $JF(x)$ is positive definite for all $x \in K$ then F is strictly monotone on K.*

(c) *F is strongly monotone on K if and only if a positive constant α exists such that $JF(x) - \alpha I$ is positive semidefinite for every $x \in K$.*

Note that in case (b) we only have a sufficient condition for strict monotonicity. To see that the reverse implication does not hold in general, consider the function x^3, which is strictly monotone on $\mathbb{R}$. Its Jacobian (the derivative in this case) is given by x^2 and it is clear that at the origin $x^2 = 0$, so that the Jacobian is not everywhere positive definite.

Computation of the solutions of a VI is a complex issue. In the main text we describe some projection-type algorithms that, although quite simple, are, at least theoretically, very interesting. Although there exist other types of algorithms – see [39] – that might practically outperform projection-type algorithms, these latter algorithms are most interesting, since they are, on the one hand, simple, and on the other hand, they match well with the decomposition schemes that are of paramount interests in the solution of games. The use of more complex VI solution methods in the solution of GNEPs is certainly a topic that should be addressed in the near future.

Bibliography

[1] Adida E. and Perakis G.: *Dynamic pricing and inventory control: Uncertainty and competition.* Part A: *Existence of Nash equilibrium.* Operations Research Center, Sloan School of Management, MIT, Technical Report (2006).

[2] Adida E. and Perakis G.: *Dynamic pricing and inventory control: Uncertainty and competition.* Part B: *An algorithm for the normalized Nash equilibrium.* Operations Research Center, Sloan School of Management, MIT, Technical Report (2006).

[3] Altman, E. and Altman, Z.: *S-modular games and power control in wireless networks.* IEEE Transactions on Automatic Control **48(5)** (2003).

[4] Altman E., Boulogne T., El-Azouzi R., Jiménez T., and Wynter L.: *A survey on networking games in telecommunications.* Computers & Operations Research **33** (2006), 286–311.

[5] Altman E. and Wynter L.: *Equilibrium, games, and pricing in transportation and telecommunication networks.* Networks and Spatial Economics **4** (2004), 7–21.

[6] Antipin A.S.: *Solution methods for variational inequalities with coupled constraints.* Computational Mathematics and Mathematical Physics **40** (2000), 1239–1254.

[7] Antipin A.S.: *Solving variational inequalities with coupling constraints with the use of differential equations.* Differential Equations **36** (2000), 1587–1596.

[8] Antipin A.S.: *Differential equations for equilibrium problems with coupled constraints.* Nonlinear Analysis **47** (2001), 1833–1844.

[9] Arrow K.J. and Debreu G.: *Existence of an equilibrium for a competitive economy.* Econometrica **22** (1954), 265–290.

[10] Aubin J.-P. and Frankowska H.: *Set-Valued Analysis.* Birkhäuser, Boston (1990).

[11] Başar T. and Olsder G.J.: *Dynamic noncooperative game theory.* Academic Press London/New York, Second edition (1989) (Reprinted in SIAM Series "Classics in Applied Mathematics", 1999).

[12] Bassanini A., La Bella A., and Nastasi A.: *Allocation of railroad capacity under competition: A game theoretic approach to track time pricing.* In Trans-

portation and Networks Analysis: Current Trends (edited by M. Gendreau and P. Marcotte), 1–17, Kluwer Academic Publisher, Dordrecht (2002).

[13] Baye M.R., Tian G., and Zhou J.: *Characterization of existence of equilibria in games with discontinuous and non-quasiconcave payoffs.* The Review of Economic Studies **60** (1993), 935–948.

[14] Bensoussan A.: *Points de Nash dans le cas de fonctionnelles quadratiques et jeux différentiels linéaires à N personnes.* SIAM Journal of Control **12** (1974), 460–499.

[15] Berman A. and Plemmons R.J.: *Nonnegative matrices.* In the Mathematical Sciences SIAM Classics in Applied Mathematics No. 9, Philadelphia (1994).

[16] Berridge S. and Krawczyk J.B.: *Relaxation algorithms in finding Nash equilibria.* Economic working papers archives (1997), http://econwpa.wustl.edu/eprints/comp/papers/9707/9707002.abs.

[17] Bertrand J.: Review of "*Théorie mathématique de la richesse sociale*" by Léon Walras and "*Recherches sur les principes mathématiques de la théorie des richesses*" by Augustin Cournot. Journal des Savants, 499–508 (1883).

[18] Bertsekas D.P. and Tsitsiklis J.N.: *Parallel and distributed computation: Numerical methods.* Athena Scientific, Cambridge (1997).

[19] Breton M., Zaccour G., and Zahaf M.: *A game-theoretic formulation of joint implementation of environmental projects.* European Journal of Operational Research **168** (2005), 221–239.

[20] Chung M., Hande, P., Lan T., and Tan C.W.: *Power control.* In Wireless Cellular Networks, Now Publisher Inc. (2008).

[21] Contreras J., Klusch M.K., and Krawczyk J.B.: *Numerical solution to Nash-Cournot equilibria in coupled constraints electricity markets.* IEEE Transactions on Power Systems **19** (2004), 195–206.

[22] Cottle R.W., Pang J.-S., and Stone R.E.: *The linear complementarity problem.* Academic Press (1992).

[23] Cournot, A.A.: *Recherches sur les principes mathématiques de la théorie des richesses.* Hachette, Paris (1838).

[24] Dasgupta P. and Maskin E.: *The existence of equilibrium in discontinuous economic games, I: Theory.* Rev. Econ. Stud. **53** (1986), 1–26.

[25] Dasgupta P. and Maskin E.: *The existence of equilibrium in discontinuous economic games, II: Applications.* Rev. Econ. Stud. **53** (1986), 27–41.

[26] Debreu G.: *A social equilibrium existence theorem.* Proceedings of the National Academy of Sciences **38** (1952), 886–893.

[27] Debreu G.: *Theory of values.* Yale University Press, New Haven, USA (1959).

[28] Demyanov V.F., Di Pillo G, and Facchinei F.: *Exact penalization via Dini and Hadamard conditional derivatives.* Optimization Methods and Software **9** (1998), 19–36.

[29] Di Pillo G. and Facchinei F.: *Exact barrier function methods for Lipschitz programs.* Applied Mathematics and Optimization **32** (1995), 1–31.

[30] Dreves A., Facchinei F., Kanzow C., and Sagratella S.: *On the solution of the KKT conditions of generalized Nash equilibrium problems.* In preparation (2010).

[31] Dreves A. and Kanzow C.: *Nonsmooth optimization reformulations characterizing all solutions of jointly convex generalized Nash equilibrium problems.* Technical report, Institute of Mathematics, University of Würzburg, Würzburg, Germany (2009).

[32] Drouet L., Haurie A., Moresino F., Vial J.-P., Vielle M., and Viguier L.: *An oracle based method to compute a coupled equilibrium in a model of international climate policy.* Computational Management Science **5** (2008), 119–140.

[33] Ehrenmann, A.: *Equilibrium problems with equilibrium constraints and their application to electricity markets.* Ph.D. Dissertation, Judge Institute of Management, The University of Cambridge, Cambridge, UK, Cambridge, UK (2004).

[34] Facchinei F., Fischer A., and Piccialli V.: *On generalized Nash games and variational inequalities.* Operations Research Letters **35** (2007), 159–164.

[35] Facchinei F., Fischer A., and Piccialli V.: *Generalized Nash equilibrium problems and Newton methods.* Mathematical Programming **117** (2009), 163–194.

[36] Facchinei F. and Kanzow C.: *Generalized Nash equilibrium problems.* 4OR **5** (2007), 173–210.

[37] Facchinei F. and Kanzow C.: *Penalty methods for the solution of generalized Nash equilibrium problems.* SIAM Journal on Optimization **20(5)** (2010), 2228–2253.

[38] Facchinei F. and Lampariello L.: *Partial penalization for the solution of generalized Nash equilibrium problems.* Journal of Global Optimization **50** (2011), 39–57.

[39] Facchinei F. and Pang J.-S.: *Finite-dimensional variational inequalities and complementarity problems.* Springer, New York (2003).

[40] Facchinei F. and Pang J.-S.: *Exact penalty functions for generalized Nash problems.* In G. Di Pillo and M. Roma, editors, Large-Scale Nonlinear Optimization, 115–126, Springer (2006).

[41] Facchinei F. and Pang J.-S.: *Nash equilibria: The variational approach.* In "Convex optimization in signal processing and communications" (Palomar D.P. and Eldar Y. eds.), 443–493. Cambridge University Press (2010).

[42] Facchinei F., Piccialli V., and Sciandrone M.: *Decomposition algorithms for generalized potential games.* Computational Optimization and Applications **50** (2011), 237–262.

[43] Facchinei F. and Sagratella S.: *On the computation of all solutions of jointly convex generalized Nash equilibrium problems.* Optimization Letters **5** (2011), 531–547.

[44] Flåm S.D.: *Paths to constrained Nash equilibria.* Applied Mathematics and Optimization **27** (1993), 275–289.

[45] Flåm S.D. and Ruszczyński A.: *Noncooperative convex games: Computing equilibrium by partial regularization.* IIASA Working Paper 94–42 (1994).

[46] Fukushima, M.: *Restricted generalized Nash equilibria and controlled penalty algorithm.* Computational Management Science **8** (2011), 201–218.

[47] Fukushima M. and Pang J.-S.: *Quasi-variational inequalities, generalized Nash equilibria, and multi-leader-follower games.* Computational Management Science **2** (2005), 21–56.

[48] Gabriel S.A., Kiet S., and Zhuang J.: *A mixed complementarity-based equilibrium model of natural gas markets.* Operations Research **53** (2005), 799–818.

[49] Gabriel S. and Smeers Y.: *Complementarity problems in restructured natural gas markets.* In Recent Advances in Optimization, Lectures Notes in Economics and Mathematical Systems **563** (2006), 343–373.

[50] Garcia C.B. and Zangwill W.I.: *Pathways to solutions, fixed points, and equilibria.* Prentice-Hill, New Jersey (1981).

[51] Gürkan G. and Pang J.-S.: *Approximations of Nash equilibria.* Mathematical Programming **117** (2009), 223–253.

[52] Harker P.T.: *Generalized Nash games and quasi-variational inequalities.* European Journal of Operational Research **54** (1991), 81–94.

[53] Harker P.T. and Hong S.: *Pricing of track time in railroad operations: An internal market approach.* Transportation Research B-Meth. **28** (1994), 197–212.

[54] Haurie A. and Krawczyk J.-B.: *Optimal charges on river effluent from lumped and distributed sources.* Environmental Modelling and Assessment **2** (1997), 93–106.

[55] von Heusinger A. and Kanzow C.: *SC^1 optimization reformulations of the generalized Nash equilibrium problem.* Optimization Methods and Software **23** (2008), 953–973.

[56] von Heusinger A. and Kanzow C.: *Optimization reformulations of the generalized Nash equilibrium problem using Nikaido-Isoda-type functions.* Computational Optimization and Applications **43** (2009), 353–377.

[57] von Heusinger A. and Kanzow C.: *Relaxation methods for generalized Nash equilibrium problems with inexact line search.* Journal of Optimization Theory and Applications **143** (2009), 159–183.

[58] von Heusinger A., Kanzow C., and Fukushima M.: *Newton's method for computing a normalized equilibrium in the generalized Nash game through fixed point formulation.* Technical report, Institute of Mathematics, University of Würzburg, Würzburg, Germany (2009).

[59] Hobbs B., Helman U., and Pang J.-S.: *Equilibrium market power modeling for large scale power systems.* IEEE Power Engineering Society Summer Meeting 2001. 558–563. (2001).

[60] Hobbs B. and Pang J.-S.: *Nash-Cournot equilibria in electric power markets with piecewise linear demand functions and joint constraints.* Operations Research **55** (2007), 113–127.

[61] Hotelling H.: *Game theory for economic analysis.* The Econometrics Journal **39** (1929), 41–47.

[62] Huang J., Berry R., and Honig M.L.: *Distributed interference compensation for wireless networks.* IEEE Journal on Selected Areas in Communications **24(5)** (2006), 1074–1084.

[63] Ichiishi T.: *Game theory for economic analysis.* Academic Press, New York (1983).

[64] Jiang H.: Network capacity management competition. Technical report, Judge Business School at University of Cambridge, UK (2007).

[65] Jorswieck E.A., Larsson E.G., Luise M., and Poor H.V. (guest editors): *Game theory in signal processing and communications.* Special issue of the IEEE Signal Processing Magazine, 26 (2009).

[66] Kesselman A., Leonardi S., and Bonifaci V.: *Game-theoretic analysis of internet switching with selfish users.* Proceedings of the First International Workshop on Internet and Network Economics, WINE 2005, Lectures Notes in Computer Science **3828** (2005), 236–245.

[67] Krawczyk J.-B.: *An open-loop Nash equilibrium in an environmental game with coupled constraints.* In "Proceedings of the 2000 Symposium of the International Society of Dynamic Games", 325–339, Adelaide, South Australia (2000).

[68] Krawczyk J.-B.: *Coupled constraint Nash equilibria in environmental games.* Resource and Energy Economics **27** (2005), 157–181.

[69] Krawczyk J.-B.: *Numerical solutions to coupled-constraint (or generalised Nash) equilibrium problems.* Computational Management Science **4** (2007), 183–204.

[70] Krawczyk J.-B. and Uryasev S.: *Relaxation algorithms to find Nash equilibria with economic applications.* Environmental Modelling and Assessment **5** (2000), 63–73.

[71] Kubota, K. and Fukushima, M.: *Gap function approach to the generalized Nash equilibrium problem.* Journal of Optimization Theory and Applications **144** (2010), 511–531

[72] Kulkarni A.A. and Shanbhag U.V.: *On the variational equilibrium as a refinement of the generalized Nash equilibrium.* Manuscript, University of Illinois at Urbana-Champaign, Department of Industrial and Enterprise Systems Engineering.

[73] Leshem A. and Zehavi E.: *Game theory and the frequency selective interference channel – A tutorial.* IEEE Signal Processing Magazine **26** (2009), 28–40.

[74] Luo Z.-Q. and Pang J.-S.: *Analysis of iterative waterfilling algorithm for multiuser power control in digital subscriber lines.* EURASIP Journal on Applied Signal Processing (2006), 1–10.

[75] Luo Z.-Q. and Zhang S.: *Spectrum management: Complexity and duality.* Journal of Selected Topics in Signal Processing **2(1)** (2008), 57–72.

[76] MacKenzie A.B. and Wicker S.B.: *Game theory and the design of self-configuring, adaptive wireless networks.* IEEE Communication Magazine **39(11)** (2001), 126–131.

[77] Monderer D. and Shapley L.S.: *Potential games.* Games and Economic Behavior **14** (1996), 124–143.

[78] Monteiro R.D. C. and Pang J.-S.: *A potential reduction Newton method for constrained equations.* SIAM Journal on Optimization **9** (1999), 729–754.

[79] Morgan J. and Scalzo V.: *Existence of equilibria in discontinuous abstract economies.* Preprint 53-2004, Dipartimento di Matematica e Applicazioni R. Caccioppoli, Napoli (2004).

[80] Nabetani K., Tseng P., and Fukushima M.: *Parametrized variational inequality approaches to generalized Nash equilibrium problems with shared constraints.* Computational Optimization and Applications **48** (2011), 423–452.

[81] Nash J.F.: *Equilibrium points in n-person games.* Proceedings of the National Academy of Sciences **36** (1950), 48–49.

[82] Nash J.F.: *Non-cooperative games.* Annals of Mathematics **54** (1951), 286–295.

[83] von Neumann J.: *Zur Theorie der Gesellschaftsspiele.* Mathematische Annalen **100** (1928), 295–320.

[84] von Neumann J. and Morgenstern O.: *Theory of games and economic behavior.* Princeton University Press, Princeton (1944).

[85] Nisan N., Roughgarden T., Tardos E., and Vazirani V.V. (Edited by): *Algorithmic game theory.* Cambridge University Press (2007).

[86] O'Neill D., Julian D., and Boyd S.: *Seeking Foschini's genie: optimal rates and powers in wireless networks.* IEEE Transactions on Vehicular Technology, 2004.

[87] Outrata J.V., Kočvara M., and Zowe J.: *Nonsmooth approach to optimization problems with equilibrium constraints.* Kluwer Academic Publishers, Dordrecht (1998).

[88] Pang J.-S.: *Computing generalized Nash equilibria.* Manuscript, Department of Mathematical Sciences, The Johns Hopkins University (2002).

[89] Pang J.-S. and Fukushima M.: *Quasi-variational inequalities, generalized Nash equilibria, and multi-leader-follower games.* Computational Management Science **2** (2005), 21–56 (erratum: ibid. **6** (2009), 373–375.)

[90] Pang J.-S., Scutari G., Facchinei F., and Wang C.: *Distributed power allocation with rate constraints in gaussian parallel interference channels.* IEEE Transactions on Information Theory **54** (2008), 3471–3489.

[91] Pang J.-S., Scutari G., Palomar D.P., and Facchinei F.: *Design of cognitive radio systems under temperature-interference constraints: A variational inequality approach.* IEEE Transactions on Signal Processing **58** (2010), 3251–3271.

[92] Puerto J., Schöbel A., and Schwarze S.: *The path player game: Introduction and equilibria.* Preprint 2005-18, Göttingen Georg-August University, Göttingen, Germany (2005).

[93] Rao S.S., Venkayya V.B., and Khot N.S.: *Game theory approach for the integrated design of structures and controls.* AIAA Journal **26** (1988), 463–469.

[94] Reny P.J.: *On the existence of pure and mixed strategy Nash equilibria in discontinuous games.* Econometrica **67** (1999), 1026–1056.

[95] Robinson S.M.: *Shadow prices for measures of effectiveness. I. Linear Model.* Operations Research **41** (1993), 518–535.

[96] Robinson S.M.: *Shadow prices for measures of effectiveness. II. General Model.* Operations Research **41** (1993), 536–548.

[97] Rockafellar R.T. and Wets R.J.-B.: *Variational analysis.* Springer Verlag, Berlin (1998).

[98] Rosen J.B.: *Existence and uniqueness of equilibrium points for concave n-person games.* Econometrica **33** (1965), 520–534.

[99] Schmit L.A.: *Structural synthesis – Its genesis and development.* AIAA Journal **19** (1981), 1249–1263.

[100] Scotti S.J.: *Structural design using equilibrium programming formulations.* NASA Technical Memorandum 110175 (1995).

[101] Scutari G., Facchinei F., Pang J.-S., and Palomar D.P.: *Monotone communication games.* In preparation.

[102] Scutari A., Palomar D.P., and Barbarossa S.: *Asynchronous iterative water-filling for Gaussian frequency-selective interference channels.* IEEE Transactions on Information Theory **54** (2008), 2868–2878.

[103] Scutari A., Palomar D.P., and Barbarossa S.: *Optimal linear precoding strategies for wideband noncooperative systems based on game theory* – Part I: *Nash equilibria.* IEEE Transactions on Signal Processing **56** (2008), 1230–1249.

[104] Scutari A., Palomar D.P., and Barbarossa S.: *Optimal linear precoding strategies for wideband noncooperative systems based on game theory* – Part II: *Algorithms.* IEEE Transactions on Signal Processing **56** (2008), 1250–1267.

[105] Scutari A, Palomar D.P., and Barbarossa S.: *Competitive optimization of cognitive radio MIMO systems via game theory.* In "Convex optimization in signal processing and communications" (Palomar D.P. and Eldar Y. eds.), 387–442. Cambridge University Press (2010).

[106] Scutari G., Palomar D.P., Pang J.-S., and Facchinei F.: *Flexible design of cognitive radio wireless systems: From game theory to variational inequality theory.* IEEE Signal Processing Magazine **26(5)** (2009), 107–123.

[107] Sun L.-J. and Gao Z.-Y.: *An equilibrium model for urban transit assignment based on game theory.* European Journal of Operational Research **181** (2007), 305–314.

[108] Tian G. and Zhou J.: *Transfer continuities, generalizations of the Weierstrass and maximum theorems: a full characterization.* Journal of Mathematical Economics **24** (1995), 281–303.

[109] Tidball M. and Zaccour G.: *An environmental game with coupling constraints.* Environmental Modeling Assessment **10**, 153–158 (2005).

[110] Topkis D.: *Supermodularity and complementarity.* Princeton University Press (1998).

[111] Uryasev S. and Rubinstein R.Y.: *On relaxation algorithms in computation of noncooperative equilibria.* IEEE Transactions on Automatic Control **39** (1994), 1263–1267.

[112] Vincent T.L.: *Game theory as a design tool.* ASME Journal of Mechanisms, Transmissions, and Automation in Design **105** (1983), 165–170.

[113] Vives X.: *Nash equilibrium with strategic complementarities.* Journal of Mathematical Economics **19** (1994), 305–321.

[114] Walras L.: *Éléments d'économie politique pure.* Lausanne (1900).

[115] Wei J.-Y. and Smeers Y.: *Spatial oligopolistic electricity models with cournot generators and regulated transmission prices.* Operations Research **47** (1999), 102–112.

[116] Yu W., Ginis G., and Cioffi J.M.: *Distributed multiuser power control for digital subscriber lines.* IEEE Journal on Selected Areas in Communications **20** (2002), 1105–1115.

[117] Zhou J., Lam W.H.K., and Heydecker B.G.: *The generalized Nash equilibrium model for oligopolistic transit market with elastic demand.* Transportation Research B-Meth. **39** (2005), 519–544.

Part III

Equilibrium and Learning in Traffic Networks

Roberto Cominetti

Foreword

To Cecilia, Andrés and Pablo

These notes provide a brief introduction to some of the mathematical aspects involved in the study of equilibrium models for congested traffic networks. They were written to support a series of lectures delivered by the author at the Centre de Recerca Matemàtica (Barcelona, Spain) in July 2009.

Time and space limitations imposed a choice of topics to be covered and the level of detail in their treatment. After reviewing the classical concepts of Wardrop and stochastic user equilibrium, we provide a detailed treatment of the notion of Markovian traffic equilibrium. From these static equilibrium concepts, we move on to describe recent work on an adaptive procedure that models travel behavior, and its asymptotic convergence towards equilibrium. While this choice reflects the author's bias, I hope it serves as an introduction to the subject and to motivate a deeper study of an area that has still many interesting open questions.

I would like to express my gratitude to the organizers of the Advanced Courses at CRM, and particularly to Aris Daniilidis and Juan Enrique Martínez-Legaz, for their invitation and for making this course possible. I am also deeply indebted to Vianney Perchet and Guillaume Vigeral for their careful and thoughtful revision that greatly contributed to improve these notes.

Santiago de Chile, September 2010

Roberto Cominetti

Introduction and Overview

Congestion is a common characteristic of large urban areas, especially at peak hours when transport demands approach the saturation capacities of the roads. Urban planning and traffic management policies require quantitative models to forecast traffic in order to evaluate and compare alternative designs. These notes provide a brief introduction to some of the mathematical tools involved in the modeling of traffic flows in congested networks.

Traffic is often described as a steady state that emerges from the adaptive behavior of selfish drivers who strive to minimize travel times while competing for limited road capacity. Such equilibrium models provide a static description of how traffic demands flow through a congested network. In 1952, Wardrop [76] proposed a first model using continuous variables to represent aggregate flows. Drivers are assumed to behave rationally by selecting shortest paths according to the prevailing traffic conditions, while congestion is modeled by travel times that increase with the total flow carried by each route. In this setting, any given traffic pattern induces route loads and travel times that determine in turn which routes are optimal. An equilibrium is defined as a consistent traffic pattern for which the routes that are actually used are the shortest ones according to the induced travel times. Soon after the model was introduced, Beckman et al. [11] realized that these equilibria could be characterized as the optimal solutions of an equivalent convex minimization problem. In the 1980s, these results were supplemented by Daganzo [28] and Fukushima [43] who described a dual characterization that led to alternative numerical methods to compute the equilibrium. In Chapter 1 we present a brief overview of *Wardrop equilibrium* and its characterization.

The relatively poor agreement between the traffic patterns predicted by Wardrop equilibrium and the ones observed in real networks led to a critical revision of the assumptions that supported the model. One criticism focused on the assumption of homogeneity in driver behavior. As a matter of fact, drivers' decisions are subject to random fluctuations that arise from the difficulty in discriminating routes with similar costs, as well as from the intrinsic variability of travel times that induce different perceptions of the routes depending on the individual past experience. The idea of a random selection of routes is captured by discrete choice models based on random utility theory, leading to the notion of *stochastic user equilibrium* (SUE) [27, 28, 30, 35, 58, 62, 70, 74]. In Chapter 1 we give a short overview of this concept. More detailed accounts can be found in the book by Ben-Akiva & Lerman [16] and the surveys by Florian & Hearn [36, 37].

As the network size increases, the number of routes grows exponentially and the assumption that drivers are able to compare all the routes becomes less and less plausible. The exponential growth also makes route-based models computationally intractable, so that methods to circumvent path enumeration were developed in [2, 3, 17, 51, 55]. These ideas evolved into the notion of *Markovian equilibrium* [9] which looks at route choice as a stochastic dynamic process: drivers proceed towards their destination by a sequential process of arc selection using a discrete choice model at every intermediate node in their trip. The route is no longer fixed at the origin but it is the outcome of this sequential process, so that driver movements are governed by a Markov chain and the network flows are the corresponding invariant measures. In Chapter 2 we discuss this model in detail.

Equilibrium models assume implicitly the existence of an underlying mechanism in travel behavior that stabilizes traffic flows at a steady state. However, the models are stated directly as equilibrium equations at an aggregate population level and are not tied to a specific adaptive mechanism of individual drivers. Empirical evidence of adaptive behavior based on experiments and simulations has been reported in the literature [8, 31, 49, 57, 68, 77], though it was observed that the resulting steady states may differ from the standard notions of equilibria. Also, several continuous time dynamics describing plausible adaptive mechanisms that converge to Wardrop equilibrium were studied in [41, 67, 73] while a class of finite-lag discrete time adjustment procedures was considered in [23, 24, 29, 47]. However, these dynamics are again of an aggregate nature and are not explicitly tied to the behavior of individual drivers. An alternative to Wardrop's model in which drivers are considered as individual players in a game was studied by Rosenthal [64]. The main result established the existence of a Nash equilibrium in pure strategies by exploiting a potential function which is a discrete analog of Beckman et al. [11]. This motivated a number of extensions to the so-called *congestion games* and the more general class of *potential games*. Although we will not review these contributions, in Chapter 3 we reconsider the atomic framework with finitely many drivers but in a dynamic setting that models their adaptive behavior.

Learning and adaptation are fundamental issues in the study of repeated games with boundedly rational players. They have been intensively explored in the last decades [42, 79], though most results apply to games with a small number of players. The most prominent adaptive procedure is *fictitious play*, which assumes that at each stage players choose a best reply to the observed empirical distribution of past moves by their opponents [22, 63]. The assumption that players are able to record the past moves of their opponents is very stringent for games involving many players with limited observation capacity. A milder assumption is that players observe only the outcome vector, namely the payoff obtained at every stage and the payoff that would have resulted if a different move had been played. Procedures such as *no-regret* [44, 45], *exponential weight* [40], and *calibration* [38], deal with such limited information contexts assuming that players adjust their behavior based on payoff statistics. Eventually, adaptation leads to configurations where no player regrets the choices he makes. Although these procedures are flexible and robust, the underlying rationality may still be too demanding for games with a large number of strategies and poorly informed players. This is the case for traffic where a multitude of small players make routing decisions with little information about the strategies of other drivers nor the actual congestion in the network.

A simpler adaptive rule that relies only on the sequence of realized moves and payoffs is *reinforcement* [7, 19, 33], where players select moves proportionally to the cumulative payoff of each alternative. In Chapter 3 we describe a similar approach to model the adaptive behavior of drivers in a simple network with parallel links: each player observes only the travel time of the specific route chosen on any given day, and future decisions are adjusted based on past observations. Specifically, each player has a prior estimate of the average payoff of each route and makes a decision based on this rough information using a random choice rule. The payoff of the chosen route is then observed and is used to update the perception for that particular move. This procedure is repeated day after day, generating a discrete time stochastic process called the *learning process*. Since travel times depend on the congestion of routes imposed collectively by all the players' decisions, the process progressively reveals to each player the congestion conditions on all the routes. In the long run these dynamics lead the system to coordinate on a steady state that can be characterized as a Nash equilibrium for a particular limit game. Convergence depends on a viscosity parameter that represents the amount of noise in players' choices. If noise is large enough, the adaptive dynamics have a unique global attractor which almost surely attracts the sequences generated by the learning process. This approach proceeds bottom-up: a simple and explicit discrete time stochastic model for individual behavior gives rise to an associated continuous time deterministic dynamics which leads ultimately to an equilibrium of a particular limit game. The equilibrium is not postulated a priori but it is derived from basic assumptions on player behavior, providing a microeconomic foundation for equilibrium.

Chapter 1

Wardrop and Stochastic User Equilibrium

This chapter provides a short overview of two classical models for traffic equilibrium: *Wardrop equilibrium* and *stochastic user equilibrium.*

The traffic network is represented by a directed graph $G = (N, A)$ on which every arc $a \in A$ has an associated travel time $t_a = s_a(w_a)$ that depends on the total flow w_a on the link. Congestion is captured by the fact that these travel times increase with the flow, namely $s_a : \mathbb{R} \to (0, \infty)$ is supposed strictly increasing and continuous. Given a subset of destinations $D \subseteq N$ and traffic demands $g_i^d \geq 0$ from every node $i \neq d$ to each destination $d \in D$, a traffic equilibrium model aims to predict how these demands flow through the network.

1.1 Wardrop equilibrium

The basic principle used by Wardrop [76] to model traffic is that users behave rationally by choosing the shortest available paths. The model considers a situation in which all users perceive the same travel times and hence they agree on which are the shortest routes. Since travel times depend on the flow that is actually observed, an equilibrium is defined as a flow pattern that is consistent with the shortest paths determined by the induced travel times.

More precisely, let R_i^d denote the set of simple paths connecting i to d, which is assumed nonempty, and let R be the union of all the R_i^d's. An equilibrium is a path-flow assignment $h = (h_r)_{r \in R}$ with nonnegative entries $h_r \geq 0$ that satisfies the flow conservation constraints

$$g_i^d = \sum_{r \in R_i^d} h_r, \quad \forall d \in D,\ i \neq d, \tag{1.1}$$

and such that $h_r > 0$ can only occur when r is a shortest path connecting the corresponding origin-destination pair (i, d). The total cost of the path $r \in R$ is

$$c_r = \sum_{a \in r} t_a = \sum_{a \in r} s_a(w_a), \tag{1.2}$$

where $t_a = s_a(w_a)$ is the *travel time* or *generalized cost* of the arc a expressed as a function of the total flow w_a on that arc,

$$w_a = \sum_{r \ni a} h_r. \tag{1.3}$$

Definition 1.1. Let H be the set of all pairs (w, h) satisfying (1.1) and (1.3) with $h \geq 0$. A pair $(w, h) \in H$ is called a *Wardrop equilibrium* if for each origin-destination pair $(i, d) \in N \times D$ and every route $r \in R_i^d$ one has

$$h_r > 0 \implies c_r = \tau_i^d,$$

where the route costs c_r are given by (1.2) and $\tau_i^d = \min_{r \in R_i^d} c_r$ is the shortest travel time from i to d.

Wardrop equilibrium can be reformulated in a number of equivalent ways, either as a fixed point of an appropriate set-valued map or as a solution to a variational inequality. However, the most powerful description was discovered by Beckman et al. [11], which characterizes the equilibria as the optimal solutions of a specific convex program, namely

Theorem 1.2 ([11]). *A pair $(w, h) \in H$ is a Wardrop equilibrium if and only if it solves the convex minimization problem*

$$\min_{(w,h) \in H} \sum_{a \in A} \int_0^{w_a} s_a(z)\, dz. \tag{P_H}$$

Proof. Since (P_H) is a convex program, a pair $(w, h) \in H$ is optimal if and only if it satisfies the first-order optimality condition

$$\sum_{a \in A} s_a(w_a)(w'_a - w_a) \geq 0, \quad \forall (w', h') \in H.$$

Replacing the expressions $w'_a = \sum_{r \ni a} h'_r$ and $w_a = \sum_{r \ni a} h_r$ into this equation, and rearranging the summation, this may be equivalently stated as

$$\sum_{(i,d)} \left[\sum_{r \in R_i^d} c_r (h'_r - h_r) \right] \geq 0, \quad \forall (w', h') \in H.$$

Using the flow conservation constraints (1.1) satisfied by h' and h, it follows that the latter is equivalent to the fact that for each OD pair (i, d) and each route $r \in R_i^d$ the cost c_r must be minimal whenever $h_r > 0$, which is precisely Wardrop's condition. □

Corollary 1.3. *There exists a Wardrop equilibrium (w, h). Furthermore, the vector of total arc flows w is the same in all equilibria.*

Proof. Since the feasible set H is compact, it follows that (P_H) has optimal solutions and therefore there exist equilibria. The strict convexity of the objective function then implies that the optimal w is unique. □

We denote by w^* the vector of total arc flows common to all Wardrop equilibria. Although there may be several path-flow assignments h corresponding to w^*, we loosely refer to w^* as *the* equilibrium flow.

1.1.1 Arc-flow formulation

When the number of paths becomes large, problem (P_H) may not be amenable to direct computation. An equivalent arc-flow formulation is obtained by considering variables $v_a^d \geq 0$ that represent the flow on arc a with destination d, and the set V of feasible pairs (w, v) with $v \geq 0$ satisfying flow conservation

$$g_i^d + \sum_{a \in A_i^-} v_a^d = \sum_{a \in A_i^+} v_a^d, \quad \forall d \in D,\, i \neq d \tag{1.4}$$

and the total flow relations

$$w_a = \sum_{d \in D} v_a^d, \quad \forall a \in A. \tag{1.5}$$

To each feasible assignment $(w, h) \in H$ it corresponds a pair $(w, v) \in V$ by setting $v_a^d = \sum\{h_r : r \in \cup_i R_i^d,\, a \in r\}$. Note that not all elements in V are of this form, since the latter allow flow along cycles that may be forbidden in H. However, as long as we deal with positive costs, cyclic flows are automatically excluded in any optimal solution and therefore the convex program

$$\min_{(w,v) \in V} \sum_{a \in A} \int_0^{w_a} s_a(z)\, dz \tag{P_V}$$

is a relaxation of (P_H) with the same optimal w^*. More precisely, any v with $(w^*, v) \in V$ is optimal for (P_V) and the sets $\{a \in A : v_a^d > 0\}$ are cycle-free so that v may be decomposed into path-flows $h_r \geq 0$ satisfying (1.1) and (1.3) (see, e.g., [1, Theorem 2.1]). Any such decomposition gives an equilibrium and, conversely, every equilibrium is of this form.

1.1.2 Dual formulation

As noted above, the basic issue when computing a traffic equilibrium is to find the vector of total arc flows w^*. A dual problem which provides an alternative for this was described in Daganzo [28] and Fukushima [43]. The dual variables are the arc travel times t_a for $a \in A$ and the time-to-destination variables τ_i^d for $d \in D$ and $i \neq d$. For notational convenience we set $\tau_d^d = 0$ (considered as a constant rather than a constrained variable) and we let T denote the set of all vectors (t, τ) satisfying

$$\tau_{i_a}^d \leq t_a + \tau_{j_a}^d, \quad \forall a \in A,\, d \in D, \tag{1.6}$$

where i_a and j_a are the initial and terminal nodes of arc a. Setting $t_a^0 = s_a(0)$ it turns out that

Theorem 1.4 ([28, 43]). *The convex program*

$$\min_{(t,\tau)\in T} \sum_{a\in A} \int_{t_a^0}^{t_a} s_a^{-1}(y)\,dy - \sum_{\substack{d\in D\\ i\neq d}} g_i^d \tau_i^d \tag{D}$$

has optimal solutions, and any such solution (t^*, τ^*) satisfies $t_a^* = s_a(w_a^*) > 0$.

Proof. Let (t, τ) be an optimal solution for (D). Since $s_a(\cdot)$ is strictly increasing and continuous, the map $t_a \mapsto \int_{t_a^0}^{t_a} s_a^{-1}(y)\,dy$ is strictly convex and differentiable on its domain, which is an open interval. Moreover, since the constraints of (D) are linear, there exist multipliers $v_a^d \geq 0$ satisfying the KKT conditions

$$\begin{aligned}
&\text{(a)}\quad v_a^d\left[\tau_{i_a}^d - t_a - \tau_{j_a}^d\right] = 0, && \forall a \in A;\\
&\text{(b)}\quad s_a^{-1}(t_a) = \sum_{d\in D} v_a^d, && \forall a \in A;\\
&\text{(c)}\quad g_i^d + \sum_{a\in A_i^-} v_a^d = \sum_{a\in A_i^+} v_a^d, && \forall d \in D,\, i \neq d.
\end{aligned}$$

Setting $w_a = \sum_{d\in D} v_a^d$, the pair (w, v) turns out to be feasible for (P_V), while conditions (a), (b) and (1.6) imply that t_a and τ_i^d are Lagrange multipliers for the constraints (1.5) and (1.4) respectively. Hence (w, v) is a stationary point for (P_V) and, since the latter is a convex program, it follows that (w, v) is optimal. Therefore $w = w^*$ and the equality $w_a^* = s_a^{-1}(t_a)$ results from (b).

The existence of optimal solutions for (D) follows by proceeding in the other direction: take (w^*, v) optimal for (P_V) and choose corresponding multipliers t_a and τ_i^d. Then (t, τ) is a stationary point for (D), hence an optimal solution. □

Note that, for any given $t > 0$, a corresponding optimal $\tau = \bar{\tau}(t)$ may be obtained by solving shortest paths for each OD pair (i, d). Indeed, the objective function in (D) is minimized by making the τ_i^d's as large as possible. Taking into account the constraints (1.6), it follows that the optimal choice is to take

$$\tau_i^d = \min_{a\in A_i^+} \left\{t_a + \tau_{j_a}^d\right\}, \tag{1.7}$$

which are the Bellman's equations that define the shortest travel times $\tau_i^d = \bar{\tau}_i^d(t)$. Hence problem (D) is equivalent to the unconstrained, non-smooth and strictly convex program

$$\min_t \; \bar{\Phi}(t) \triangleq \sum_{a\in A} \int_{t_a^0}^{t_a} s_a^{-1}(y)\, dy - \sum_{\substack{d\in D\\ i\neq d}} g_i^d \bar{\tau}_i^d(t) \qquad (\bar{D})$$

with unique optimal solution $t_a^* = s_a(w_a^*) \geq t_a^0$.

1.2 Stochastic user equilibrium

Wardrop equilibrium conveys the assumption that users are homogeneous and all of them perceive the same costs c_r. This was singled out as one of the main factors explaining the relatively poor agreement between the predictions of the model and the flows observed in real urban networks. Actually, since travel times are subject to some inherent stochasticity, it is natural to expect that drivers have only an approximate perception of the average travel time to be experienced on any given route. These individual perceptions depend on past experience and observations that will most likely differ from one driver to another. On the other hand, different users might have different goals when choosing a route: some might be risk neutral and be essentially concerned with average travel times, while others may be risk averse and penalize those routes that exhibit more variability. These observations pointed out the necessity of extending the concept of Wardrop equilibrium to a stochastic setting in which the travel time perceptions, and hence the corresponding shortest paths, depend on the particular user considered. The idea was formalized by looking at the route selection process as a discrete choice model based on random utility theory. In Appendix A we summarize the main facts that we will require from the theory of discrete choice models.

Stochastic models assume that users are randomly drawn from a large population having a variable perception of routes. Specifically, the travel time of a route $r \in R$ is modeled as a random variable $\tilde{c}_r = c_r + \epsilon_r$ where $c_r = \sum_{a\in r} t_a$ as before, and ϵ_r is a random noise with $\mathbb{E}(\epsilon_r) = 0$ that accounts for the variability of route perceptions among the population. The demand g_i^d from origin i to destination d splits among the routes $r \in R_i^d$ according to

$$h_r = g_i^d \, \mathbb{P}(\tilde{c}_r \leq \tilde{c}_p, \, \forall p \in R_i^d). \qquad (1.8)$$

Note that these route flows are nonnegative and satisfy the constraints (1.1).

For instance, in the multinomial Logit assignment model proposed by Dial [30] in an uncongested setting with constant travel times t_a (see also [70, 74]), the $\tilde{c}_r$'s are taken as independent Gumbel variables with expectation c_r and dispersion parameter β, leading to

$$h_r = g_i^d \frac{\exp(-\beta c_r)}{\sum_{p\in R_i^d} \exp(-\beta c_p)}.$$

To reduce the burden of path enumeration involved in this formula, Dial restricted the analysis to the set of *efficient routes*, namely those for which each arc leads away from the origin and closer to the destination. The number of efficient routes still grows exponentially with network size, so that the applicability of the model is limited to small networks. On the other hand, since independence of route travel times is an unlikely assumption when dealing with overlapping routes, Daganzo and Sheffi [27] proposed an alternative model based on a Probit stochastic assignment in which the random vector $\epsilon = (\epsilon_r)_{r\in R}$ is normally distributed $\mathcal{N}(0,\Sigma)$. In this case there is no close form expression for the probabilities in (1.8), which must be estimated using Monte Carlo simulation, reducing once again the applicability to small networks.

Stochastic assignment was extended by Fisk [35] to an equilibrium model for the case in which the travel times are flow-dependent. Namely,

Definition 1.5. A pair (w,h) is called a *stochastic user equilibrium* (SUE) if it satisfies the flow equations (1.3) and for each origin-destination pair (i,d) the flow h_r on route $r \in R_i^d$ is given by (1.8), where the travel times $\tilde{c}_r = c_r + \epsilon_r$ are random variables with $\mathbb{E}(\epsilon_r) = 0$ and c_r given by (1.2).

Fisk [35] considered the Logit case in which the perturbations $-\epsilon_r$ are i.i.d. Gumbel, and showed that SUE could be characterized by an equivalent convex optimization problem. Namely,

Theorem 1.6. *A pair (w,h) is a stochastic user equilibrium if and only if it is an optimal solution for the problem*

$$\min_{(w,h)\in H} \sum_{a\in A} \int_0^{w_a} s_a(z)\,dz + \frac{1}{\beta} \sum_{r\in R} h_r \ln h_r. \tag{S}$$

Proof. We begin by noting that if $g_i^d = 0$ for a given OD pair, the flow constraints in H impose $h_r = 0$ for all $r \in R_i^d$. Hence the minimization can be restricted to the remaining flows. Let $OD^+ = \{(i,d) : g_i^d > 0\}$ and $R^+ = \cup\{R_i^d : (i,d) \in OD^+\}$. We observe that Slater's constraint qualification condition holds for this restricted optimization problem: take $h_r = g_i^d/|R_i^d| > 0$ for all $r \in R_i^d$, and $w_a = \sum_{r\ni a} h_r$.

Now we proceed by considering the Lagrangian

$$\begin{aligned}\mathcal{L} = \sum_{a\in A} \int_0^{w_a} s_a(z)\,dz + \frac{1}{\beta} \sum_{r\in R^+} h_r \ln h_r + \sum_{a\in A} t_a \left[\sum_{r\ni a} h_r - w_a\right] \\ + \sum_{(i,d):g_i^d>0} \tau_i^d \left[g_i^d - \sum_{r\in R_i^d} h_r\right] - \sum_{r\in R^+} \lambda_r h_r.\end{aligned}$$

A feasible point $(w,h) \in H$ is optimal in (S) if and only if there exist multipliers $t_a \in \mathbb{R}$, $\tau_i^d \in \mathbb{R}$ and $\lambda_r \geq 0$ satisfying the KKT conditions

$$s_a(w_a) - t_a = 0, \quad \forall a \in A, \tag{1.9}$$

$$\frac{1}{\beta}[1 + \ln h_r] + \sum_{a \in r} t_a - \tau_i^d - \lambda_r = 0, \quad \forall (i,d) \in OD^+, r \in R_i^d, \tag{1.10}$$

$$\lambda_r h_r = 0, \quad \forall r \in R^+. \tag{1.11}$$

Equation (1.10) yields $h_r = \mu_i^d \exp(-\beta[c_r - \lambda_r])$ for a certain constant $\mu_i^d > 0$. In particular $h_r > 0$, so that (1.11) implies $\lambda_r = 0$, and then using the flow constraints (1.1) we deduce that (1.10)–(1.11) are equivalent to

$$h_r = g_i^d \frac{\exp(-\beta c_r)}{\sum_{p \in R_i^d} \exp(-\beta c_p)},$$

from which the result follows easily. □

Since problem (S) is strictly convex and coercive it has a unique optimal solution so that we immediately get the following

Corollary 1.7. *There exists a unique stochastic user equilibrium.*

The method of successive averages was proposed in [35] as a practical computational scheme. A variant of this scheme, with an adaptive rule for selecting a subset of efficient routes in the spirit of Dial's, was implemented by Leurent [52]. Since these methods iterate in the high-dimensional space of route flows, alternative methods based on a dual formulation were investigated by Daganzo [28] and Miyagi [58], providing at the same time a unified framework encompassing simultaneously the stochastic and deterministic models. In order to state the result, let us consider the *expected travel time* functions

$$\varphi_i^d(c_r : r \in R_i^d) = \mathbb{E}\Big(\min_{r \in R_i^d} \{c_r + \epsilon_r\}\Big),$$

which, according to Proposition A.1 in Appendix A, are concave and smooth with

$$\frac{\partial \varphi_i^d}{\partial c_r}(c_r : r \in R_i^d) = \mathbb{P}(\tilde{c}_r \le \tilde{c}_p, \forall p \in R_i^d).$$

Theorem 1.8 ([28, 58]). *The strictly convex program*

$$\min_t \Phi(t) = \sum_{a \in A} \int_{t_a^0}^{t_a} s_a^{-1}(y)\, dy - \sum_{\substack{d \in D \\ i \neq d}} g_i^d \varphi_i^d \left(\sum_{a \in r} t_a : r \in R_i^d \right) \tag{SD}$$

has a unique optimal solution t^* and $w_a^* = s_a^{-1}(t_a^*)$ is the unique SUE.

Proof. To prove the existence of solutions we show that $\Phi(\cdot)$ is coercive. We must show that $\Phi^\infty(t) > 0$ for all $t \neq 0$, where $\Phi^\infty(\cdot)$ is the recession function

$$\begin{aligned}\Phi^\infty(t) &= \lim_{\lambda \to \infty} \frac{1}{\lambda} \Big[\Phi(t^0 + \lambda t) - \Phi(t^0) \Big] \\ &= \sum_{a \in A} \lim_{\lambda \to \infty} t_a s_a^{-1}(\lambda t_a) - \sum_{i,d} g_i^d \min\Big\{ \sum_{a \in r} t_a : r \in R_i^d \Big\}.\end{aligned}$$

Any $t \neq 0$ has a nonzero component $t_a \neq 0$. Since $\lim_{z\to\pm\infty} s_a^{-1}(z) = \pm\infty$, the corresponding terms in the first sum are equal to $+\infty$, which implies $\Phi^\infty(t) = +\infty$. This proves that $\Phi(\cdot)$ is coercive and therefore (SD) admits at least one optimal solution t^*. Since $\Phi(\cdot)$ is strictly convex, this solution must be unique.

The optimality conditions that characterize t^* are

$$s_a^{-1}(t_a^*) = \sum_{i,d} g_i^d \sum_{r\ni a} \frac{\partial \varphi_i^d}{\partial c_r}(c_r : r \in R_i^d)$$

so that defining

$$h_r^* = g_i^d \frac{\partial \varphi_i^d}{\partial c_r}(c_r : r \in R_i^d) = g_i^d \, \mathbb{P}(\tilde{c}_r \leq \tilde{c}_p, \forall p \in R_i^d)$$

and letting $w_a^* = \sum_{r\ni a} h_r^*$, it follows that $t_a^* = s_a(w_a^*)$ and (w^*, h^*) is therefore the unique SUE, as claimed. □

1.3 Remarks on Wardrop and SUE

Wardrop equilibrium and SUE are *route-based* models: each driver on a given OD pair selects an optimal path by comparing all the available routes. From a modelling perspective this has several drawbacks. Firstly, partial information and limited discrimination capacity of the users suggest that the choice should be restricted to a few efficient routes as proposed by Dial. However, defining which routes should then be considered may not be clear, especially when travel times are flow-dependent. A discussion of this issue and possible remedies were given in [3, 52]. On the other hand, Logit-based models assume the independence of route costs, which is not adequate when dealing with overlapping routes. A further drawback specific to the Logit models in the literature is that they consider a uniform dispersion parameter throughout the network, conveying the implicit assumption that the variance of travel time perception is constant and does not scale with distance. In addition, both Logit and Probit models share the counter-intuitive feature of assigning flows to all the available routes, no matter how large their travel times might be. Finally, we note that models based on path enumeration not only become computationally impractical for large networks, but one may argue that drivers do not have enough information nor are capable of evaluating an exponentially increasing number of alternative routes. Methods that avoid path enumeration were described in [3, 17, 51, 55]. In these works, a Markovian property of the Logit model is exploited to derive an equivalent reformulation of Fisk's optimization problem in the space of arc flows. Yet, the analysis relies heavily on the assumption that the dispersion parameter of the Logit model is constant throughout the network.

Chapter 2

Markovian Traffic Equilibrium

As pointed out at the end of the previous chapter, route-based equilibrium models present several drawbacks from the modeling and computational viewpoints. In particular, the independence of route travel times is not well suited when dealing with overlapping paths. For instance, in the small example of Figure 2.1 with 3 routes from i to d, all of them with the same expected cost $c_r = 1$, the Logit rule assigns one third of the demand to each route. However, since both routes using the lower arc are almost identical, the assignment $(\frac{1}{2}, \frac{1}{4}, \frac{1}{4})$ seems more natural. The latter is the solution obtained if we focus on *arc choices* rather than *path choices:* at node i there are only two arc options (upper and lower), both offering the same travel time so that one may expect that each one gets roughly one half of the demand. The half taking the lower arc faces a second choice at the intermediate node j where it splits again giving $\frac{1}{4}$ of the demand on each of the two lower routes. Although the arc-choice approach alleviates the independence issue, it does not solve it completely: when small deviations occur next to the origin, as in a mirror image of the network of Figure 2.1 with the direction of links reversed from d towards i, arc-choices and path-choices coincide, yet one may not expect the two lower paths to be independent.

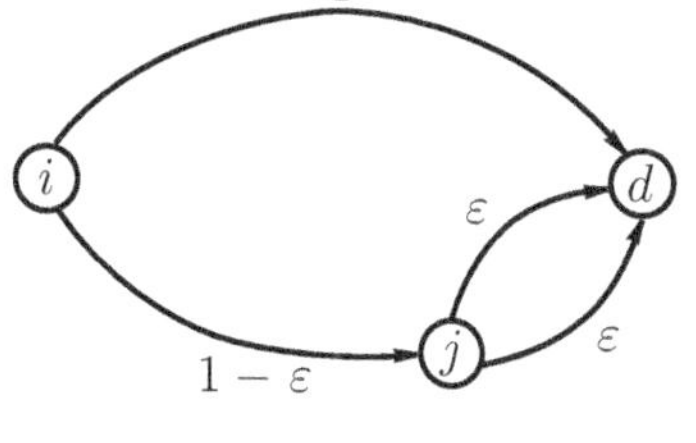

Figure 2.1: Path vs. arc choices on a small network.

This chapter, based on [9], explores an arc-based approach that looks at route selection as a recursive decision process based on arc choices: route is no longer fixed at the origin before departure but is built along the way in a dynamic programming manner. More precisely, drivers move towards their destinations by a sequential process of arc selection using a discrete choice model at every intermediate node in their trip. Route selection is the outcome of this sequential process while passenger movements are governed by an embedded Markov chain so that network flows are given by the corresponding invariant measures. The resulting equilibrium concept is called *Markovian traffic equilibrium* (MTE).

We stress that the Markovian character of MTE concerns the *spatial* dimension and not the *inter-temporal* dynamics (for models describing the day-to-day variability of flows, see [23, 24, 47, 77]). The idea is similar to the Logit Markovian model in Akamatsu [2], allowing for general discrete choice models. The resulting model is very flexible and can handle different choice models at every node, and even to mix deterministic and stochastic assignment rules at different nodes. This flexibility can be exploited to model the simultaneous selection of transportation mode and route by considering multi-modal networks. The random route costs are not required to be independent and the number of alternative arcs to be considered at each node remains within the discrimination capabilities of the users with no need for arbitrarily reducing the set of possible routes. Moreover, by considering discrete choices based on distributions with compact support, the resulting assignments use only nearly optimal routes and not every route as in the Logit or Probit settings. Still, MTE admits a simple characterization through a strictly convex minimization problem that avoids path enumeration and leads to computational procedures that are effective even for large networks.

The chapter is structured as follows. Section 2.1 discusses the preliminaries on Markovian route choice required to define the notion of Markovian equilibrium. The latter is introduced in Section 2.2 showing that it solves an unconstrained strictly convex minimization problem. This variational characterization yields the existence and uniqueness of an equilibrium, providing a unifying framework for stochastic and deterministic equilibrium. In a purely stochastic setting, this problem turns out to be smooth, and since it is also low-dimensional (one variable per arc), it opens the way to solve large networks. In Section 2.3 we report some numerical examples comparing the performance of the method of successive averages (MSA) and Newton's method on one small and one large network, providing a formal convergence proof for MSA.

2.1 Markovian route choice

Suppose that the arc travel times are random variables $\tilde{t}_a = t_a + \nu_a$ expressed as the sum of an expected value t_a and a random noise ν_a that accounts for the variability of perceptions within the population and which is assumed to have a continuous distribution with $\mathbb{E}(\nu_a) = 0$. Then, the route travel times and optimal costs are also random variables

$$\tilde{c}_r = \sum_{a \in r} \tilde{t}_a,$$
$$\tilde{\tau}_i^d = \min_{r \in R_i^d} \tilde{c}_r.$$

Note that the costs of two overlapping routes are dependent. Let $c_r = \mathbb{E}(\tilde{c}_r)$ and $\tau_i^d = \mathbb{E}(\tilde{\tau}_i^d)$ denote the expected values, and observe that $\tau_i^d = \mathbb{E}(\min_{r \in R_i^d} \tilde{c}_r)$ is different from $\min_{r \in R_i^d} c_r$.

Now consider a driver heading toward destination d and reaching an intermediate node i in his trip. To exit from i, he selects an arc $a \in A_i^+$ according to his own evaluation of the available arcs. Namely, the cost of an arc $a \in A$ in regard to destination d is defined by the random variable

$$\tilde{z}_a^d = \tilde{t}_a + \tilde{\tau}_{j_a}^d = z_a^d + \epsilon_a^d,$$

where $z_a^d = \mathbb{E}(\tilde{z}_a^d)$ and $\mathbb{E}(\epsilon_a^d) = 0$. Each driver observes the variables $\tilde{z}_a^d$ for $a \in A_i^+$ and then selects the arc with the lowest cost. Since ϵ_a^d has continuous distribution, the probability of a tie is zero. The process is repeated at each subsequent node giving rise to a recursive discrete choice model. Thus, for each destination $d \in D$ we have an underlying Markov chain on the graph G where, for $i \neq d$, the transition probabilities are given by

$$P_{ij}^d = \begin{cases} \mathbb{P}(\tilde{z}_a^d \leq \tilde{z}_b^d,\ \forall b \in A_i^+) & \text{if } (i,j) = a \in A_i^+, \\ 0 & \text{otherwise,} \end{cases}$$

while the destination d is an absorbing state with

$$P_{dj}^d = \begin{cases} 1 & \text{if } j = d, \\ 0 & \text{otherwise.} \end{cases}$$

According to Proposition A.1 in Appendix A, these transition probabilities can be expressed as

$$\mathbb{P}(\tilde{z}_a^d \leq \tilde{z}_b^d,\ \forall b \in A_i^+) = \frac{\partial \varphi_i^d}{\partial z_a^d}(z^d), \tag{2.1}$$

where $\varphi_i^d(\cdot)$ are the expected travel time functions

$$\varphi_i^d(z^d) \triangleq \mathbb{E}\Big(\min_{a \in A_i^+} \{z_a^d + \epsilon_a^d\} \Big).$$

These functions – which are componentwise non-decreasing and concave – are determined by the random variables ϵ_a^d, which are themselves tied to the original arc travel time random variables ν_a. Notice also that, since the ϵ_a^d's have continuous distribution, the maps $\varphi_i^d(\cdot)$ are smooth. These maps convey all the information required to state the model and may be considered as the primary modeling objects. It is worth noting that the class $\mathcal{E}$ of functions that can be expressed in this form admits an analytic characterization (see Proposition A.2). In the sequel we assume that the model is specified directly in terms of a given family of functions $\varphi_i^d \in \mathcal{E}$ with $\varphi_d^d \equiv 0$.

Remark. Since driver movements are governed by a Markov chain, cycling may occur so that additional conditions are required to ensure that drivers reach the destination with probability one. A simple condition is to assume that a driver at an intermediate node i will not consider all the arcs in A_i^+ but only those which take him closer to his destination d, so that the corresponding Markov chain is supported over an acyclic graph (N, A^d). To deal with this case it suffices to redefine

$$\varphi_i^d(z^d) \triangleq \mathbb{E}\Big(\min_{a \in A_i^{d+}} \{z_a^d + \epsilon_a^d\} \Big),$$

so that $P_{ij}^d = \frac{\partial \varphi_i^d}{\partial z_a^d}(z^d) = 0$ for all $a = (i,j) \in A_i^+ \setminus A_i^{d+}$ and no flow heading towards d uses those arcs. A more general condition will be considered later.

2.1.1 Expected travel times

Since $\tilde{z}_a^d = \tilde{t}_a + \tilde{\tau}_{j_a}^d$, we may write Bellman's equations of dynamic programming in the form $\tilde{\tau}_i^d = \min_{a \in A_i^+} \tilde{z}_a^d$. Taking expectation we get

$$\begin{cases} z_a^d = t_a + \tau_{j_a}^d \\ \tau_i^d = \varphi_i^d(z^d), \end{cases}$$

which can be expressed solely in terms of the variables τ_i^d as

$$\tau_i^d = \varphi_i^d\Big((t_a + \tau_{j_a}^d)_{a \in A} \Big). \tag{2.2}$$

Note that the functions $\varphi_i^d(z^d)$ can be interpreted as smooth approximations of the polyhedral functions $\bar{\varphi}_i^d(z^d) = \min_{a \in A_i^+} z_a^d$, and in fact we have $\varphi_i^d(z^d) \leq \bar{\varphi}_i^d(z^d)$. The maps $\bar{\varphi}_i^d(\cdot)$ correspond to the deterministic case in which the noises ϵ_a^d have zero variance, so that (2.2) may be interpreted as the stochastic analog of the Bellman's equations (1.7).

2.1.2 Expected flows

For each destination d, driver movements are governed by a Markov chain. Hence, the *route-based* distribution rule (1.8) is replaced by an *arc-based* recursive scheme where the expected throughput flow x_i^d entering a node i with destination d splits among the outgoing arcs $a \in A_i^+$ according to

$$v_a^d = x_i^d \, \mathbb{P}(\tilde{z}_a^d \leq \tilde{z}_b^d, \, \forall b \in A_i^+) = x_i^d \, \frac{\partial \varphi_i^d}{\partial z_a^d}(z^d).$$

These equations can be written in matrix form as

$$v^d = \widehat{Q}^d(z^d)' x^d, \tag{2.3}$$

where $\widehat{Q}^d(z^d) = (Q_{ia}^d)_{i \neq d, \, a \in A}$ with $Q_{ia}^d = \frac{\partial \varphi_i^d}{\partial z_a^d}(z^d)$ if $i = i_a$ and $Q_{ia}^d = 0$ otherwise.

The expected throughputs are computed as follows. Let $\widehat{P}^d(z^d) = (P_{ij}^d)_{i,j \neq d}$ denote the restriction of the transition matrix P^d to the set of nodes $N \setminus \{d\}$, and let $g^d = (g_i^d)_{i \neq d}$. Then the expected throughput vector $x^d = (x_i^d)_{i \neq d}$ for destination d is given by $x^d = \sum_{k=0}^{\infty} [\widehat{P}^d(z^d)']^k g^d$. This may also be expressed as

$$x^d = g^d + \widehat{P}^d(z^d)' x^d, \tag{2.4}$$

which corresponds to the familiar flow conservation equations

$$x_i^d = g_i^d + \sum_{a \in A_i^-} v_a^d.$$

2.1.3 Implicit functions

As mentioned earlier, additional conditions are required to ensure that drivers reach their destination in finite time with probability one. We will prove that this is the case if the expected travel times t belong to the set

$$\mathcal{C} = \left\{ t \in \mathbb{R}^{|A|} \; : \; \exists \widehat{\tau} \text{ with } \widehat{\tau}_i^d < \varphi_i^d \left(\left(t_a + \widehat{\tau}_{j_a}^d \right)_{a \in A} \right) \text{ for all } i \neq d \right\}.$$

More precisely, we will prove that the equations (2.2)–(2.4) uniquely define τ, v and x as implicit functions of t over the domain $\mathcal{C}$. Notice that, since the $\varphi_i^d(\cdot)$'s are continuous, concave and componentwise non-decreasing, it follows that the domain $\mathcal{C}$ is open, convex, and for each $t \in \mathcal{C}$ we have $t' \in \mathcal{C}$ for all $t' \geq t$. The next proofs are based on the following technical remark.

Lemma 2.1. *Suppose that τ^d solves (2.2) for a given $t \in \mathcal{C}$ and let $\delta_i^d = \tau_i^d - \widehat{\tau}_i^d$ with $\widehat{\tau}$ such that $\widehat{\tau}_i^d < \varphi_i^d((t_a + \widehat{\tau}_{j_a}^d)_{a \in A})$ for $i \neq d$. Then for each $i \neq d$ there exists $j \in N$ with $P_{ij}^d > 0$ and $\delta_j^d < \delta_i^d$.*

Proof. Let $\widehat{z}_a^d = t_a + \widehat{\tau}_{j_a}^d$ and $z_a^d = t_a + \tau_{j_a}^d$. Since φ_i^d is concave and smooth, we get

$$\varphi_i^d(\widehat{z}) \leq \varphi_i^d(z) + \sum_{a \in A} \frac{\partial \varphi_i^d}{\partial z_a^d}(z^d)(\widehat{z}_a^d - z_a^d)$$

$$= \varphi_i^d(z) + \sum_{a \in A_i^+} P_{ij_a}^d (\widehat{\tau}_{j_a}^d - \tau_{j_a}^d)$$
$$= \varphi_i^d(z) - \sum_j P_{ij}^d \delta_j^d$$

and, since $\widehat{\tau}_i^d < \varphi_i^d(\widehat{z})$ and $\tau_i^d = \varphi_i^d(z)$, we deduce that $\sum_j P_{ij}^d \delta_j^d < \delta_i$, from which the conclusion follows directly. □

Proposition 2.2. *Suppose that τ^d solves (2.2) for a given $t \in C$ and let $z_a^d = t_a + \tau_{j_a}^d$. Then the matrix $[I - \widehat{P}^d(z^d)]$ is invertible and therefore the equations (2.3)–(2.4) have a unique solution $v^d = \widehat{Q}^d(z^d)' x^d$ with $x^d = [I - \widehat{P}^d(z^d)']^{-1} g^d$.*

Proof. Using Lemma 2.1 inductively, for each $i \neq d$ we may find a sequence of nodes $i_0, i_1, \ldots, i_m$ with $i_0 = i$, $i_m = d$ and $P_{i_k i_{k+1}}^d > 0$. Hence the Markov chain started from any i has a positive probability of reaching the absorbing state d. It follows that $\widehat{P}^d(z^d)^k$ is strictly sub-Markovian for k large and therefore $[I - \widehat{P}^d(z^d)]$ is invertible. The claim about equations (2.3)–(2.4) follows directly from this. □

Proposition 2.3. *If $t \in C$ then for each destination $d \in D$ the equations (2.2) have unique solutions $\tau^d = \tau^d(t)$. The maps $t \mapsto \tau_i^d(t)$ are smooth, concave, componentwise non-decreasing, and we have $\tau_i^d(t) \leq \bar{\tau}_i^d(t)$ with $\bar{\tau}_i^d(t)$ the shortest travel times defined by (1.7).*

Proof. EXISTENCE. Fix any $\widehat{\tau}$ with $\widehat{\tau}_i^d \leq \varphi_i^d((t_a + \widehat{\tau}_{j_a}^d)_{a \in A})$. We will prove the existence of a solution τ_i^d for (2.2) with $\widehat{\tau}_i^d \leq \tau_i^d \leq \bar{\tau}_i^d(t)$. To this end, consider the recursion $\tau_i^{k+1} = \varphi_i^d((t_a + \tau_{j_a}^k)_{a \in A})$ started from $\tau^0 = \widehat{\tau}^d$. The choice of $\widehat{\tau}$ gives $\tau^0 \leq \tau^1$ and, since the φ_i^d's are componentwise non-decreasing, we inductively get that $\tau^k \leq \tau^{k+1}$.

Now, since $\varphi_i^d(z^d) \leq \min_{a \in A_i^+} z_a^d$, we also have $\widehat{\tau}_i^d \leq \min_{a \in A_i^+} \{t_a + \widehat{\tau}_{j_a}^d\}$, and then the iteration $\bar{\tau}_i^{k+1} = \min_{a \in A_i^+} \{t_a + \bar{\tau}_{j_a}^k\}$ started from $\bar{\tau}^0 = \widehat{\tau}^d$ is increasing and converges after at most $|V|$ iterations to the shortest travel times $\bar{\tau}_i^d = \bar{\tau}_i^d(t)$. In particular, $\widehat{\tau}^d = \bar{\tau}^0 \leq \bar{\tau}^d$.

Going back to the sequence τ^k, since $\tau^0 = \widehat{\tau}^d \leq \bar{\tau}^d$, it follows inductively that $\tau^k \leq \bar{\tau}$ for all k. Indeed, if $\tau^k \leq \bar{\tau}^d$ then

$$\begin{aligned}
\tau_i^{k+1} &= \varphi_i^d\left(\left(t_a + \tau_{j_a}^k\right)_{a \in A}\right) \\
&\leq \varphi_i^d\left(\left(t_a + \bar{\tau}_{j_a}^d\right)_{a \in A}\right) \\
&\leq \min_{a \in A_i^+} \left\{t_a + \bar{\tau}_{j_a}^d\right\} = \bar{\tau}_i^d.
\end{aligned}$$

Thus, the non-decreasing sequences τ_i^k are bounded from above by $\bar{\tau}_i^d$ and therefore they have a limit $\tau_i^d \in [\widehat{\tau}_i^d, \bar{\tau}_i^d]$ which satisfies $\tau_i^d = \varphi_i^d((t_a + \tau_{j_a}^d)_{a \in A})$.

UNIQUENESS. Let τ^1, τ^2 be two solutions and denote $\alpha = \max_{i\in N}[\tau_i^1 - \tau_i^2]$. Let N^* be the set of nodes i where the maximum is attained. For each $i \in N^*$ we have

$$\begin{aligned}\tau_i^1 = \varphi_i^d\Big(\big(t_a + \tau_{j_a}^1\big)_{a\in A}\Big) &\leq \varphi_i^d((t_a + \tau_{j_a}^2 + \alpha)_{a\in A}) \\ &= \varphi_i^d\Big(\big(t_a + \tau_{j_a}^2\big)_{a\in A}\Big) + \alpha \\ &= \tau_i^2 + \alpha \\ &= \tau_i^1.\end{aligned}$$

Strict monotonicity then shows that whenever $\frac{\partial \varphi_i^d}{\partial z_a^d}((t_a + \tau_{j_a}^1)_{a\in A}) > 0$ one must also have $\tau_{j_a}^1 = \tau_{j_a}^2 + \alpha$, that is to say $j_a \in N^*$. Using Lemma 2.1 we deduce that for all $i \in N^* \setminus \{d\}$ there exists $j \in N^*$ with $\delta_j^1 < \delta_i^1$, and proceeding inductively we eventually reach $j = d \in N^*$, so that $\alpha = \tau_d^1 - \tau_d^2 = 0$, which proves $\tau^1 \leq \tau^2$. Exchanging the roles of τ^1 and τ^2 we get $\tau^2 \leq \tau^1$; hence $\tau^1 = \tau^2$.

CONCAVITY. Let $\tau^1 = \tau^d(t^1)$ and $\tau^2 = \tau^d(t^2)$. Set $t^\alpha = \alpha t^1 + (1-\alpha)t^2$ with $\alpha \in (0,1)$ and consider $\widehat{\tau} = \alpha\tau^1 + (1-\alpha)\tau^2$. Since the φ_i^d's are concave, we obtain

$$\begin{aligned}\varphi_i^d\Big(\big(t_a^\alpha + \widehat{\tau}_{j_a}^d\big)_{a\in A}\Big) &\geq \alpha\varphi_i^d\Big(\big(t_a^1 + \tau_{j_a}^1\big)_{a\in A}\Big) + (1-\alpha)\varphi_i^d\Big(\big(t_a^2 + \tau_{j_a}^2\big)_{a\in A}\Big) \\ &= \alpha\tau_i^1 + (1-\alpha)\tau_i^2 \\ &= \widehat{\tau}_i^d\end{aligned}$$

and therefore the existence proof above implies $\widehat{\tau}_i^d \leq \tau_i^d(t^\alpha)$, which is precisely the concavity inequality

$$\alpha\tau_i^d(t^1) + (1-\alpha)\tau_i^d(t^2) \leq \tau_i^d(t^\alpha).$$

MONOTONICITY. Let $\widehat{\tau} = \tau(t^1)$. Since the $\varphi_i^d(\cdot)$'s are componentwise non-decreasing, for $t^2 \geq t^1$ we have

$$\widehat{\tau}_i^d = \varphi_i^d\Big(\big(t_a^1 + \widehat{\tau}_{j_a}^d\big)_{a\in A}\Big) \leq \varphi_i^d\Big(\big(t_a^2 + \widehat{\tau}_{j_a}^d\big)_{a\in A}\Big)$$

and then the existence proof above implies $\tau_i^d(t^2) \geq \widehat{\tau}_i^d = \tau_i^d(t^1)$.

SMOOTHNESS. This is a direct consequence of the Implicit Function Theorem. Indeed, noting that $\tau_d^d = 0$ we may reduce (2.2) to a system in the variables $(\tau_i^d)_{i\neq d}$. The Jacobian of this reduced system is $[I - \widehat{P}^d]$ which is invertible according to Proposition 2.2, so the conclusion follows. □

Combining the previous two propositions it results that, for each $t \in \mathcal{C}$ and for each destination $d \in D$, the equations (2.2)–(2.4) define unique implicit functions

$v^d = v^d(t)$, $x^d = x^d(t)$, and $\tau^d = \tau^d(t)$. In the sequel we denote $\widehat{P}^d(t) = \widehat{P}^d(z^d(t))$ and $\widehat{Q}^d(t) = \widehat{Q}^d(z^d(t))$ with $z_a^d(t) = t_a + \tau_{j_a}^d(t)$, so that

$$x^d(t) = \left[I - \widehat{P}^d(t)'\right]^{-1} g^d,$$

$$v^d(t) = \widehat{Q}^d(t)' x^d(t).$$

2.2 Markovian traffic equilibrium

We proceed to formalize the notion of Markovian equilibrium. In this setting we assume that, in addition to the maps $\varphi_i^d \in \mathcal{E}$, we are given the family of strictly increasing and continuous arc travel time functions $s_a\colon \mathbb{R} \to (0, \infty)$.

Definition 2.4. A vector $w \in \mathbb{R}^{|A|}$ is a *Markovian traffic equilibrium* (MTE) if $w_a = \sum_{d \in D} v_a^d$ where the v_a^d's satisfy the flow distribution equations (2.3)–(2.4) with $z_a^d = t_a + \tau_{j_a}^d$ where τ^d solves (2.2) with $t_a = s_a(w_a)$.

Using the implicit functions defined in Section 2.1.3, we may restate this definition by saying that w is an MTE if and only if $w_a = \sum_{d \in D} v_a^d(t)$ with $t = s_a(w_a)$. This is equivalent to saying that $w_a = s_a^{-1}(t_a)$ where t solves the system of equations $s_a^{-1}(t_a) = \sum_{d \in D} v_a^d(t)$. Although this notion might look rather different from Wardrop equilibrium, we will show that they are in fact closely related by proving that the MTE admits a variational characterization similar to $(\bar{D})$.

Theorem 2.5. *Assume that $t^0 \in \mathcal{C}$ where $t_a^0 = s_a(0)$ for all $a \in A$. Then there exists a unique* MTE *given by $w_a^* = s_a^{-1}(t_a^*)$ where t^* is the unique optimal solution of the smooth strictly convex program*

$$\min_{t \in \mathcal{C}} \Phi(t) \triangleq \sum_{a \in A} \int_{t_a^0}^{t_a} s_a^{-1}(y)\, dy - \sum_{\substack{d \in D \\ i \neq d}} g_i^d \tau_i^d(t). \tag{MD}$$

Proof. Let us begin by noting that the map $t_a \mapsto \int_{t_a^0}^{t_a} s_a^{-1}(y)\, dy$ is decreasing for $t_a < t_a^0$ and increasing for $t_a > t_a^0$. Then, since the functions $t \mapsto \tau_i^d(t)$ are also componentwise non-decreasing, it follows that the optimum in (MD) must be attained in the region $\{t \geq t^0\} \subseteq \mathcal{C}$. Since the $\tau_i^d(\cdot)$'s are concave and the $s_a^{-1}(\cdot)$'s are strictly increasing with $s_a^{-1}(t_a) \to \infty$ when $t_a \to \infty$, we infer that $\Phi(\cdot)$ is strictly convex and coercive in this region and therefore (MD) has a unique optimal solution t^*.

Now we already noted that w is an MTE if and only if $w_a = s_a^{-1}(t_a)$ where t solves $s_a^{-1}(t_a) = \sum_{d \in D} v_a^d(t)$. To complete the proof it suffices to show that the latter corresponds to $\nabla \Phi(t) = 0$, so that $t = t^*$.

Let us compute $\nabla \Phi(t)$. An implicit differentiation of the system (2.2) yields $\frac{\partial \tau^d}{\partial t_a} = \widehat{Q}_{\cdot a}^d(t) + \widehat{P}^d(t) \frac{\partial \tau^d}{\partial t_a}$, from which we get $\frac{\partial \tau^d}{\partial t_a} = [I - \widehat{P}^d(t)]^{-1} \widehat{Q}_{\cdot a}^d(t)$. Then, considering the functions $\psi^d(t) = \sum_{i \neq d} g_i^d \tau_i^d(t)$, we have

$$\frac{\partial \psi^d}{\partial t_a} = \left(\frac{\partial \tau^d}{\partial t_a}\right)' g^d = \widehat{Q}^d_{\cdot a}(t)' \left[I - \widehat{P}^d(t)'\right]^{-1} g^d = \widehat{Q}^d_{\cdot a}(t)' x^d(t) = v_a^d(t)$$

and therefore

$$\frac{\partial \Phi}{\partial t_a} = s_a^{-1}(t_a) - \sum_{d \in D} v_a^d(t), \tag{2.5}$$

which yields the desired conclusion $w_a^* = s_a^{-1}(t^*)$. □

Although problem (MD) includes the constraint $t \in \mathcal{C}$, it is essentially unconstrained since $\mathcal{C}$ is open and the optimum is attained in its interior. Observe also that, in contrast with the deterministic case, the previous results show that in the case of MTE not only w^* but also the arc-flows $v_a^d(t^*)$ are uniquely determined. Moreover (MD) is smooth and strictly convex, and the dimension of the space $\mathbb{R}^{|A|}$ is moderate compared to the models based on route flows. These nice features are balanced by the need of computing the implicit functions $\tau_i^d(\cdot)$. Nevertheless, the existence proof in Proposition 2.3 provides an algorithm which can be used to devise numerical schemes for computing the MTE.

Problem (MD) may be dualized in order to obtain a *primal* characterization of MTE similar to (P_V) for the deterministic equilibrium, namely

Theorem 2.6. *If $t^0 \in \mathcal{C}$ then the MTE is the unique optimal solution w^* of*

$$\min_{(w,v) \in V} \sum_{a \in A} \int_0^{w_a} s_a(z)\, dz + \sum_{d \in D} \chi^d(v^d) \tag{MP}$$

where $\chi^d(v^d) = \sup_{z^d} \sum_{a \in A} (\varphi^d_{i_a}(z^d) - z_a^d) v_a^d$.

Proof. Let w^* be the MTE and $v_a^d = v_a^d(t^*)$. Since (MP) is strictly convex with respect to w, it suffices to check that (w^*, v) is a stationary point of the Lagrangian

$$\mathcal{L} = \sum_{a \in A} \int_0^{w_a} s_a(z)\, dz + \sum_{d \in D} \chi^d(v^d) + \sum_{\substack{d \in D \\ i \neq d}} \mu_i^d \left[g_i^d + \textstyle\sum_{a \in A_i^-} v_a^d - \sum_{a \in A_i^+} v_a^d\right]$$
$$+ \sum_{a \in A} \nu_a \left[\textstyle\sum_{d \in D} v_a^d - w_a\right] - \sum_{\substack{d \in D \\ a \in A}} \lambda_a^d v_a^d .$$

The multipliers $\mu_i^d \in \mathbb{R}$, $\nu_a \in \mathbb{R}$ and $\lambda_a^d \geq 0$ correspond to the constraints (1.4), (1.5) and $v_a^d \geq 0$ respectively, and stationarity amounts to $s_a(w_a^*) = \nu_a$, $\lambda_a^d v_a^d = 0$, and $\zeta^d \in \partial \chi^d(v^d)$ where $\zeta_a^d = \mu_{i_a}^d - \mu_{j_a}^d - \nu_a + \lambda_a^d$. The first condition imposes $\nu_a = t_a^*$ while for the second we simply set $\lambda_a^d = 0$. To check the last condition, take $\mu_i^d = \tau_i^d$ with $\tau_i^d = \tau_i^d(t^*)$, and let $z_a^d = t_a^* + \tau_i^d$. Combining (2.3) and (2.4), we get, for all $i \neq d$ and $a \in A_i^+$,

$$v_a^d = \frac{\partial \varphi_i^d}{\partial z_a^d}(z^d) \sum_{a \in A_i^+} v_a^d.$$

This shows that z^d is an optimal solution for $\chi^d(v^d)$ and, setting $\gamma_a^d = \varphi_{i_a}^d(z^d) - z_a^d$, we get $\gamma^d \in \partial \chi^d(v^d)$. Since $\varphi_{i_a}^d(z^d) = \tau_{i_a}^d$ and $z_a^d = t_a^* + \tau_{j_a}^d$, we deduce that $\zeta^d = \gamma^d \in \partial \chi^d(v^d)$, as required. □

We remark that $\chi^d(v^d)$ may be expressed in terms of the Fenchel conjugates of the smooth convex functions $-\varphi_i^d(\cdot)$ as

$$\chi^d(v^d) = \sum_{i \neq d} \chi_i^d\left((v_a^d)_{a \in A_i^+}, \sum_{b \in A_i^+} v_b^d \right)$$

with $\chi_i^d(v, x) = x\,(-\varphi_i^d)^*(-\frac{v}{x})$. Problem (MP) is the stochastic analog of (P_V). As a matter of fact, in the deterministic case with $\varphi_i^d(z^d) = \min\{z_a^d : a \in A_i^+\}$ we get $\chi^d(v^d) \equiv 0$, so that (MP) coincides with (P_V). In the case of a Logit Markovian model where

$$\varphi_i^d(z^d) = -\frac{1}{\beta_i^d} \ln \left(\sum_{a \in A_i^+} e^{-\beta_i^d z_a^d} \right),$$

a straightforward computation yields

$$\chi^d(v^d) = \sum_{i \neq d} \frac{1}{\beta_i^d} \left[\sum_{a \in A_i^+} v_a^d \ln (v_a^d) - \left(\sum_{a \in A_i^+} v_a^d \right) \ln \left(\sum_{a \in A_i^+} v_a^d \right) \right],$$

so that (MP) also extends Akamatsu [2], where the case $\beta_i^d \equiv \theta$ was considered.

2.2.1 Extensions

Arc capacities and saturation

So far the volume-delay functions $s_a(\cdot)$ have been assumed finite. Since often the arcs have a maximal capacity, it is useful to extend the model by considering strictly increasing continuous functions $s_a\colon (-\infty, \bar{w}_a) \to (0, \infty)$ with $s_a(w_a) \to \infty$ when $w_a \to \bar{w}_a$. The saturation capacity $\bar{w}_a$ is supposed strictly positive and may be infinite for those arcs which are not subject to saturation. To ensure feasibility we must assume that these capacities are large enough for the given demands.

Theorem 2.7. *Suppose $t^0 \in C$ and assume that there exists a vector $(\widehat{w}, \widehat{v})$ satisfying* (1.4) *and* (1.5) *with $\widehat{v} \geq 0$ and $\widehat{w}_a < \bar{w}_a$ for all $a \in A$. Then the conclusions of Theorems* 2.5 *and* 2.6 *remain valid.*

Proof. The analysis in Sections 2.1 and 2.2 remains unchanged except for the existence of an MTE which requires to show that $\Phi(t)$ is coercive on the region $\{t \geq t^0\}$. To this end we prove that the recession map $\Phi^\infty(t) \triangleq \lim_{\lambda\to\infty}\Phi(\lambda t)/\lambda$ satisfies $\Phi^\infty(t) > 0$ for all nonzero $t \geq 0$. For such a t, the recession function of $\sum_{a\in A}\int_{t_a^0}^{t_a} s_a^{-1}(z)\,dz$ is equal to $\sum_{a\in A}\bar{w}_a t_a$. In order to compute the recession functions of the $\tau_i^d(\cdot)$'s, we exploit equation (2.2) to obtain

$$\left(\tau_i^d\right)^\infty(t) = \left(\varphi_i^d\right)^\infty\left(\left(t_a + \left(\tau_{j_a}^d\right)^\infty(t)\right)_{a\in A}\right). \tag{2.6}$$

Now the definition of the class $\mathcal{E}$ and Lebesgue's theorem imply that

$$(\varphi_i^d)^\infty(z^d) = \min\{z_a^d \,:\, a \in A_i^+\}$$

and therefore (2.6) implies that $(\tau_i^d)^\infty(t)$ is the shortest distance from i to d, namely $(\tau_i^d)^\infty(t) = \bar{\tau}_i^d(t)$. Combining these facts, we obtain

$$\Phi^\infty(t) = \sum_{a\in A}\bar{w}_a t_a - \sum_{\substack{d\in D\\ i\neq d}} g_i^d\bar{\tau}_i^d(t).$$

Multiplying the inequalities $\bar{\tau}_{i_a}^d \leq t_a + \bar{\tau}_{j_a}^d$ by $\widehat{v}_a^d$ and adding over all the arcs $a \notin A_d^+$, we get $\sum_{i\neq d} g_i^d\bar{\tau}_i^d \leq \sum_{a\in A} t_a\widehat{v}_a^d$. Then, summing over all $d \in D$ and assuming that $t \geq 0$, $t \neq 0$, we obtain

$$\sum_{\substack{d\in D\\ i\neq d}} g_i^d\bar{\tau}_i^d(t) \leq \sum_{a\in A} t_a\widehat{w}_a < \sum_{a\in A} t_a\bar{w}_a,$$

which implies that $\Phi^\infty(t) > 0$, as claimed. □

Mixed deterministic/stochastic assignment

The functions $\varphi_i^d \in \mathcal{E}$ in the stochastic model are used to describe the flow distribution rule (2.3), and may differ from node to node: at some nodes one could consider Logit distribution, while other nodes may be governed by probit or other discrete choice models. On the other hand, the deterministic model assumes that the flow entering each node is distributed among optimal arcs, which may also be written in the form (2.3) by taking $\varphi_i^d(z^d) = \max\{z_a^d \,:\, a \in A_i^+\}$ and replacing the gradient by the subdifferential. This further explains the analogy between the characterizations ($\bar{D}$) and (MD), and leads naturally to consider the possibility of a hybrid model where some nodes have a stochastic distribution rule while other nodes are deterministic. The analysis carries over to this more general setting with (MD) characterizing the traffic equilibrium, though $\Phi(\cdot)$ will no longer be smooth.

Simultaneous mode/route choice and elastic demands

Noting that the graph $G = (N, A)$ need not be limited to a single transportation mode, the previous framework turns out to be flexible enough to handle more complex decision processes such as the simultaneous choice of *mode* and *route*. To this end it suffices to apply the model over a multi-modal graph built by connecting every origin and destination to one or several corresponding nodes on the subgraphs representing the basic transportation modes (car, bus, metro, walk, etc.). Combined modes such as car-metro-walk may be easily included by connecting the nodes on the corresponding subgraphs through additional *transfer arcs*. At every origin node one may adopt a particular distribution rule based on a Logit or probit model, while at other nodes (e.g., the metro sub-network) one may use a deterministic rule. A further extension concerns the modelling of elastic demands. The option of not making a trip for a given OD pair may be simulated by adding a no-trip arc which connects directly the origin to the destination, with cost equal to the inverse of the demand function (see [36]).

2.3 Numerical methods

In this section we discuss some numerical methods to solve (MD) and we report a few numerical tests. According to (2.5), the derivatives of the objective function $\Phi(t)$ are given by

$$\frac{\partial \Phi}{\partial t_a} = s_a^{-1}(t_a) - \widehat{w}_a(t)$$

with $\widehat{w}_a(t) = \sum_{d \in D} v_a^d(t)$. The computation of these derivatives requires solving first the system (2.2) in order to find the functions $\tau_i^d(t)$ and then solving (2.3)–(2.4) to find $v_a^d(t)$. As seen from the existence proof of Proposition 2.3, this may be done for each destination $d \in D$ by a fixed point iteration:

$$FP(t) \begin{cases} \text{– Iterate } \tau_i^{d,n+1} = \varphi_i^d(t_a + \tau_{j_a}^{d,n}) \text{ to find estimates } \tau_i^d \sim \tau_i^d(t). \\ \text{– Compute } \frac{\partial \varphi_i^d}{\partial z_a^d}(z^d) \text{ with } z_a^d = t_a + \tau_{j_a}^d. \\ \text{– Build the matrices } \widehat{P}^d \text{ and } \widehat{Q}^d. \\ \text{– Compute } x^d = \left[I - (\widehat{P}^d)'\right]^{-1} g^d \text{ and } v^d = (\widehat{Q}^d)' x^d. \\ \text{– Aggregate } \widehat{w} = \sum_{d \in D} v^d. \end{cases}$$

A gradient method for minimizing $\Phi(\cdot)$ would use $FP(t^k)$ in order to estimate $\widehat{w}^k \sim \widehat{w}(t^k)$ and then use the update

$$t_a^{k+1} = t_a^k - \alpha_k \left[s_a^{-1}(t_a^k) - \widehat{w}_a^k\right] \tag{G$_t$}$$

with α_k a suitably chosen stepsize. Since performing a linesearch is too expensive, we implemented a normalized gradient stepsize by taking $\alpha_k = \lambda_k / \|h^k\|$ with $h_a^k = [s_a^{-1}(t_a^k) - \widehat{w}_a^k]$ and $\lambda_k > 0$ such that $\lambda_k \to 0$ and $\sum_k \lambda_k = \infty$. Although [72, Theorem 2.3] guarantees that $t^k \to t^*$, our numerical tests confirmed the slow convergence expected from a gradient method.

An alternative method is obtained by the change of variables $t_a = s_a(w_a)$, which transforms (MD) into a non-convex optimization problem in the total flow variables w_a, that is to say, minimizing the function $\Psi(w) = \Phi(s(w))$. The derivatives of the latter are given by $\frac{\partial \Psi}{\partial w_a} = [w_a - \widehat{w}_a(s(w))]s_a'(w_a)$, so that the gradient iteration now becomes

$$w_a^{k+1} = w_a^k - \alpha_k \big[w_a^k - \widehat{w}_a^k\big] s_a'\big(w_a^k\big). \tag{$\mathrm{G_w}$}$$

This was also implemented with a normalized stepsize, but the convergence was again too slow. A more efficient variant is the method of successive averages which slightly modifies the latter as

$$w_a^{k+1} = w_a^k - \alpha_k \big[w_a^k - \widehat{w}_a^k\big] \tag{MSA}$$

and which may be interpreted as a variable metric descent method.

In our tests we took $\alpha_k = \max\left\{\frac{1}{k+1}, 0.125\right\}$, which uses the standard MSA stepsize in the early stages after which the stepsize becomes constant. A Matlab implementation was tested on the traffic network of Sioux Falls, a small network of 24 nodes and 76 arcs with travel times $s_a(w_a) = t_a^0[1 + b_a(\frac{w_a}{c_a})^{p_a}]$ for $w_a \geq 0$, where $t_a^0 > 0$ represents the free-flow travel time of the link, b_a is a scaling parameter, c_a is the link capacity, and the exponent p_a is usually taken equal to 4 (data are available for download from `www.bgu.ac.il/~bargera`).

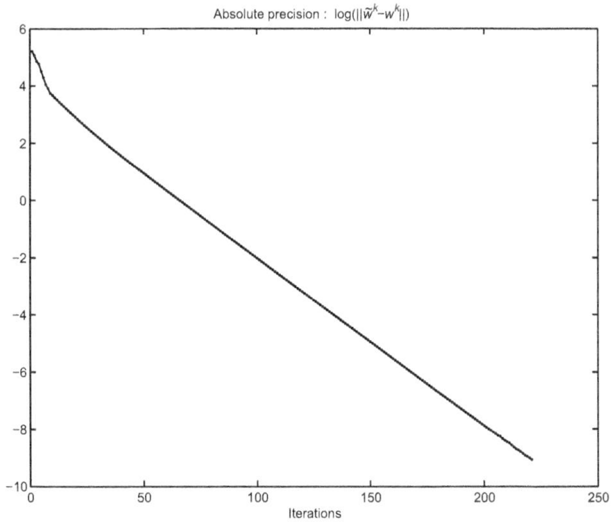

Figure 2.2: MSA iterations.

Note that the expression of $s_a(w_a)$ for $w_a < 0$ is irrelevant since MSA iterates remain nonnegative. A Logit model was used by taking

$$\varphi_i^d(z^d) = -\ln\left[\sum_{a\in A_i^+} \exp(-\beta_i^d z_a^d)\right] / \beta_i^d$$

with $\beta_i^d = 0.5$.

Figure 2.2 plots the absolute precision $\log(\|\widehat{w}^k - w^k\|)$ along the iterations, which exhibits a linear rate of convergence. The method attains high accuracy[1] but the number of iterations is large. However, the cost per iteration is low and the overall CPU time on a 1.6 Ghz processor was 3.7[s]. In order to speed up convergence, the MSA iteration was combined with a Newton method which was activated after MSA reached a 10% relative precision, i.e., $\|\widehat{w}^k - w^k\| \leq 0.1\|w^k\|$. Figure 2.3 illustrates the performance of this variant. The faster convergence compensates the additional work involved in computing Newton's direction, reducing the total CPU time to 0.7[s].

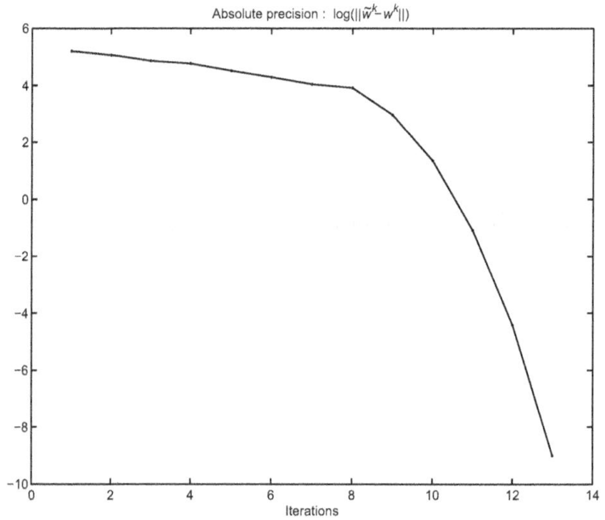

Figure 2.3: MSA-Newton iterations.

An additional test was performed on the larger network of Chicago which comprises 2950 arcs and 933 nodes, among which 387 correspond to destinations. The performance of both methods is very similar to the case of Sioux Falls reaching $\|\widehat{w}^k - w^k\| \leq 10^{-9}$ after 234 iterations and 29[min] of CPU for the case of MSA, and 14 iterations and 11[min] of CPU for Newton. The advantage of Newton's method is less clear for medium precision computations. Furthermore, in contrast

[1]The relative accuracy $\|\widehat{w}^k - w^k\|/\|w^k\|$ is even higher, reaching 10^{-14}.

with Newton's method, MSA may be used to solve mixed stochastic-deterministic models in which the objective function of (MD) is non-smooth.

The most time-consuming part in each iteration $FP(t)$ is the first step, which involves solving approximately the system of equations (2.2). This is done by a standard fixed point iteration which is convergent but may require several steps, especially in the early stages, when we are still far from the equilibrium. In this inner fixed point iteration as well as in the solution of the linear systems that determine x^d and v^d, we fully exploit the sparsity structure of the underlying graph. Sparsity is also crucial when computing Newton's direction, for which we use a conjugate gradient method. The latter does not require explicitly the Hessian H but only to evaluate the product Hx, which can exploit sparsity in order to gain efficiency both in terms of processing time and storage requirements. Indeed, according to the formulas in Section 2.3.2, the product Hx may be computed by solving $2|D|$ sparse linear systems of dimension $|N| \times |N|$ (two systems per destination $d \in D$), while we avoid building the Hessian, whose computation and storage becomes very expensive for large networks.

2.3.1 Convergence of MSA

The following results provide sufficient conditions for the convergence of MSA. We assume that the averaging sequence $\alpha_k > 0$ satisfies $\sum_k \alpha_k = \infty$ and $\sum_k \alpha_k^2 < \infty$, while the functions $s_a \colon \mathbb{R} \to (0, \infty)$ are of class C^1 with $t^0 \in \mathcal{C}$.

Theorem 2.8. *Suppose that the sequence $(w^k)_{k \in \mathbb{N}}$ generated by the MSA iteration $w^{k+1} = (1 - \alpha_k) w^k + \alpha_k \widehat{w}^k$ remains bounded. Then w^k converges to the MTE w^*.*

Proof. Since $(w^{k+1} - w^k)/\alpha_k = -\nabla\Phi(t^k)$, we have the subgradient inequality

$$\Phi(t^k) + \Big\langle \frac{w^{k+1} - w^k}{\alpha_k}, t^k - t^* \Big\rangle \leq \Phi(t^*),$$

which can be written in the form

$$\big[\Phi(t^k) - \Phi(t^*)\big]\alpha_k + \langle \nabla h(w^k), w^{k+1} - w^k \rangle \leq 0 \tag{2.7}$$

by just noting that $t^k - t^* = \nabla h(w^k)$, where $h(w) = \sum_{a \in A} \int_{w_a^*}^{w_a} [s_a(z) - s_a(w_a^*)]\, dz$. We observe that $h(w^*) = 0$ and $h(w) > 0$ for all $w \neq w^*$.

The mean value theorem gives the existence of $\xi^k \in [w^k, w^{k+1}]$ such that

$$h(w^{k+1}) = h(w^k) + \langle \nabla h(w^k), w^{k+1} - w^k \rangle + \tfrac{1}{2} \langle \nabla^2 h(\xi^k)(w^{k+1} - w^k), w^{k+1} - w^k \rangle$$

and, since the w^k's are bounded, we may find $\alpha \geq 0$ such that

$$h(w^{k+1}) \leq h(w^k) + \langle \nabla h(w^k), w^{k+1} - w^k \rangle + \alpha \|w^{k+1} - w^k\|^2.$$

Now, since $t_a^k = s_a(w_a^k)$ is also bounded, we get $\|w^{k+1} - w^k\| = \alpha_k \|\nabla\Phi(t^k)\| \leq \beta\alpha_k$ for some constant $\beta \geq 0$, and using (2.7) we deduce that

$$\big[\Phi(t^k) - \Phi(t^*)\big]\alpha_k + \big[h(w^{k+1}) - h(w^k)\big] \leq \alpha\beta^2\alpha_k^2.$$

In particular, $0 \leq h(w^{k+1}) \leq h(w^k) + \alpha\beta^2\alpha_k^2$, which implies that $h(w^k)$ converges. On the other hand, summing these inequalities we obtain

$$\sum_{k=0}^{\infty} \big[\Phi(t^k) - \Phi(t^*)\big]\alpha_k \leq h(w^0) + \alpha\beta^2 \sum_{k=0}^{\infty} \alpha_k^2 < \infty$$

and then the condition $\sum \alpha_k = \infty$ implies that $\liminf \Phi(t^k) \leq \Phi(t^*)$. Since $\Phi(\cdot)$ is convex with a unique minimum at t^*, it has bounded level sets and then we may extract a subsequence $t^{k_j} \to t^*$. It follows that $w^{k_j} \to w^*$, and therefore $\lim_k h(w^k) = \lim_j h(w^{k_j}) = 0$. As observed above, $h(w) > 0$ for all $w \neq w^*$, from which it follows that $w^k \to w^*$, as claimed. □

Corollary 2.9. *Suppose that the Markov chain for each destination d is supported over an acyclic graph (N, A^d) (see the remark before* Section 2.1.1*). Then $w^k \to w^*$.*

Proof. Since the flows $v_a^d(t)$ satisfy flow conservation and are supported over an acyclic graph, they are bounded. Hence the aggregate flows $\widehat{w}_a(t)$ are also bounded, and then so is the sequence w^k which is obtained by averaging the $\widehat{w}^k$'s. The conclusion then follows from the previous result. □

Theorem 2.10. *Suppose that $s_a(\cdot)$ and $s_a'(\cdot)$ are bounded from above. Then the sequence $(w^k)_{k\in\mathbb{N}}$ generated by the* MSA *iteration $w^{k+1} = (1-\alpha_k)w^k + \alpha_k\widehat{w}^k$ converges to the* MTE *w^*.*

Proof. The proof is similar to the one of Theorem 2.8. In fact, the boundedness of w^k was only used to ensure that $\nabla^2 h(\xi^k)$ and t^k are bounded in order to derive the existence of the constants α and β. Now, since $\nabla^2 h(\xi^k)$ is a diagonal matrix with entries $s_a'(\xi_a^k)$ while $t_a^k = s_a(w_a^k)$, these bounds follow directly from the alternative assumptions of this theorem. □

2.3.2 Hessian calculation

This final section provides analytic expressions for the Hessians of the functions $\Phi(t)$ and $\Psi(w) = \Phi(s(w))$ required to implement Newton's method. We denote

$$B^d(t) = I + J^d \frac{\partial \tau^d}{\partial t},$$

where $J^d = (J_{aj})_{a\in A,\, j\neq d}$ with $J_{aj}^d = 1$ if $j = j_a$ and $J_{aj}^d = 0$ otherwise. Recall from the proof of Theorem 2.5 that

$$\frac{\partial \tau^d}{\partial t} = \big[I - \widehat{P}^d(t)\big]^{-1}\widehat{Q}^d(t).$$

Proposition 2.11. *If the functions $\varphi_i^d(\cdot)$ are of class C^2, then for all $t \in \mathcal{C}$ we have*

$$\nabla^2\Phi(t) = \mathrm{diag}\big[1/s_a'(s_a^{-1}(t_a))\big] - \sum_{d\in D} B^d(t)' \left[\sum_{i\neq d} x_i^d(t)\nabla^2\varphi_i^d(z^d(t))\right] B^d(t).$$

Proof. Setting $\psi^d(t) = \sum_{i\neq d} g_i^d\tau_i^d(t)$ as in the proof of Theorem 2.5, the formula for $\nabla^2\Phi(t)$ follows easily if we show that

$$\nabla^2\psi^d(t) = (B^d)' \left[\sum_{i\neq d} x_i^d\nabla^2\varphi_i^d(z^d)\right] B^d. \tag{2.8}$$

To prove the latter, we exploit the equality $[I-\widehat{P}^d]\frac{\partial\tau^d}{\partial t_a} = \widehat{Q}_{\cdot a}^d$, which gives

$$\frac{\partial^2\tau^d}{\partial t_b\partial t_a} = \big[I-\widehat{P}^d\big]^{-1}\left\{\frac{\partial\widehat{Q}_{\cdot a}^d}{\partial t_b} + \frac{\partial\widehat{P}^d}{\partial t_b}\frac{\partial\tau^d}{\partial t_a}\right\}.$$

Then, using the fact that $x^d = [I-(\widehat{P}^d)']^{-1}g^d$ and the chain rule, we get

$$\begin{aligned}
\frac{\partial^2\psi^d}{\partial t_b\partial t_a} &= (g^d)'\big[I-\widehat{P}^d\big]^{-1}\left\{\frac{\partial\widehat{Q}_{\cdot a}^d}{\partial t_b} + \frac{\partial\widehat{P}^d}{\partial t_b}\frac{\partial\tau^d}{\partial t_a}\right\}\\
&= (x^d)'\left\{\frac{\partial\widehat{Q}_{\cdot a}^d}{\partial t_b} + \frac{\partial\widehat{P}^d}{\partial t_b}\frac{\partial\tau^d}{\partial t_a}\right\}\\
&= \sum_{i\neq d} x_i^d\left\{\frac{\partial Q_{ia}^d}{\partial t_b} + \sum_{j\neq d}\frac{\partial P_{ij}^d}{\partial t_b}\frac{\partial\tau_j^d}{\partial t_a}\right\}\\
&= x_{i_a}^d\frac{\partial}{\partial t_b}\left(\frac{\partial\varphi_{i_a}^d}{\partial z_a^d}\right) + \sum_{i\neq d} x_i^d\sum_{c\in A_i^+}\frac{\partial}{\partial t_b}\left(\frac{\partial\varphi_i^d}{\partial z_c^d}\right)\frac{\partial\tau_{j_c}^d}{\partial t_a}\\
&= \sum_{i\neq d} x_i^d\sum_{c\in A_i^+}\frac{\partial}{\partial t_b}\left(\frac{\partial\varphi_i^d}{\partial z_c^d}\right)\left[\delta_{ca} + \frac{\partial\tau_{j_c}^d}{\partial t_a}\right]\\
&= \sum_{i\neq d} x_i^d\sum_{c,e\in A_i^+}\frac{\partial^2\varphi_i^d}{\partial z_e^d\partial z_c^d}\left[\delta_{eb} + \frac{\partial\tau_{j_e}^d}{\partial t_b}\right]\left[\delta_{ca} + \frac{\partial\tau_{j_c}^d}{\partial t_a}\right],
\end{aligned}$$

where $\delta_{ca} = 1$ if $c = a$ and $\delta_{ca} = 0$ otherwise. This corresponds precisely to (2.8), as was to be proved. □

Using the chain rule and the equality $\nabla\left[\sum_{d\in D}\psi^d(t)\right] = \sum_{d\in D} v^d(t) = \widehat{w}(t)$, we get

Proposition 2.12. *If the functions $\varphi_i^d(\cdot)$ are of class C^2 then we have*

$$\nabla^2\Psi(w) = M(w) + \Gamma(w)\nabla^2\Phi\big(s(w)\big)\Gamma(w)$$

with

$$\begin{aligned} M(w) &= \operatorname{diag}\big[s_a''(w_a)(w_a - \widehat{w}_a(s(w)) : a \in A\big], \\ \Gamma(w) &= \operatorname{diag}\big[s_a'(w_a) : a \in A\big]. \end{aligned}$$

Chapter 3

Adaptive Dynamics in Traffic Games

The equilibrium concepts discussed in the previous chapters were eminently static, with traffic modeled as a steady state that results from the adaptive behavior of drivers. These models assume implicitly the existence of a hidden mechanism that leads to equilibrium, though they are stated directly in the form of equilibrium equations which are not tied to a specific adaptive dynamics. Furthermore, they consider a non-atomic framework that ignores individual drivers and uses continuous variables to represent aggregate flows. In this chapter we discuss a dynamical model for the adaptive behavior of finitely many drivers in a simple network and we study its convergence towards a steady state.

Empirical evidence of some form of adaptive behavior has been reported in [8, 49, 68], based on experiments and simulations of different discrete time adaptive dynamics, though it has been observed that the steady states attained may differ from the standard equilibria and may even depend on the initial conditions. Also, empirical support for the use of discrete choice models in the context of games was given in [57], while the convergence of a class of finite-lag adjustment procedures was established in [23, 24, 29]. On a different direction, several continuous time dynamics describing plausible adaptive mechanisms that converge to Wardrop equilibrium were studied in [41, 67, 73], though these models are of an aggregate nature and are not directly linked to the behavior of individual drivers.

Learning and adaptive behavior in repeated games have been intensively explored in the last decades (see [42, 79]). Instead of studying aggregate dynamics in which the informational and strategic aspects are unspecified, the question is considered from the perspective of individual players that make decisions day after day based on information derived from past observations. The accumulated experience is summarized in a *state variable* that determines the strategic behavior of players through a certain stationary rule. At this level, the most prominent adaptive procedure is *fictitious play*, studied by Brown [22] in the early 50s (see also [63]), which assumes that at each stage players choose a best reply to the observed empirical distribution of past moves of their opponents. A variant called *smooth fictitious play* for games with perturbed payoffs and reminiscent of Logit random choice is discussed in [42, 48].

The assumption that players are able to record the past moves of their opponents is very stringent for games involving many players with limited information. A milder assumption is that players observe only the outcome vector, namely the payoff obtained at every stage and the payoff that would have resulted if a different move had been played. Procedures such as *no-regret* [44, 45], *exponential weight* [40], and *calibration* [38], deal with such limited information contexts assuming that players adjust their behavior based on statistics of past performance. Eventually, adaptation leads to configurations where no player regrets the choices he makes. These procedures, initially conceived for the case when the complete outcome vector is available, were adapted to the case where only the realized payoff can be observed [6, 38, 42, 45]. The idea is to use the observed payoffs to build an unbiased estimator of the outcome vector to which the initial procedure is applied.

Although these procedures are flexible and robust, the underlying rationality may still be too demanding for games with a large number of strategies and poorly informed players. This is the case for traffic where a multitude of *small* players make routing decisions with little information about the strategies of other drivers nor the actual congestion in the network. A simpler adaptive rule which relies only on the past sequence of realized moves and payoffs is *reinforcement dynamics* [7, 10, 19, 33, 53, 61]: players select moves proportionally to a propensity vector built by a cumulative procedure in which the current payoff is added to the component played, while the remaining components are kept unchanged. This mechanism is related to the replicator dynamics [61] and its convergence has been established for zero-sum 2-player games as well as for some games with unique equilibria [10, 53]. We should also mention here the mechanism proposed in [19], which uses an averaging rule with payoff dependent weights.

A similar idea is considered in this chapter. The model assumes that each player has a prior perception or estimate of the payoff performance for each possible move and makes a decision based on this rough information by using a random choice rule such as Logit. The payoff of the chosen alternative is then observed and is used to update the perception for that particular move. This procedure is repeated day after day, generating a discrete time stochastic process which we call the *learning process.* The basic ingredients are therefore: a state parameter; a decision rule from states to actions; and an updating rule on the state space. This structure is common to many procedures in which the incremental information leads to a change in a state parameter that determines the current behavior through a given stationary map. However, the specific dynamics considered here are structurally different from those studied previously, while preserving the qualitative features of *probabilistic choice* and *sluggish adaptation* [79, Section 2.1]. Although players observe only their own payoffs, these values are affected by everybody else's choices revealing information on the game as a whole. The basic question is whether a simple learning mechanism based on such a minimal piece of information may be sufficient to induce coordination and make the system stabilize to an equilibrium.

The chapter follows closely the presentation of [25] and is structured as follows. Section 3.1 describes the *learning process* and the *adaptive dynamics* in the general setting of repeated games, providing sufficient conditions for this process to converge almost surely towards a stationary state which is characterized as an equilibrium for a related limit game. The analysis relies on techniques borrowed from stochastic algorithms (see, e.g., [12, 50]), and proceeds by studying an associated continuous deterministic dynamical system. Under suitable assumptions, the latter has a unique rest point which is a global attractor, from which the convergence of the learning process follows. Section 3.2 particularizes the general model to the case of a simple traffic game on a network with parallel links. In this restricted setting, the convergence of the dynamics depend on a *"viscosity parameter"* that represents the amount of noise in players' choices. If noise is large enough then the learning process and the associated adaptive dynamics have a unique global attractor. A potential function is used to derive an equivalent Lagrangian description of the dynamics, together with alternative characterizations of the rest points. Finally, in the symmetric case when all players are identical it is shown that a unique symmetric equilibrium exists which is moreover a local attractor for the dynamics under a weaker assumption on the viscosity parameter.

This approach proceeds bottom-up from a simple and explicit behavioral rule to equilibrium: a discrete time stochastic model for individual behavior gives rise to an associated continuous time deterministic dynamics which leads ultimately to an equilibrium of a particular limit game. The equilibrium is not postulated a priori but it is derived from basic postulates on player behavior, providing a microeconomic foundation for equilibrium.

3.1 Payoff-based adaptive dynamics

3.1.1 The model

We introduce a dynamical model of adaptive behavior in repeated games, where each player adjusts iteratively her strategy as a function of past payoffs.

Let $\mathcal{P} = \{1, \ldots, N\}$ denote the set of players. Each $i \in \mathcal{P}$ is characterized by a finite set S^i of pure strategies and a payoff function $G^i \colon S^i \times S^{-i} \to \mathbb{R}$ where $S^{-i} = \prod_{j \neq i} S^j$. We denote by Δ^i the set of mixed strategies or probability vectors over S^i, and we set $\Delta = \prod_{i \in \mathcal{P}} \Delta^i$. As usual, we keep the notation G^i for the multilinear extension of payoffs to the set of mixed strategies.

The game is played repeatedly. At stage n, each player $i \in \mathcal{P}$ selects a random move $s^i_n \in S^i$ according to the mixed strategy

$$\pi^i_n = \sigma^i(x^i_n) \in \Delta^i, \tag{3.1}$$

which depends on a vector $x_n^i = (x_n^{is})_{s\in S^i}$ that represents her perception of the payoff performance of the pure strategies available. Here $\sigma^i : \mathbb{R}^{S^i} \to \Delta^i$ is a continuous map from the space of perceptions to the space of mixed strategies, which describes the stationary behavior of player i. We assume throughout that $\sigma^{is}(\cdot)$ is *strictly positive* for all $s \in S^i$.

At the end of the stage, player i observes her own payoff $g_n^i = G^i(s_n^i, s_n^{-i})$, with no additional information about the moves or payoffs of the opponents, and uses this value to adjust her perception of the pure strategy just played and keeping unchanged the perceptions of the remaining strategies, namely

$$x_{n+1}^{is} = \begin{cases} (1-\gamma_n)x_n^{is} + \gamma_n g_n^i & \text{if } s = s_n^i, \\ x_n^{is} & \text{otherwise.} \end{cases}$$

Here γ_n is a sequence of averaging factors with $\gamma_n \in (0,1)$, $\sum_n \gamma_n = \infty$, and $\sum_n \gamma_n^2 < \infty$ (a simple choice is $\gamma_n = \frac{1}{n+1}$). This iteration may be written in vector form as

$$x_{n+1} - x_n = \gamma_n \big[w_n - x_n\big] \tag{3.2}$$

with

$$w_n^{is} = \begin{cases} g_n^i & \text{if } s = s_n^i, \\ x_n^{is} & \text{otherwise.} \end{cases}$$

The distribution of the payoffs and therefore of the random vector w_n is determined by the current perceptions x_n, so that (3.1)–(3.2) yield a Markov process for the evolution of perceptions. It may be interpreted as a process in which players probe the different pure strategies to *learn* about their payoffs, and adapt their behavior according to the accumulated information. The iteration from x_n to x_{n+1} can be decomposed into a chain of elementary steps: the prior perceptions give rise to mixed strategies that lead to moves, which determine the payoffs that are finally used to update the perceptions. Schematically, the procedure looks like: $x_n^{is} \rightsquigarrow \pi_n^{is} \rightsquigarrow s_n^i \rightsquigarrow g_n^i \rightsquigarrow x_{n+1}^{is}$. The information gathered at every stage by each player is very limited – only the payoff of the specific move played at that stage – but it conveys implicit information on the behavior of the rest of the players. The basic question we address is whether this iterative procedure based on a minimal piece of information can lead players to coordinate on a steady state.

Informally, dividing (3.2) by the small parameter γ_n, the iteration may be interpreted as a finite difference Euler scheme for a related differential equation, except that the right-hand side is not deterministic but a random field. Building on this observation, the theory of stochastic algorithms (see, e.g., [12, 50] or Appendix B) establishes a connection between the asymptotics of the discrete time random process (3.2) for $n \to \infty$ and the behavior as $t \to \infty$ of the continuous-time deterministic *averaged* dynamics

$$\frac{dx}{dt} = \mathbb{E}(w|x) - x, \tag{3.3}$$

where $\mathbb{E}(\cdot\,|x)$ stands for the expectation on the distribution of moves induced by the mixed strategies $\sigma^i(x^i)$. In particular, if (3.3) admits a global attractor x^* (in the sense of dynamical systems), then the discrete process (3.2) will also converge to x^* with probability one. This point will be further developed in Section 3.1.3 using Lyapunov function techniques to establish the existence of such an attractor.

Equation (3.3) can be made more explicit. To this end we consider the state space of perceptions $\Omega = \prod_{i\in\mathcal{P}} \mathbb{R}^{S^i}$ and the map $\Sigma\colon \Omega \to \Delta$ that gives the profile of mixed strategies for the players as a function of the state x,

$$\Sigma(x) = \left(\sigma^i(x^i)\right)_{i\in\mathcal{P}}. \tag{3.4}$$

Let also $F\colon \Delta \to \Omega$ be given by $F(\pi) = (F^i(\pi))_{i\in\mathcal{P}}$, where $F^i(\pi) = (F^{is}(\pi))_{s\in S^i}$ is the expected payoff vector of player i, namely

$$F^{is}(\pi) = G^i\left(s, \pi^{-i}\right). \tag{3.5}$$

The latter represents the expected payoff for player i when she chooses $s \in S^i$ and the other players use mixed strategies $\{\pi^j\}_{j\neq i}$. Note that F^i does not depend on π^i and that the expected payoff of player i can be expressed as

$$G^i(\pi) = \left\langle \pi^i, F^i(\pi)\right\rangle.$$

Finally, define the map $C\colon \Omega \to \Omega$ of expected vector payoffs as a function of the state $x \in \Omega$ as

$$C(x) = F(\Sigma(x)). \tag{3.6}$$

Proposition 3.1. *The continuous dynamics* (3.3) *may be expressed as*

$$\frac{dx^{is}}{dt} = \sigma^{is}(x^i)\left[C^{is}(x) - x^{is}\right]. \tag{3.7}$$

Proof. Denoting $\pi = \Sigma(x)$ and using the definition of the random vector w, we compute the expected value $\mathbb{E}(w|x)$ by conditioning on player i's move as

$$\begin{aligned}\mathbb{E}(w^{is}|x) &= \pi^{is}\, G^i\left(s, \pi^{-i}\right) + \left(1 - \pi^{is}\right) x^{is}\\ &= \sigma^{is}\left(x^i\right) C^{is}(x) + \left(1 - \sigma^{is}(x^i)\right) x^{is}\end{aligned}$$

which plugged into (3.3) yields (3.7). □

We call (3.7) the *adaptive dynamics* associated to the *learning process* (3.2). Note that (3.7) is not directly postulated as a mechanism of adaptive behavior, but instead it is an auxiliary construction to help analyze (3.2). These dynamics describe the evolution of perceptions and are not stated in the space of mixed strategies as in other adaptive procedures like fictitious play, nor in the space of correlated strategies such as for no-regret or reinforcement dynamics. Moreover, we recall that fictitious play requires the knowledge of all the past moves of the opponents, while no-regret procedures rely on the knowledge of $G^i(a, s^{-i})$ for all $a \in S^i$. Concerning the cumulative proportional reinforcement rule, the state variable is a vector indexed over the set of pure strategies where each component is the sum of the payoffs obtained when that pure strategy was played, while the decision rule is the normalized state variable, so it corresponds to a different dynamics.

3.1.2 Rest points and perturbed game

According to the general results on stochastic algorithms, the rest points of the continuous dynamics (3.7) are natural candidates to be limit points for the stochastic process (3.2). Since $\sigma^{is}(x^i) > 0$, these rest points are the fixed points of the map $x \mapsto C(x)$. The existence of such fixed points follows easily from Brouwer's theorem if one notes that this map is continuous with bounded range. In this subsection we describe the nature of these rest points focusing on the case where the profile map $\Sigma(x)$ is given by a Logit discrete choice model.

Let $\mathcal{B}$ denote the set of rest points for (3.7). We note that a point $x \in \mathcal{B}$ is completely characterized by its image $\pi = \Sigma(x)$. To see this, it suffices to restate the fixed point equation $x = C(x)$ as a coupled system in (x, π),

$$\begin{cases} \pi = \Sigma(x) \\ x = F(\pi), \end{cases} \tag{3.8}$$

so that for $x \in \mathcal{B}$ the map $x \mapsto \Sigma(x)$ has an inverse given by $\pi \mapsto F(\pi)$. We state this observation in the following

Proposition 3.2. *The map $x \mapsto \Sigma(x)$ is one-to-one over the set $\mathcal{B}$.*

A particular choice for the map $\sigma^i(\cdot)$ is given by the Logit rule

$$\sigma^{is}(x^i) = \frac{\exp(\beta_i x^{is})}{\sum_{a \in S^i} \exp(\beta_i x^{ia})}, \tag{3.9}$$

where the parameter $\beta_i > 0$ has a smoothing effect with $\beta_i \downarrow 0$ leading to a uniform choice while for $\beta_i \uparrow \infty$ the probability concentrates on the pure strategies with higher perceptions. We notice the formal similarity with smooth fictitious play (as well as with the exponential weight algorithm) in which x^{is} is replaced by the average past payoff that would have been produced by the move s and which is unknown in the current framework. A closer connection may be established with the notion of *quantal response equilibria* introduced in [57], which are exactly the π's corresponding to system (3.8), that is to say, the solutions of $\pi = \Sigma(F(\pi))$. As a matter of fact, under the Logit choice formula, the one-to-one correspondence between rest points x and the associated π's allows to establish a link between the set $\mathcal{B}$ and the Nash equilibria for a related N-person game $\mathcal{G}$ defined by strategy sets Δ^i for $i \in \mathcal{P}$ and payoff functions $\mathcal{G}^i : \Delta^i \times \Delta^{-i} \to \mathbb{R}$ given by

$$\mathcal{G}^i(\pi) = \langle \pi^i, F^i(\pi) \rangle - \frac{1}{\beta_i} \sum_{s \in S^i} \pi^{is} \left[\ln \pi^{is} - 1 \right],$$

which is a perturbation of the original game by an entropy term.

Proposition 3.3. *If the maps $\sigma^i(\cdot)$ are given by the Logit rule (3.9) then $\Sigma(\mathcal{B})$ is the set of Nash equilibria of the perturbed game $\mathcal{G}$.*

Proof. A well-known characterization of the Logit probabilities gives

$$\sigma^i(x^i) = \operatorname{argmax}_{\pi^i \in \Delta^i} \langle \pi^i, x^i \rangle - \frac{1}{\beta_i} \sum_{s \in S^i} \pi^{is} \left[\ln \pi^{is} - 1 \right].$$

Setting $x^i = F^i(\pi)$ and since this expression depends only on π^{-i}, Nash equilibria of $\mathcal{G}$ are characterized by $\pi^i = \sigma^i(x^i)$ with $x^i = F^i(\pi)$. But this is precisely (3.8), so that $\Sigma(\mathcal{B})$ is the set of Nash equilibria of $\mathcal{G}$. □

Remark. The previous characterization extends to the case where the maps $\sigma^i(\cdot)$ are given by more general discrete choice models. Namely, suppose that player i selects an element $s \in S^i$ that yields a maximal utility $x^{is} + \varepsilon^{is}$ where ε^{is} are non-atomic random variables (the Logit model corresponds to the case when $\{-\varepsilon^{is}\}_{s \in S^i}$ are independent Gumbel variables with shape parameter β_i). In this more general framework, the probabilities

$$\sigma^{is}(x^i) = \mathbb{P}\big(s \text{ maximizes } x^{is} + \varepsilon^{is}\big)$$

can be expressed as $\sigma^i(x^i) = \nabla \varphi^i(x^i)$ with $\varphi^i(x^i) = \mathbb{E}[\max_{s \in S^i}\{x^{is} + \varepsilon^{is}\}]$, which is smooth and convex. The perturbed payoff functions are now given by

$$\mathcal{G}^i(\pi) = \langle \pi^i, F^i(\pi) \rangle - (\varphi^i)^*(\pi^i),$$

where $(\varphi^i)^*$ denotes the Fenchel conjugate of $\varphi^i(\cdot)$.

3.1.3 Asymptotic convergence of the dynamics

As mentioned earlier, the asymptotics of (3.2) and (3.7) are intimately linked. More precisely, since payoffs are bounded, the same holds for any sequence x_n generated by (3.2), and therefore combining [12, Propositions 4.1 and 4.2] and the Limit Set Theorem [12, Theorem 5.7] it follows that the set of accumulation points of the discrete time random process x_n is almost surely an *internally chain transitive set* (ICT) for the deterministic continuous time dynamics (3.7). The latter is a strong notion of invariant set for dynamical systems, which allows asymptotically vanishing shocks in the dynamics. For the precise definitions and results, which are somewhat technical, we refer to [12] or to Appendix B below. For our purposes it suffices to mention that, according to [12, Corollary 5.4], if (3.7) has a unique rest point $\bar{x}$ which is a global attractor then it is the only ICT and x_n converges to $\bar{x}$ almost surely.

Theorem 3.4. *If $C\colon \Omega \to \Omega$ is a $\|\cdot\|_\infty$-contraction then its unique fixed point $\bar{x} \in \Omega$ is a global attractor for the adaptive dynamics* (3.7) *and the learning process* (3.2) *converges almost surely towards $\bar{x}$.*

Proof. Let $\ell \in [0,1)$ be a Lipschitz constant for $C(\cdot)$. The existence and uniqueness of $\bar{x}$ is clear, while almost sure convergence of (3.2) will follow from the previously cited results in stochastic approximation together with [12, Corollary 6.6], provided that we exhibit a strict Lyapunov function with a unique minimum at $\bar{x}$. We claim that $\Phi(x) = \|x - \bar{x}\|_\infty$ has this property.

Since $\Phi(x(t))$ is the maximum of the smooth functions $\pm(x^{is}(t) - \bar{x}^{is})$, it is absolutely continuous and its derivative coincides with the derivative of one the functions attaining the max. Specifically, let $i \in \mathcal{P}$ and $s \in S^i$ be such that $\Phi(x(t)) = |x^{is}(t) - \bar{x}^{is}|$. If $x^{is}(t) \geq \bar{x}^{is}$, the fixed point property $\bar{x}^{is} = C^{is}(\bar{x})$ gives

$$\begin{aligned}\frac{d}{dt}\left[x^{is}(t) - \bar{x}^{is}\right] &= \sigma^{is}(x^i)\left[C^{is}(x) - C^{is}(\bar{x}) + \bar{x}^{is} - x^{is}\right]\\ &\leq \sigma^{is}(x^i)\left[\ell\|x - \bar{x}\|_\infty + \bar{x}^{is} - x^{is}\right]\\ &= -\sigma^{is}(x^i)\left[1 - \ell\right]\Phi(x),\end{aligned}$$

while a similar argument for the case $x^{is}(t) < \bar{x}^{is}$ then yields

$$\frac{d}{dt}\Phi(x(t)) \leq -\min_{is} \sigma^{is}(x^i(t))\ [1 - \ell]\ \Phi(x(t)).$$

Now, since $C^{is}(x)$ is bounded it follows from (3.7) that the same holds for $x(t)$ and then $\sigma^{is}(x^i(t))$ stays bounded away from 0 so that $\frac{d}{dt}\Phi(x(t)) \leq -\varepsilon\,\Phi(x(t))$ for some $\varepsilon > 0$. This implies that Φ is a Lyapunov function which decreases to 0 exponentially fast along the trajectories of (3.7), and since $\bar{x}$ is the unique point with $\Phi(\bar{x}) = 0$ the conclusion follows. □

It is worth noting that the convergence of the state variables $x_n \to \bar{x}$ for the learning dynamics (3.2) entails the convergence of the corresponding mixed strategies $\pi_n^i = \sigma^i(x_n^i)$ and therefore of the behavior of players.

An explicit condition for $C(\cdot)$ to be a contraction is obtained as follows. Let $\omega = \max_{i\in\mathcal{P}} \sum_{j\neq i} \beta_j$ and take θ an upper bound for the impact over a player's payoff when a single player changes her move, namely

$$\left|G^i(s,u) - G^i(s,v)\right| \leq \theta$$

for each player $i \in \mathcal{P}$, every pure strategy $s \in S^i$, and all pairs $u, v \in S^{-i}$ such that $u^j = v^j$ except for one $j \neq i$.

Proposition 3.5. *Under the Logit rule* (3.9), *if $2\omega\theta < 1$ then $C(\cdot)$ is a $\|\cdot\|_\infty$-contraction.*

Proof. Consider the difference $C^{is}(x)-C^{is}(y) = F^{is}(\Sigma(x))-F^{is}(\Sigma(y))$ for a fixed player $i \in \mathcal{P}$ and pure strategy $s \in S^i$. We may write this difference as a telescopic sum of the terms $\Lambda_j = F^{is}(\pi_j) - F^{is}(\pi_{j-1})$ for $j = 1, \ldots, N$, where

$$\pi_j = \big(\sigma^1(x^1), \ldots, \sigma^j(x^j), \sigma^{j+1}(y^{j+1}), \ldots, \sigma^N(y^N)\big).$$

Since $F^{is}(\pi)$ does not depend on π^i, we have $\Lambda_i = 0$. For the remaining terms, we note that they can be expressed as $\Lambda_j = \langle A_j, \sigma^j(x^j) - \sigma^j(y^j)\rangle$ where for $t \in S^j$ we denote

$$A_j^t = \sum_{u \in S^{-i},\, u^j = t} G^i(s,u) \prod_{k \neq i,j} \pi_j^{ku^k}.$$

Moreover, since both $\sigma^j(x^j)$ and $\sigma^j(y^j)$ belong to the unit simplex Δ^j, we may also write $\Lambda_j = \langle A_j - A_j^r \mathbb{1},\, \sigma^j(x^j) - \sigma^j(y^j)\rangle$ for any fixed $r \in S^j$. It is now easy to see that $|A_j^t - A_j^r| \leq \theta$, and therefore we deduce

$$\big|C^{is}(x) - C^{is}(y)\big| \leq \theta \sum_{j \neq i} \big\|\sigma^j(x^j) - \sigma^j(y^j)\big\|_1. \tag{3.10}$$

Take $w_j \in \mathbb{R}^{S^j}$ with $\|w_j\|_\infty = 1$ and $\|\sigma^j(x^j) - \sigma^j(y^j)\|_1 = \langle w_j, \sigma^j(x^j) - \sigma^j(y^j)\rangle$, so that using the mean value theorem we may find $z^j \in [x^j, y^j]$ with

$$\begin{aligned}\big\|\sigma^j(x^j) - \sigma^j(y^j)\big\|_1 &= \sum_{t \in S^j} w_j^t \big\langle \nabla\sigma^{jt}(z^j), x^j - y^j\big\rangle \\ &\leq \sum_{t \in S^j} \big\|\nabla\sigma^{jt}(z^j)\big\|_1 \big\|x^j - y^j\big\|_\infty.\end{aligned}$$

Using Lemma 3.6 below, together with the fact that $\sigma^j(z^j) \in \Delta^j$, we get

$$\big\|\sigma^j(x^j) - \sigma^j(y^j)\big\|_1 \leq \sum_{t \in S^j} 2\beta_j \sigma^{jt}(z^j) \big\|x^j - y^j\big\|_\infty \leq 2\beta_j \big\|x - y\big\|_\infty,$$

which combined with (3.10) gives finally

$$\big|C^{is}(x) - C^{is}(y)\big| \leq 2\omega\theta \big\|x - y\big\|_\infty. \qquad \square$$

Lemma 3.6. *For each $i \in \mathcal{P}$ and $s \in S^i$ the Logit rule (3.9) satisfies*

$$\big\|\nabla\sigma^{is}(x^i)\big\|_1 = 2\beta_i \sigma^{is}(x^i)\big(1 - \sigma^{is}(x^i)\big).$$

Proof. Let $\pi^i = \sigma^i(x^i)$. A direct computation gives $\frac{\partial\sigma^{is}}{\partial x^{it}} = \beta_i \pi^{is}(\delta_{st} - \pi^{it})$ with $\delta_{st} = 1$ if $s = t$ and $\delta_{st} = 0$ otherwise, from which we get

$$\big\|\nabla\sigma^{is}(x^i)\big\|_1 = \beta_i \pi^{is} \sum_{t \in S^i} \big|\delta_{st} - \pi^{it}\big| = 2\beta_i \pi^{is}\big(1 - \pi^{is}\big). \qquad \square$$

3.2 Application to traffic games

In this section we use the previous framework to model the adaptive behavior of drivers in a congested traffic network. The setting for the *traffic game* is as follows. Every day a set of N users choose one among M alternative routes from a set $\mathcal{R}$. The combined choices of all players determine the total route loads and the corresponding travel times. Each user experiences only the cost of the route chosen on that day and uses this information to adjust the perception for that particular route, affecting the mixed strategy to be played in the next stage.

More precisely, a route $r \in \mathcal{R}$ is characterized by an increasing sequence $c_1^r \leq \cdots \leq c_N^r$ where c_u^r represents the average travel time of the route when it carries a load of u users. The set of pure strategies for each player $i \in \mathcal{P}$ is $S^i = \mathcal{R}$ and if $r_n^j \in \mathcal{R}$ denotes the route chosen by each player j at stage n, then the payoff to player i is given as the negative of the experienced travel time $g_n^i = G^i(r_n) = -c_u^r$ with $r = r_n^i$ and $u = |\{j \in \mathcal{P} : r_n^j = r\}|$.

The evolution of perceptions is governed by (3.2) as in section Section 3.1, while the route r_n^i is randomly chosen according to a mixed strategy $\pi_n^i = \sigma^i(x_n^i)$ which depends on prior perceptions about route payoffs through a Logit model

$$\sigma^{ir}(x^i) = \frac{\exp(\beta_i x^{ir})}{\sum_{a \in \mathcal{R}} \exp(\beta_i x^{ia})}. \tag{3.11}$$

In the model we assume that all users on route r experience exactly the same travel time c_u^r, though the analysis remains unchanged if we merely suppose that each $i \in \mathcal{P}$ observes a random time $\tilde{c}^{ir}$ with conditional expected value c_u^r given the number u of users that choose r. On the other hand, the network topology here is very simple with only a set of parallel routes, and a natural extension is to consider more general networks. Note however that the parallel link structure allows to model more complex decision problems such as the simultaneous choice of route and departure time, by modeling a physical route by a set of parallel links that represent the route at different time windows.

3.2.1 Potential function and global attractor

In the traffic game setting, the vector payoff map $F(\cdot)$ defined by (3.5) can be expressed as the gradient of a potential function[1] which is inspired from Rosenthal [64]. Namely, consider the map $H\colon [0,1]^{\mathcal{P}\times\mathcal{R}} \to \mathbb{R}$ defined by

$$H(\pi) = -\mathbb{E}_\pi^B \left[\sum_{r \in \mathcal{R}} \sum_{u=1}^{U^r} c_u^r \right], \tag{3.12}$$

[1]Although the traffic game is also a *potential game* in the sense of Monderer–Shapley [59], the notion of potential used here simply means that the payoff vector is obtained as the gradient of a real-valued smooth function.

where $\mathbb{E}_\pi^B$ denotes the expectation with respect to the variables $U^r = \sum_{i\in\mathcal{P}} X^{ir}$ with X^{ir} independent non-homogeneous Bernoulli random variables such that $\mathbb{P}(X^{ir} = 1) = \pi^{ir}$.

A technical remark is in order here. We observe that $H(\pi)$ was defined for $\pi \in [0,1]^{\mathcal{P}\times\mathcal{R}}$ and not only for $\pi \in \Delta$, which allows to differentiate H with respect to each variable π^{ir} independently, ignoring the constraints $\sum_{r\in\mathcal{R}} \pi^{ir} = 1$. As a result of this independence assumption, the random variable X^{ir} *cannot* be identified with the indicator Y^{ir} of the event "player i chooses route r", for which we do have these constraints: each player $i \in \mathcal{P}$ must choose one and only one route $r \in \mathcal{R}$, so that the family $\{Y^{ir}\}_{r\in\mathcal{R}}$ is not independent. However, since player i chooses his route independently from other players, once a route $r \in \mathcal{R}$ is *fixed* we do have that $\{Y^{kr}\}_{k\neq i}$ is an independent family, so that the distinction between X^{kr} and Y^{kr} becomes superfluous when computing expected payoffs as shown next.

Lemma 3.7. *For any given $r \in \mathcal{R}$ and $i \in \mathcal{P}$ let $U_i^r = \sum_{k\neq i} X^{kr}$ with X^{kr} independent non-homogeneous Bernoulli's such that $\mathbb{P}(X^{kr} = 1) = \pi^{kr}$. Then*

$$F^{ir}(\pi) = -\mathbb{E}_\pi^B\big[c^r_{U^r} \mid X^{ir} = 1\big] = -\mathbb{E}_\pi^B\big[c^r_{U_i^r+1}\big]. \tag{3.13}$$

Proof. By definition we have $F^{ir}(\pi) = -\mathbb{E}[c^r_{V_i^r+1}]$ with the expectation taken with respect to the random variable $V_i^r = \sum_{k\neq i} Y^{kr}$ where Y^{kr} denotes the indicator of the event "player k chooses route r". Since r is fixed, the variables $\{Y^{kr}\}_{k\neq i}$ are independent Bernoulli's with $\mathbb{P}(Y^{kr} = 1) = \pi^{kr}$, so we may replace them by X^{kr}, that is to say

$$F^{ir}(\pi) = -\mathbb{E}_\pi^B\big[c^r_{U_i^r+1}\big] = -\mathbb{E}_\pi^B\big[c^r_{U^r} \mid X^{ir} = 1\big]. \qquad \square$$

We deduce from this property that H is a potential.

Proposition 3.8. $F(\pi) = \nabla H(\pi)$ *for all* $\pi \in \Delta$.

Proof. We note that $H(\pi) = -\sum_{r\in\mathcal{R}} \mathbb{E}_\pi^B[\sum_{u=1}^{U^r} c_u^r]$, so that conditioning on the variables $\{X^{ir}\}_{r\in\mathcal{R}}$ we get

$$H(\pi) = -\sum_{r\in\mathcal{R}} \left[\pi^{ir}\,\mathbb{E}_\pi^B\left(\sum_{u=1}^{U_i^r+1} c_u^r\right) + (1-\pi^{ir})\,\mathbb{E}_\pi^B\left(\sum_{u=1}^{U_i^r} c_u^r\right)\right],$$

which combined with (3.13) yields

$$\frac{\partial H}{\partial \pi^{ir}}(\pi) = \mathbb{E}_\pi^B\left[\sum_{u=1}^{U_i^r} c_u^r\right] - \mathbb{E}_\pi^B\left[\sum_{u=1}^{U_i^r+1} c_u^r\right] = -\mathbb{E}_\pi^B\big[c^r_{U_i^r+1}\big] = F^{ir}(\pi). \qquad \square$$

Using formula (3.13), or equivalently Proposition 3.8, we may obtain the Lipschitz estimates required to study the convergence of the learning process. In particular we note that $F^{ir}(\pi)$ turns out to be a symmetric polynomial in the variables $(\pi^{kr})_{k\neq i}$ only, and does not depend on the probabilities with which the users choose other routes. This allows us to obtain sufficient conditions that are tighter than the one derived from Proposition 3.5. The following results are expressed in terms of a parameter that measures the congestion effect produced by an additional user, namely

$$\delta = \max\left\{c_u^r - c_{u-1}^r : r \in \mathcal{R}; u = 2, \ldots, N\right\}. \tag{3.14}$$

Lemma 3.9. *The second derivatives of H are all zero except for*

$$\frac{\partial^2 H}{\partial \pi^{jr} \partial \pi^{ir}}(\pi) = \mathbb{E}_\pi^B\left[c_{U_{ij}^r+1}^r - c_{U_{ij}^r+2}^r\right] \in \left[-\delta, 0\right] \tag{3.15}$$

with $j \neq i$, where $U_{ij}^r = \sum_{k\neq i,j} X^{kr}$.

Proof. We just noted that $\frac{\partial H}{\partial \pi^{ir}}(\pi) = \mathbb{E}_\pi^B[-c_{U_i^r+1}^r]$ depends only on $(\pi^{kr})_{k\neq i}$. Also, conditioning on X^{jr} we get

$$\frac{\partial H}{\partial \pi^{ir}}(\pi) = \pi^{jr}\mathbb{E}_\pi^B\left[-c_{U_{ij}^r+2}^r\right] + (1-\pi^{jr})\mathbb{E}_\pi^B\left[-c_{U_{ij}^r+1}^r\right],$$

from which (3.15) follows at once. □

As a corollary of these estimates and Theorem 3.4, we obtain the following global convergence result. Recall that $\omega = \max_{i\in\mathcal{P}} \sum_{j\neq i} \beta_j$.

Theorem 3.10. *Assume in the traffic game that $\omega\delta < 2$. Then the corresponding adaptive dynamics* (3.7) *has a unique rest point $\bar{x}$ which is a global attractor and the process* (3.2) *converges almost surely to $\bar{x}$.*

Proof. Proposition 3.8 gives $C^{ir}(x) = F^{ir}(\Sigma(x)) = \frac{\partial H}{\partial \pi^{ir}}(\Sigma(x))$, and since $\frac{\partial H}{\partial \pi^{ir}}(\pi)$ depends only on $\{\pi^{kr}\}_{k\neq i}$, using Lemma 3.9 we deduce that

$$\left|C^{ir}(x) - C^{ir}(y)\right| = \left|\frac{\partial H}{\partial \pi^{ir}}(\Sigma(x)) - \frac{\partial H}{\partial \pi^{ir}}(\Sigma(y))\right| \leq \delta \sum_{j\neq i} \left|\sigma^{jr}(x^j) - \sigma^{jr}(y^j)\right|.$$

Now Lemma 3.6 and the inequality $\sigma(1-\sigma) \leq \frac{1}{4}$ give us

$$\left|\sigma^{jr}(x^j) - \sigma^{jr}(y^j)\right| \leq \frac{1}{2}\beta_j\left\|x^j - y^j\right\|_\infty \leq \frac{1}{2}\beta_j\left\|x - y\right\|_\infty,$$

which, combined with the previous estimate, yields

$$\left|C^{ir}(x) - C^{ir}(y)\right| \leq \frac{1}{2}\delta\sum_{j\neq i}\beta_j \left\|x-y\right\|_\infty \leq \frac{1}{2}\omega\delta\left\|x-y\right\|_\infty.$$

Thus $C(\cdot)$ is a $\|\cdot\|_\infty$-contraction and we conclude using Theorem 3.4. □

To interpret this result, we note that the β_i's in the Logit formula are inversely proportional to the standard deviation of the random terms in the discrete choice model. Thus, the condition $\omega\delta < 2$ requires either a weak congestion effect (small δ) or a sufficiently large noise (small ω). Although this condition is sharper than the one obtained in Proposition 3.5, the parameter ω involves sums of β_i's, so that it becomes more and more stringent as the number of players increases. In the sequel we show that uniqueness of the rest point still holds under the much weaker condition $\beta_i\delta < 1$ for all $i \in \mathcal{P}$, and even for $\beta_i\delta < 2$ in the case of linear costs ($c_u^r = a^r + \delta^r u$) or when players are symmetric ($\beta_i \equiv \beta$).

It is important to note that at lower noise levels (large β_i's) player behavior becomes increasingly deterministic and multiple pure equilibria will coexist, as in the case of the market entry game proposed by Selten and Güth [69], which in our setting corresponds to the special case of two roads with linear costs and symmetric players. For this game, two alternative learning dynamics were analyzed by Duffy and Hopkins [31]: a proportional reinforcement rule and a Logit rule based on hypothetical reinforcement (which requires further information about the opponents' moves). In the first case, convergence to a pure Nash equilibrium is established, while in the second the analysis is done at small noise levels proving convergence towards a perturbed pure Nash equilibrium.

3.2.2 Lagrangian description of the dynamics

The potential function $H(\cdot)$ allows to rewrite the dynamics (3.7) in several alternative forms. A straightforward substitution yields

$$\dot{x}^{ir} = \sigma^{ir}(x^i)\left[\frac{\partial H}{\partial \pi^{ir}}(\Sigma(x)) - x^{ir}\right], \tag{3.16}$$

so that defining

$$\Psi(\pi) = H(\pi) - \sum_{i\in\mathcal{P}} \frac{1}{\beta_i} \sum_{r\in\mathcal{R}} \pi^{ir}\left[\ln\left(\pi^{ir}\right) - 1\right],$$

$$\lambda^i(x^i) = \frac{1}{\beta_i}\ln\left(\sum_{r\in\mathcal{R}} \exp\left(\beta_i x^{ir}\right)\right),$$

one may also put it as

$$\dot{x}^{ir} = \sigma^{ir}(x^i)\left[\frac{\partial \Psi}{\partial \pi^{ir}}(\Sigma(x)) - \lambda^i(x^i)\right]. \tag{3.17}$$

Now, setting $\mu^i = \lambda^i(x^i)$ we have $\sigma^{ir}(x^i) = \bar{\pi}^{ir}(x^{ir}, \mu^i) \triangleq \exp[\beta_i(x^{ir} - \mu^i)]$. If instead of considering μ^i as a function of x^i we treat it as an independent variable, we find $\frac{\partial \bar{\pi}^{ir}}{\partial x^{ir}} = \beta_i \bar{\pi}^{ir}$, and then, introducing the Lagrangians

$$\mathcal{L}(\pi; \mu) = \Psi(\pi) - \sum_{i\in\mathcal{P}} \mu^i \left[\sum_{r\in\mathcal{R}} \pi^{ir} - 1\right], \qquad L(x; \mu) = \mathcal{L}(\bar{\pi}(x, \mu); \mu),$$

we may rewrite the adaptive dynamics in gradient form:

$$\dot{x}^{ir} = \frac{1}{\beta_i}\frac{\partial L}{\partial x^{ir}}(x;\lambda(x)). \tag{3.18}$$

Alternatively, we may differentiate $\mu^i = \lambda^i(x^i)$ in order to get

$$\dot{\mu}^i = \sum_{r\in\mathcal{R}} \bar{\pi}^{ir}(x^{ir},\mu^i)\,\dot{x}^{ir} = \frac{1}{\beta_i}\sum_{r\in\mathcal{R}} \bar{\pi}^{ir}(x^{ir},\mu^i)\frac{\partial L}{\partial x^{ir}}(x;\mu),$$

which may be integrated back to yield $\mu^i = \lambda^i(x^i)$ as unique solution, so that (3.18) is also equivalent to the system of coupled differential equations

$$\begin{cases} \dot{x}^{ir} = \dfrac{1}{\beta_i}\dfrac{\partial L}{\partial x^{ir}}(x;\mu) \\ \dot{\mu}^i = \dfrac{1}{\beta_i}\displaystyle\sum_{r\in\mathcal{R}} \bar{\pi}^{ir}(x^{ir},\mu^i)\frac{\partial L}{\partial x^{ir}}(x;\mu). \end{cases} \tag{3.19}$$

Finally, all these dynamics may also be expressed in terms of the evolution of the probabilities π^{ir} as

$$\begin{cases} \dot{\pi}^{ir} = \beta_i\pi^{ir}\left[\pi^{ir}\dfrac{\partial\mathcal{L}}{\partial\pi^{ir}}(\pi;\mu) - \dot{\mu}^i\right] \\ \dot{\mu}^i = \displaystyle\sum_{r\in\mathcal{R}} (\pi^{ir})^2\frac{\partial\mathcal{L}}{\partial\pi^{ir}}(\pi;\mu). \end{cases} \tag{3.20}$$

We stress that (3.16), (3.17), (3.18), (3.19) and (3.20) are equivalent ways to describe the adaptive dynamics (3.7), so they provide alternative means for studying the convergence of the learning process. In particular, (3.17) may be interpreted as a gradient flow for finding critical points of the functional Ψ on the product of the unit simplices defined by $\sum_{r\in\mathcal{R}}\pi^{ir} = 1$ (even if the dynamics are in the x-space), while (3.20) can be seen as a dynamical system that searches for saddle points of the Lagrangian $\mathcal{L}$ with the variables μ^i playing the role of multipliers. We show next that these critical points are closely related to the rest points of the adaptive dynamics.

Proposition 3.11. *Let $x \in \Omega$ and $\pi = \Sigma(x)$. The following are equivalent:*

(a) $x \in \mathcal{B}$,

(b) $\nabla_x L(x,\mu) = 0$ *for* $\mu = \lambda(x)$,

(c) π *is a Nash equilibrium of the game* $\mathcal{G}$,

(d) $\nabla_\pi \mathcal{L}(\pi,\mu) = 0$ *for some* $\mu \in \mathbb{R}^N$,

(e) π *is a critical point of* Ψ *on* $\Delta(\mathcal{R})^N$, *i.e.,* $\nabla\Psi(\pi) \perp \Delta_0^N$ *where* Δ_0 *is the tangent space to* $\Delta(\mathcal{R})$, *namely* $\Delta_0 = \{z \in \mathbb{R}^M : \sum_{r\in\mathcal{R}} z^r = 0\}$.

Proof. The equivalence (a) $\Longleftrightarrow$ (b) is obvious if we note that (3.7) and (3.18) describe the same dynamics, while (a) $\Longleftrightarrow$ (c) was proved in Proposition 3.3. The equivalence (d) $\Longleftrightarrow$ (e) is also straightforward. For (a) $\Longleftrightarrow$ (d) we observe that the vector μ in (d) is uniquely determined: indeed, the condition $\nabla_\pi \mathcal{L}(\pi,\mu)=0$ gives $\frac{\partial H}{\partial \pi^{ir}}(\pi)-\frac{1}{\beta_i}\ln(\pi^{ir})=\mu^i$, so that setting $x=\nabla H(\pi)$ we get $\pi^{ir}=\exp[\beta_i(x^{ir}-\mu^i)]$ and since $\pi^i\in\Delta(\mathcal{R})$ we deduce $\mu^i=\lambda^i(x^i)$. From this observation it follows that (d) may be equivalently expressed by the equations $x=\nabla H(\pi)$ and $\pi=\Sigma(x)$, which is precisely (3.8) and therefore (a) $\Longleftrightarrow$ (d). □

When the quantities $\beta_i\delta$ are small, the function Ψ turns out to be concave so we may add another characterization of the equilibria and a weaker alternative condition for uniqueness.

Proposition 3.12. *Let $\beta=\max_{i\in\mathcal{P}}\beta_i$. If $\beta\delta<1$ then Ψ is strongly concave with parameter $\frac{1}{\beta}-\delta$ and attains its maximum at a unique point $\bar\pi\in\Delta$. This point $\bar\pi$ is the only Nash equilibrium of the game $\mathcal{G}$ while $\bar x=F(\bar\pi)$ is the corresponding unique rest point of the adaptive dynamics* (3.7).

Proof. It suffices to prove that $h'\nabla^2\Psi(\pi)h\le-(\frac{1}{\beta}-\delta)\|h\|^2$ for all $h\in\Delta_0^N$. Using Lemma 3.9, we get

$$h'\nabla^2\Psi(\pi)h=\sum_{r\in\mathcal{R}}\left[\sum_{i\neq j}h^{ir}h^{jr}\mathbb{E}_\pi^B\left[c^r_{U^r_{ij}+1}-c^r_{U^r_{ij}+2}\right]-\sum_i\frac{1}{\beta_i\pi^{ir}}\left(h^{ir}\right)^2\right]. \tag{3.21}$$

Setting $Z^{ir}=v^{ir}X^{ir}$ with $v^{ir}=h^{ir}/\pi^{ir}$, and $\delta^r_u=(c^r_u-c^r_{u-1})$ with $\delta^r_0=\delta^r_1=0$, this may be rewritten as

$$\begin{aligned}
h'\nabla^2\Psi(\pi)h&=\sum_{r\in\mathcal{R}}\left[\sum_{i\neq j}v^{ir}v^{jr}\pi^{ir}\pi^{jr}\mathbb{E}_\pi^B\left[c^r_{U^r_{ij}+1}-c^r_{U^r_{ij}+2}\right]-\sum_i\frac{\pi^{ir}}{\beta_i}\left(v^{ir}\right)^2\right]\\
&=\sum_{r\in\mathcal{R}}\mathbb{E}_\pi^B\left[\sum_{i\neq j}Z^{ir}Z^{jr}\left(c^r_{U^r-1}-c^r_{U^r}\right)-\sum_i\frac{1}{\beta_i}\left(Z^{ir}\right)^2\right]\\
&\le\sum_{r\in\mathcal{R}}\mathbb{E}_\pi^B\left[-\delta^r_{U^r}\sum_{i\neq j}Z^{ir}Z^{jr}-\frac{1}{\beta}\sum_i\left(Z^{ir}\right)^2\right]\\
&=\sum_{r\in\mathcal{R}}\mathbb{E}_\pi^B\left[-\delta^r_{U^r}\left(\sum_i Z^{ir}\right)^2-\left(\frac{1}{\beta}-\delta^r_{U^r}\right)\sum_i\left(Z^{ir}\right)^2\right].
\end{aligned}$$

The conclusion follows by neglecting the first term in the latter expectation and noting that $\delta^r_{U^r}\le\delta$ while $\mathbb{E}[(Z^{ir})^2]=(h^{ir})^2/\pi^{ir}\ge(h^{ir})^2$. □

When the costs c_u^r are linear, the following sharper estimate holds.

Proposition 3.13. *Suppose that the route costs are linear:* $c_u^r = a^r + \delta^r u$. *Let* $\beta = \max_{i\in\mathcal{P}} \beta_i$ *and* δ *given by* (3.14). *If* $\beta\delta < 2$ *then the function* $\Psi(\cdot)$ *is strongly concave on the space* $\Delta(\mathcal{R})^N$ *with parameter* $\frac{2}{\beta} - \delta$.

Proof. Under the linearity assumption, equation (3.21) gives

$$\begin{aligned} h'\nabla^2\Psi(\pi)h &= -\sum_{r\in\mathcal{R}}\left[\delta^r\sum_{i\neq j} h^{ir}h^{jr} + \sum_i \frac{1}{\beta_i\pi^{ir}}\left(h^{ir}\right)^2\right] \\ &= -\sum_{r\in\mathcal{R}}\left[\delta^r\left\{\left(\sum_i h^{ir}\right)^2 - \sum_i \left(h^{ir}\right)^2\right\} + \sum_i \frac{1}{\beta_i\pi^{ir}}\left(h^{ir}\right)^2\right] \\ &\leq \sum_{r\in\mathcal{R}}\left[\delta\sum_i \left(h^{ir}\right)^2 - \frac{1}{\beta}\sum_i \frac{1}{\pi^{ir}}\left(h^{ir}\right)^2\right]. \end{aligned}$$

Maximizing this latter expression with respect to the variables $\pi^{ir} \geq 0$ under the constraints $\sum_r \pi^{ir} = 1$, we get

$$\begin{aligned} h'\nabla^2\Psi(\pi)h &\leq \delta\sum_i\sum_r \left(h^{ir}\right)^2 - \frac{1}{\beta}\sum_i\left(\sum_r \left|h^{ir}\right|\right)^2 \\ &= \sum_i\left[\delta\|h^i\|_2^2 - \frac{1}{\beta}\|h^i\|_1^2\right]. \end{aligned}$$

Now if we restrict to vectors $h = (h^i)_{i\in\mathcal{P}}$ in the tangent space Δ_0^N, that is to say, $\sum_r h^{ir} = 0$, we may use the inequality $\|h^i\|_1 \geq \sqrt{2}\|h^i\|_2$ to conclude that

$$h'\nabla^2\Psi(\pi)h \leq \sum_i\left[\delta\|h^i\|_2^2 - \frac{2}{\beta}\|h^i\|_2^2\right] = -\left(\frac{2}{\beta} - \delta\right)\|h\|_2^2. \qquad \square$$

The characterization of $\bar{\pi}$ as a maximizer suggests that Ψ might provide an alternative Lyapunov function to study the asymptotic convergence under weaker assumptions than Theorem 3.10. Unfortunately, numerical simulations show that neither the energy $\Psi(\pi)$ nor the potential $H(\pi)$ increase along the trajectories of (3.7), at least initially. However they do decrease for large t and therefore they may eventually serve as local Lyapunov functions near $\bar{\pi}$.

3.2.3 The case of symmetric players

In this final section we consider the case in which all players are identical with $\beta_i \equiv \beta$ for all $i \in \mathcal{P}$. We denote by $\sigma(\cdot)$ the common Logit function (3.9). Under these circumstances, one might expect rest points to be also symmetric with all players sharing the same perceptions: $\bar{x}^i = \bar{x}^j$ for all $i, j \in \mathcal{P}$. This is indeed the case when $\beta\delta$ is small, but beyond a certain threshold there is a multiplicity of rest points all of which except for one are non-symmetric.

Lemma 3.14. *For all $x, y \in \Omega$, each $i, j \in \mathcal{P}$ and every $r \in \mathcal{R}$, we have*

$$\left|C^{ir}(x) - C^{jr}(x)\right| \le \tfrac{1}{2}\beta\delta \left\|x^i - x^j\right\|_\infty. \tag{3.22}$$

Proof. We observe that the only difference between F^{ir} and F^{jr} is an exchange of π^{ir} and π^{jr}. Thus, Proposition 3.8 and Lemma 3.9 combined imply that $|F^{ir}(\pi) - F^{jr}(\pi)| \le \delta|\pi^{ir} - \pi^{jr}|$ and then (3.22) follows from the equality $C(x) = F(\Sigma(x))$ and Lemma 3.6. □

Theorem 3.15. *If $\beta_i \equiv \beta$ for all $i \in \mathcal{P}$, then the adaptive dynamics (3.7) has exactly one symmetric rest point $\widehat{x} = (\widehat{y}, \ldots, \widehat{y})$. Moreover, if $\beta\delta < 2$ then every rest point is symmetric (thus unique).*

Proof. Existence. Let T be the continuous map from the cube $\prod_{r\in\mathcal{R}}[-c^r_N, -c^r_1]$ to itself that maps y to $T(y) = (T^r(y))_{r\in\mathcal{R}}$ where $T^r(y) = C^{ir}(y, \ldots, y)$. Brouwer's theorem implies the existence of a fixed point $\widehat{y}$, so that setting $\widehat{x} = (\widehat{y}, \ldots, \widehat{y})$ we get a symmetric rest point for (3.7).

Uniqueness. Suppose that $\widehat{x} = (\widehat{y}, \ldots, \widehat{y})$ and $\tilde{x} = (\tilde{y}, \ldots, \tilde{y})$ are two distinct symmetric rest points and assume with no loss of generality that the set $\mathcal{R}^+ = \{r \in \mathcal{R} : \tilde{y}^r < \widehat{y}^r\}$ is nonempty. Let $\mathcal{R}^- = \mathcal{R} \setminus \mathcal{R}^+$. The fixed point condition gives $C^{ir}(\tilde{x}) < C^{ir}(\widehat{x})$ for all $r \in \mathcal{R}^+$, and, since F^{ir} is decreasing with respect to the probabilities π^{jr}, we deduce that $\sigma^r(\tilde{y}) > \sigma^r(\widehat{y})$. Summing over all $r \in \mathcal{R}^+$ and setting $Q(z) = \left[\sum_{a\in\mathcal{R}^-} e^{\beta z^a}\right] / \left[\sum_{a\in\mathcal{R}^+} e^{\beta z^a}\right]$, we get

$$\frac{1}{1 + Q(\tilde{y})} = \sum_{r\in\mathcal{R}^+} \sigma^r(\tilde{y}) > \sum_{r\in\mathcal{R}^+} \sigma^r(\widehat{y}) = \frac{1}{1 + Q(\widehat{y})}$$

and therefore $Q(\widehat{y}) > Q(\tilde{y})$. However, $e^{\beta\widehat{y}^r} > e^{\beta\tilde{y}^r}$ for $r \in \mathcal{R}^+$ and $e^{\beta\widehat{y}^r} \le e^{\beta\tilde{y}^r}$ for $r \in \mathcal{R}^-$, so that $Q(\widehat{y}) < Q(\tilde{y})$, which yields a contradiction.

Symmetry. Suppose next that $\beta\delta < 2$ and let x be any rest point. For any two players $i, j \in \mathcal{P}$ and all routes $r \in \mathcal{R}$, property (3.22) gives

$$\left|x^{ir} - x^{jr}\right| = \left|C^{ir}(x) - C^{jr}(x)\right| \le \tfrac{1}{2}\beta\delta\left\|x^i - x^j\right\|_\infty$$

and then $\|x^i - x^j\|_\infty \le \frac{1}{2}\beta\delta\|x^i - x^j\|_\infty$, which implies $x^i = x^j$. □

Corollary 3.16. *If $\beta_i \equiv \beta$ for all $i \in \mathcal{P}$, then the game $\mathcal{G}$ has a unique symmetric equilibrium. Moreover, if $\beta\delta < 2$ then every equilibrium is symmetric (hence unique).*

The existence of a symmetric rest point requires not only that players be identical in terms of the β_i's but also with respect to payoffs. If these payoffs are given by $C^{ir}(x) = C^r(x) + \alpha^{ir}$ where $C^r(x)$ is a common value which depends only on the number of players that use route r and α^{ir} is a user specific value, then symmetry may be lost.

Going back to the stability of rest points, we observe that the condition $\omega\delta < 2$ in Theorem 3.10 becomes more and more stringent as the number of players grows: for identical players this condition reads $\beta\delta < \frac{2}{N-1}$. Now, since $\beta\delta < 2$ already guarantees a unique rest point $\widehat{x}$, one may expect that this remains a global attractor under this weaker condition. Although numerical experiments confirm this conjecture, we have only been able to prove that $\widehat{x}$ is a *local* attractor. Unfortunately this does not allow to conclude the almost sure convergence of the learning process (3.2).

Theorem 3.17. *If $\beta_i \equiv \beta$ for all $i \in \mathcal{P}$ with $\beta\delta < 2$, then the unique rest point $\widehat{x} = (\widehat{y}, \ldots, \widehat{y})$ is symmetric and a local attractor for the dynamics* (3.7).

Proof. We will prove that

$$\Phi(x) = \max\left\{\max_{i,j} \left\|x^i - x^j\right\|_\infty, \frac{1}{N-1}\max_i \left\|x^i - \widehat{y}\right\|\right\}$$

is a local Lyapunov function. More precisely, fix any $\varepsilon > 0$ and choose a lower bound $\bar{\pi} \leq \pi^{ir} := \sigma^r(x^i)$ over the compact set $S_\varepsilon = \{\Phi \leq \varepsilon\}$. This set is a neighborhood of $\widehat{x}$ since the minimum of Φ is attained at $\Phi(\widehat{x}) = 0$. Now set $\alpha = \frac{1}{2}\beta\delta$ and $b = \bar{\pi}[1-\alpha]$, and reduce ε so that $\beta|C^{jr}(x) - x^{jr}| \leq b$ for all $x \in S_\varepsilon$. We claim that S_ε is invariant for the dynamics with $\rho(t) = \Phi(x(t))$ decreasing to 0. To this end we show that $\dot{\rho}(t) \leq -\frac{b}{2}\rho(t)$. We compute $\dot{\rho}(t)$ distinguishing three cases.

Case 1: $x^{ir} - x^{jr} = \rho(t)$
A simple manipulation using (3.7) gives

$$\dot{x}^{ir} - \dot{x}^{jr} = -\pi^{ir}\rho(t) + \pi^{ir}\left[C^{ir}(x) - C^{jr}(x)\right] + \left(\pi^{ir} - \pi^{jr}\right)\left[C^{jr}(x) - x^{jr}\right],$$

so that (3.22) implies

$$\dot{x}^{ir} - \dot{x}^{jr} \leq -\pi^{ir}\left[1-\alpha\right]\rho(t) + \frac{1}{2}\beta\left|C^{jr}(x) - x^{jr}\right|\rho(t) \leq -\tfrac{b}{2}\,\rho(t).$$

Case 2: $\frac{1}{N-1}(x^{ir} - \widehat{y}^r) = \rho(t)$
In this case we have $x^{ia} - \widehat{y}^a \leq x^{ir} - \widehat{y}^r$ for all $a \in \mathcal{R}$, so that

$$\sigma^r(x^i) = \left[\sum_{a\in\mathcal{R}} e^{\beta(x^{ia} - x^{ir})}\right]^{-1} \geq \left[\sum_{a\in\mathcal{R}} e^{\beta\left(\widehat{y}^a - \widehat{y}^r\right)}\right]^{-1} = \sigma^r(\widehat{y}),$$

which then implies

$$C^{ir}(x^i, \ldots, x^i) \leq C^{ir}(\widehat{y}, \ldots, \widehat{y}) = \widehat{y}^r = x^{ir} - (N-1)\rho(t).$$

On the other hand, using Lemmas 3.9 and 3.6, we have

$$\left|C^{ir}(x) - C^{ir}(x^i, \ldots, x^i)\right| \leq (N-1)\alpha \max_{j\neq i}\left\|x^j - x^i\right\|_\infty \leq (N-1)\alpha\,\rho(t), \quad (3.23)$$

so that $C^{ir}(x) - x^{ir} \leq -(N-1)[1-\alpha]\rho(t)$ and therefore

$$\frac{d}{dt}\left[\frac{1}{N-1}\left(x^{ir} - \widehat{y}^r\right)\right] \leq -\pi^{ir}\left[1-\alpha\right]\rho(t) \leq -\tfrac{b}{2}\,\rho(t).$$

CASE 3: $\frac{1}{N-1}(\widehat{y}^r - x^{ir}) = \rho(t)$
Similarly to the previous case, we now have $\sigma^r(x^i) \leq \sigma^r(\widehat{y})$, so that

$$C^{ir}\left(x^i,\ldots,x^i\right) \geq C^{ir}\left(\widehat{y},\ldots,\widehat{y}\right) = \widehat{y}^r = x^{ir} + (N-1)\rho(t).$$

This, combined with (3.23), gives $x^{ir} - C^{ir}(x) \leq -(N-1)[1-\alpha]\rho(t)$ and then, as in the previous case, we deduce that

$$\frac{d}{dt}\left[\frac{1}{N-1}\left(\widehat{y}^r - x^{ir}\right)\right] \leq -\tfrac{b}{2}\,\rho(t).$$

□

This last result shows that $\widehat{x}$ is a local attractor when $\beta\delta < 2$. As a matter of fact, numerical simulations suggest that (3.7) always converges towards $\widehat{x}$, so we conjecture that it is a global attractor. More generally, the simulations show that even for values $\beta\delta > 2$ the continuous dynamics converge towards an equilibrium, although there is a bifurcation value beyond which the symmetric equilibrium becomes unstable and convergence occurs towards one of the multiple non-symmetric equilibria. The structure of the bifurcation is quite intricate and would require further research. The possibility of selecting the equilibrium attained by controlling the payoffs using tolls or delays to incorporate the externality that each user imposes to the rest may be also of interest in this context. Eventually one might think of a feedback mechanism in the adaptive dynamics that would lead the system to a more desirable equilibrium from the point of view of the planner.

Appendices

A Discrete choice models

This short appendix reviews the basic facts about discrete choice theory used in previous chapters. For a more thorough treatment on this theory and its many applications in economics, we refer to Ben-Akiva & Lerman [16] and Train [75].

Consider an agent who faces a choice among a finite number of alternatives $i = 1, \ldots, n$, each one incurring a random cost $\tilde{x}_i = x_i + \epsilon_i$. Here x_i is the expected cost of of the ith alternative and the random term ϵ_i satisfies $\mathbb{E}(\epsilon_i) = 0$. Suppose that the agent observes the random variables $\tilde{x}_i$ and then chooses the alternative that yields the minimal cost: $\tilde{x}_i \leq \tilde{x}_j$ for $j = 1, \ldots, n$. The expected value of the minimal cost defines a map $\varphi \colon \mathbb{R}^n \to \mathbb{R}$,

$$\varphi(x) = \mathbb{E}\big(\min\{x_1 + \epsilon_1, \ldots, x_n + \epsilon_n\}\big). \tag{A.1}$$

Denote by $\mathcal{E}$ the class of maps that can be expressed in this form, where $\epsilon = (\epsilon_i)_{i=1}^n$ is a random vector with continuous distribution and $\mathbb{E}(\epsilon) = 0$. The next proposition summarizes the basic properties of the expected utility function [26, 28, 71, 78], and characterizes the choice probability of each alternative as the derivatives of φ.

Proposition A.1. *Every function $\varphi \in \mathcal{E}$ is concave and of class C^1 with $\varphi(x) \leq \min\{x_1, \ldots, x_n\}$, and we have*

$$\frac{\partial \varphi}{\partial x_i}(x) = \mathbb{P}\big(\tilde{x}_i \leq \tilde{x}_j,\ \forall j = 1, \ldots, n\big). \tag{A.2}$$

Proof. Let us denote $m(x) = \min\{x_1, \ldots, x_n\}$. The inequality $\varphi(x) \leq m(x)$ follows at once by taking expectation in the inequality $\min_{i=1\ldots n}\{x_i + \epsilon_i\} \leq x_j + \epsilon_j$. Let $F(\epsilon)$ be the joint distribution of $\epsilon = (\epsilon_1, \ldots, \epsilon_n)$ so that

$$\varphi(x) = \int_{\mathbb{R}^n} m(x + \epsilon)\, dF(\epsilon).$$

Since m is concave, the same holds for φ. To compute $\frac{\partial \varphi}{\partial x_i}$ we consider the differential quotient

$$\frac{\varphi(x + te_i) - \varphi(x)}{t} = \int_{\mathbb{R}^n} q_t(\epsilon)\, dF(\epsilon),$$

where $q_t(\epsilon) = [m(x+\epsilon+te_i) - m(x+\epsilon)]/t$. Denoting

$$A = \{\epsilon \in \mathbb{R}^n : x_i + \epsilon_i < x_j + \epsilon_j, \forall j \neq i\},$$
$$B = \{\epsilon \in \mathbb{R}^n : x_i + \epsilon_i \leq x_j + \epsilon_j, \forall j \neq i\},$$

it follows that $\lim_{t\downarrow 0^+} q_t(\epsilon) = 1_A(\epsilon)$ and $\lim_{t\uparrow 0^-} q_t(\epsilon) = 1_B(\epsilon)$. Since the convergence is monotone, we may use Lebesgue's theorem to deduce

$$\begin{aligned} D_i^+\varphi(x) &= \int_{\mathbb{R}^n} 1_A(\epsilon)\, dF(\epsilon) = \mathbb{P}(A), \\ D_i^-\varphi(x) &= \int_{\mathbb{R}^n} 1_B(\epsilon)\, dF(\epsilon) = \mathbb{P}(B), \end{aligned} \tag{A.3}$$

and since F is non-atomic we get $\mathbb{P}(A) = \mathbb{P}(B)$, so that the partial derivative $\frac{\partial\varphi}{\partial x_i}$ exists and satisfies (A.2). The C^1 character then follows since φ is concave. □

Example. The *Logit* model assumes that the ϵ_i's are independent Gumbel variables with parameter β, which gives the expected utility function

$$\varphi(x) = -\frac{1}{\beta}\ln\left(e^{-\beta x_1} + \cdots + e^{-\beta x_n}\right)$$

and the corresponding choice probabilities

$$\mathbb{P}(\tilde{x}_i \leq \tilde{x}_j, \forall j = 1, \ldots, n) = \frac{\exp(-\beta x_i)}{\sum_{j=1}^n \exp(-\beta x_j)}.$$

In the *Probit* model with normally distributed ϵ_i's there is no simple analytical expression for $\varphi(x)$ nor the choice probabilities.

We note that in the specification of the SUE and MTE models, all the relevant information was encapsulated in the functions φ_i^d, which are precisely of the form (A.1). Thus, we could take these functions as the primary modeling objects, without expliciting the random distributions that produced them. To this end, it is useful to have an analytic characterization of the class $\mathcal{E}$. The next result from [65] provides a complete characterization of this class.

Proposition A.2. *A function $\varphi\colon \mathbb{R}^n \to \mathbb{R}$ is in $\mathcal{E}$ if and only if the following hold:*

(a) *φ is C^1 and componentwise non-decreasing;*

(b) *$\varphi(x_1 + c, \ldots, x_n + c) = \varphi(x_1, \ldots, x_n) + c$;*

(c) *$\varphi(x) \to x_i$ when $x_j \to \infty$ for all $j \neq i$;*

(d) *for x_i fixed, the mapping $\frac{\partial\varphi}{\partial x_i}(x_1, \ldots, x_n)$ is a distribution with continuous density on the remaining variables.*

Proof. For $\varphi \in \mathcal{E}$, the properties (a)–(d) are direct consequences of (A.1) and (A.2) (for a proof of (a) and (b) the reader may also refer to [26, 28, 78]). To establish the converse, let us consider a random vector $\eta = (\eta_2, \ldots, \eta_n)$ with distribution function $F_\eta(x_2, \ldots, x_n) = \frac{\partial \varphi}{\partial x_1}(0, x_2, \ldots, x_n)$. We begin by noting that property (b) implies

$$\varphi(x) = x_1 - \int_a^{x_1} \left[1 - \frac{\partial \varphi}{\partial x_1}(y, x_2, \ldots, x_n)\right] dy + \varphi(0, x_2 - a, \ldots, x_n - a),$$

so that letting $a \to -\infty$ and using (c) we get

$$\varphi(x) = x_1 - \int_{-\infty}^{x_1} \left[1 - \frac{\partial \varphi}{\partial x_1}(y, x_2, \ldots, x_n)\right] dy.$$

On the other hand, setting $Y = \min\{x_2 - \eta_2, \ldots, x_n - \eta_n\}$ and using (b) we get

$$\begin{aligned} \frac{\partial \varphi}{\partial x_1}(y, x_2, \ldots, x_n) &= \frac{\partial \varphi}{\partial x_1}(0, x_2 - y, \ldots, x_n - y) \\ &= F_\eta(x_2 - y, \ldots, x_n - y) \\ &= \mathbb{P}(y \le Y), \end{aligned}$$

so that $\varphi(x) = x_1 - \int_{-\infty}^{x_1} F_Y(y)\, dy$, and then integration by parts allows to work out this expression as

$$\begin{aligned} \varphi(x) &= x_1 - \int_{-\infty}^{x_1} [x_1 - y]\, dF_Y(y) \\ &= x_1[1 - \mathbb{P}(Y \le x_1)] + \int_{-\infty}^{x_1} y\, dF_Y(y) \\ &= \int_{-\infty}^{\infty} \min\{x_1, y\}\, dF_Y(y), \end{aligned}$$

which means $\varphi(x) = \mathbb{E}(\min\{x_1, Y\}) = \mathbb{E}(\min\{x_1, x_2 - \eta_2, \ldots, x_n - \eta_n\})$. We may then conclude by taking $\epsilon_1 = 0$ and $\epsilon_i = -\eta_i$ for $i = 2, \ldots, n$. Notice that $\mathbb{E}(\epsilon) = 0$ follows from (c) and Lebesgue's theorem, while $\mathbb{P}(\epsilon = \alpha) = 0$ follows since φ is C^1. □

Remark. Condition (d) may be weakened to "$\frac{\partial \varphi}{\partial x_1}(0, x_2, \ldots, x_n)$ *is a continuous distribution on* $\mathbb{R}^{n-1}$".

B Stochastic approximation

This section provides a brief overview of stochastic approximations of differential equations. A detailed account can be found in the books by Duflo [32] and Kushner & Yin [50] (see also [12, 13, 15]). The specific material reviewed here is taken from the recent paper by Benaïm et al. [14], which extends the results from the setting of differential equations to differential inclusions.

B.1 Differential inclusions

Consider the differential inclusion

$$\frac{dx}{dt} \in F(x(t)) \tag{I}$$

where $F\colon \mathbb{R}^m \to 2^{\mathbb{R}^m}$ is a closed set-valued map with nonempty compact convex values, satisfying the growth condition $\sup_{z\in F(x)} \|z\| \le c(1+\|x\|)$ for some $c \ge 0$. A *solution* is an absolutely continous map $x\colon \mathbb{R} \to \mathbb{R}^m$ satisfying (I) for a.e. $t \in \mathbb{R}$. Let us also consider the following notions of approximate solutions:

PERTURBED SOLUTION: An absolutely continuous map $x\colon \mathbb{R}_+ \to \mathbb{R}^m$ is called a *perturbed solution* if there is a function $t \mapsto \delta(t) \ge 0$ with $\lim_{t\to\infty} \delta(t) = 0$ and a locally integrable map $t \mapsto u(t) \in \mathbb{R}^m$ with

$$\lim_{t\to\infty} \sup_{h\in[0,T]} \left\| \int_t^{t+h} u(s)\, ds \right\| = 0, \quad \forall T \ge 0,$$

such that the following holds for a.e. $t \ge 0$:

$$\frac{dx}{dt} \in F^{\delta(t)}(x(t)) + u(t),$$

where $F^\delta(x) = \{y \in \mathbb{R}^m \,:\, \text{there exists } z \in B(x,\delta) \text{ with } d(y, F(z)) \le \delta\}$.

DISCRETE APPROXIMATION: A sequence $\{x_n\}_{n\in\mathbb{N}} \subset \mathbb{R}^m$ is called a *discrete approximation* for (I) if

$$\frac{x_{n+1} - x_n}{\gamma_{n+1}} \in F(x_n) + u_{n+1}$$

with $\gamma_n > 0$, $\gamma_n \to 0$, $\sum \gamma_n = \infty$, and $u_n \in \mathbb{R}^m$. The sequence is called a *Robbins–Monro process* with respect to a filtration $\{\mathcal{F}_n\}_{n\in\mathbb{N}}$ on $(\Omega, \mathcal{F}, \mathbb{P})$ if γ_n is a deterministic sequence and u_n is a random vector which is $\mathcal{F}_n$-measurable with $\mathbb{E}(u_{n+1}|\mathcal{F}_n) = 0$.

Proposition B.1. *Let $\{x_n\}_{n\in\mathbb{N}}$ be a bounded discrete approximation. Let us denote $\tau_n = \sum_{i=1}^n \gamma_i$ and suppose that for all $T > 0$ we have*

$$\lim_{n\to\infty} \sup_k \left\{ \left\| \sum_{n+1}^{k} u_i \gamma_i \right\| \,:\, \tau_n < \tau_k \le \tau_n + T \right\} = 0. \tag{B.1}$$

Then the linearly interpolated process $t \mapsto w(t)$ defined by

$$w(t) = x_n + \frac{x_{n+1} - x_n}{\tau_{n+1} - \tau_n}(t - \tau_n), \quad \forall t \in [\tau_n, \tau_{n+1}]$$

is a perturbed solution for (I).

Proposition B.2. *Let $\{x_n\}_{n\in\mathbb{N}}$ be a Robbins–Monro discrete approximation and suppose that for some $q \in [2,\infty)$ we have*

$$\sum_n \gamma_n^{1+q/2} < \infty \text{ and } \{u_n\}_{n\in\mathbb{N}} \text{ is bounded in } L^q. \tag{H$_\text{q}$}$$

Then (B.1) *holds almost surely.*

B.2 ω-limit sets and attractors

Recall that the ω-limit set of a map $t \mapsto x(t)$ is the set of all its accumulation points as $t \to \infty$. The ω-limit of a sequence $\{x_n\}_{n\in\mathbb{N}}$ is defined similarly.

A compact set $A \subseteq \mathbb{R}^m$ is said to be *internally chain transitive* (ICT) for the dynamics (I) if for each pair $x, y \in A$, each $\epsilon > 0$, and all $T > 0$, there is a finite sequence of solutions $x_1(\cdot), \dots, x_n(\cdot)$ and times $t_1, \dots, t_n \in [T, \infty)$ such that

(a) $x_i(t) \in A$ for all $t \in [0, t_i]$;

(b) $\|x - x_1(0)\| < \epsilon$ and $\|x_n(t_n) - y\| < \epsilon$;

(c) $\|x_{i+1}(0) - x_i(t_i)\| < \epsilon$ for $i = 1, \dots, n-1$.

Theorem B.3.

(a) *If $x(\cdot)$ is a bounded perturbed solution, its ω-limit set is ICT.*

(b) *If $\{x_n\}_{n\in\mathbb{N}}$ is a bounded Robbins–Monro discrete approximation satisfying* (H$_\text{q}$) *for some $q \geq 2$, then almost surely its ω-limit set is ICT.*

Let $\Phi_t(x)$ be the set-valued dynamical system induced by (I), namely

$$\Phi_t(x) = \big\{x(t) \,:\, x(\cdot) \text{ solution of (I) with } x(0) = x\big\},$$

and define the ω-limit set of a point $x \in \mathbb{R}^m$ and a set $U \subseteq \mathbb{R}^m$ as

$$\omega_\Phi(x) = \bigcap_{t\geq 0} \overline{\Phi_{[t,\infty)}(x)}; \qquad \omega_\Phi(U) = \bigcap_{t\geq 0} \overline{\Phi_{[t,\infty)}(U)}.$$

A set $A \subseteq \mathbb{R}^m$ is called:

FORWARD PRECOMPACT if $\Phi_{[t,\infty)}(A)$ is bounded for some $t \geq 0$.

INVARIANT if for each $x \in A$ there is a solution of (I) with $x(0) = x$ and $x(t) \in A$ for all $t \in \mathbb{R}$.

ATTRACTING if it is compact and there exists a neighborhood $U \in \mathcal{N}_A$ such that for all $\epsilon > 0$ there is $t_\epsilon > 0$ with $\Phi_t(U) \subseteq A + B(0, \epsilon)$ for $t \geq t_\epsilon$.

ATTRACTOR if it is invariant and attracting.

ATTRACTOR FREE if it is invariant and contains no proper attractor for the restricted dynamics

$$\Phi_t^A(x) = \{x(t) \,:\, x(\cdot) \text{ solves (I) with } x(0) = x \text{ and } x(t) \in A \text{ for all } t \in \mathbb{R}\}.$$

Note: attractivity of Φ^A refers to neighborhoods in the trace topology of A.

Proposition B.4. *ICT's are invariant and attractor free.*

Proposition B.5. *A compact subset $A \subset \mathbb{R}^m$ is*

(a) *attracting iff there exists $U \in \mathcal{N}_A$ forward precompact with $\omega_\Phi(U) \subseteq A$;*

(b) *attractor iff there exists $U \in \mathcal{N}_A$ forward precompact with $\omega_\Phi(U) = A$;*

(c) *attractor if and only it is invariant, Lyapunov stable (for all $U \in \mathcal{N}_A$ there exists $V \in \mathcal{N}_A$ with $\Phi_t(V) \subseteq U$ for all $t \geq 0$), and its basin of attraction $B(A) = \{x \,:\, \omega_\Phi(x) \subseteq A\}$ is a neighborhood of A.*

Proposition B.6.

(a) *If A is attracting, L is invariant, and $\omega_\Phi(x) \subseteq A$ for some $x \in L$, then $L \subseteq A$.*

(b) *If A is attractor then $\Phi_t(A) \subseteq A$ for all $t \geq 0$.*

B.3 Lyapunov functions

Theorem B.7. *Let $\Lambda \subseteq \mathbb{R}^m$ be a compact set and U an open neighborhood of Λ which is forward invariant: $\Phi_t(U) \subseteq U$ for all $t \geq 0$. Let $V\colon U \to [0, \infty)$ be a continuous map such that $V(x) = 0$ for all $x \in \Lambda$, and $V(y) < V(x)$ for all $x \in U \setminus \Lambda$, $y \in \Phi_t(x)$, $t > 0$. Then Λ contains an attractor A with $U \subseteq B(A)$.*

Theorem B.8. *Let $\Lambda \subseteq \mathbb{R}^m$ and U an open neighborhood of Λ. Suppose that $V\colon U \to \mathbb{R}$ is a continuous Lyapunov function: $V(y) \leq V(x)$ for all $x \in U$, $y \in \Phi_t(x)$ and $t \geq 0$, with strict inequality if $x \notin \Lambda$. If $V(\Lambda)$ has empty interior then every ICT set A is contained in Λ and V is constant over A.*

Corollary B.9. *Under the assumptions of* Theorem B.8, *let $\{x_n\}_{n\in\mathbb{N}}$ be a bounded Robbins–Monro discrete approximation satisfying* $(\mathrm{H_q})$ *for some $q \geq 2$. Then x_n converges almost surely to Λ.*

Bibliography

[1] Ahuja R.K., Magnanti Th.L., Orlin J.B., *Network flows*, in: G.L. Nemhauser et al. (eds.), Handbooks in Operations Research and Management Science, Vol. 1 Optimization, 211–369, North-Holland, Amsterdam, 1989.

[2] Akamatsu T., *Cyclic flows, Markov processes and stochastic traffic assignment*, Transportation Research Part B **30(5)** (1996), 369–386.

[3] Akamatsu T., *Decomposition of path choice entropy in general transportation networks*, Transportation Science **31** (1997), 349–362.

[4] Akamatsu T., Matsumoto Y., *A stochastic network equilibrium model with elastic demand and its solution method*, JSCE Journal of Infrastructure Planning and Management **IV-10** (1989), 109–118.

[5] Alvarez F., Bolte J., Brahic O., *Hessian Riemannian gradient flows in convex programming*, SIAM J. Control Optim. **43(2)** (2004), 477–501.

[6] Auer P., Cesa-Bianchi N., Freund Y., Schapire R.E., *The non-stochastic multiarmed bandit problem*, SIAM J. Computing **32** (2002), 48–77.

[7] Arthur W.B., *On designing economic agents that behave like human agents*, J. Evolutionary Econ. **3** (1993), 1–22.

[8] Avinieri E., Prashker J., *The impact of travel time information on travellers' learning under uncertainty*, paper presented at the 10th International Conference on Travel Behaviour Research, Lucerne, 2003.

[9] Baillon J.-B., Cominetti R., *Markovian traffic equilibrium*, Mathematical Programming 111(1-2), Ser. B (2008), 35–36.

[10] Beggs A., *On the convergence of reinforcement learning*, J. Economic Theory **122** (2005), 1–36.

[11] Beckman M., McGuire C., Winsten C., *Studies in Economics of Transportation*, Yale University Press, New Haven, 1956.

[12] Benaïm M., *Dynamics of stochastic approximation algorithms*, in: Séminaire de Probabilités, Lecture Notes in Math. 1709, Springer, Berlin, 1–68, 1999.

[13] Benaïm M., Hirsch M.W., *Mixed equilibria and dynamical systems arising from fictitious play in perturbed games*, Games and Economic Behavior **29** (1999), 36–72.

[14] Benaïm M., Hofbauer J., Sorin S., *Stochastic approximations and differential inclusions*, SIAM J. Control Optim. **44** (2005), 328–348.

[15] Benaïm M., Hofbauer J., Sorin S., *Stochastic approximations and differential inclusions; Part II: Applications*, Mathematics of Operations Research **31(4)** (2006), 673–695.

[16] Ben-Akiva M.E., Lerman S.R., *Discrete Choice Analysis: Theory and Application to Travel Demand*, MIT Press, Cambridge, Massachussetts, 1985.

[17] Bell M.G.H., *Alternatives to Dial's Logit assignment algorithm*, Transportation Research **29B(4)** (1995), 287–296.

[18] Blackwell D., *An analog of the minmax theorem for vector payoffs*, Pacific J. Math. **6** (1956), 1–8.

[19] Börgers T., Sarin R., *Learning through reinforcement and replicator dynamics*, J. Economic Theory **77** (1997), 1–14.

[20] Borm P., Facchini G., Tijs S., Megen F.V., Voorneveld M., *Congestion games and potentials reconsidered*, International Game Theory Review **1** (2000), 283–299.

[21] Boulogne Th., Nonatomic strategic games and network applications, Ph.D. Thesis, Université Paris 6, 2004.

[22] Brown G., *Iterative solution of games by fictitious play*, in: Activity Analysis of Production and Allocation, Cowles Commission Monograph No. 13, John Wiley & Sons, Inc., New York, 374–376, 1951.

[23] Cantarella G., Cascetta E., *Dynamic processes and equilibrium in transportation networks: towards a unifying theory*, Transportation Science **29** (1995), 305–329.

[24] Cascetta E., *A stochastic process approach to the analysis of temporal dynamics in transportation networks*, Transportation Research **23B** (1989), 1–17.

[25] Cominetti R., Melo E., Sorin S., *A payoff-based learning procedure and its application to traffic games*, Games and Economic Behavior, **70** (2010), 71–83.

[26] Daganzo C.F., *Multinomial Probit: The Theory and its Applications to Demand Forecasting*, Academic Press, New York, 1979.

[27] Daganzo C., Sheffi Y., *On stochastic models of traffic assignment*, Transportation Science **11** (1977), 253–274.

[28] Daganzo C.F., *Unconstrained extremal formulation of some transportation equilibrium problems*, Transportation Science **16** (1982), 332–360.

[29] Davis G.A., Nihan N.L., *Large population approximations of a general stochastic traffic model*, Operations Research **41** (1993), 170–178.

[30] Dial R.B., *A probabilistic multipath traffic assignment model which obviates path enumeration*, Transportation Research **5** (1971), 83–111.

[31] Duffy J., Hopkins E., *Learning, information, and sorting in market entry games: theory and evidence*, Games and Economic Behavior **51** (2005), 31–62.

[32] Duflo M., *Algorithmes stochastiques*, Mathématiques & Applications 23, Springer-Verlag, Berlin, 1996.

[33] Erev I., Roth A.E., *Predicting how people play games: Reinforcement learning in experimental games with unique, mixed strategy equilibria*, American Economic Review **88** (1998), 848–881.

[34] Erev I., Rapoport A., *Coordination, "magic", and reinforcement learning in a market entry game*, Games and Economic Behavior **23** (1998), 146–175.

[35] Fisk C., *Some developments in equilibrium traffic assignment*, Transportation Research **14B** (1980), 243–255.

[36] Florian M., Hearn D., *Network equilibrium models and algorithms*, in: M.O. Hall et al. (eds.), Handbooks in Operations Research and Management Science, Vol. 8, Ch. 6, North-Holland, Amsterdam, 1995.

[37] Florian M., Hearn D., *Network equilibrium and pricing*, in: R.W. Hall (ed.), Handbook of Transportation Science, Kluwer, Norwell, 361–393, 1999.

[38] Foster D., Vohra R.V., *Calibrated learning and correlated equilibria*, Games and Economic Behavior **21** (1997), 40–55.

[39] Foster D., Vohra R.V., *Asymptotic calibration*, Biometrika **85** (1998), 379–390.

[40] Freund Y., Schapire R.E., *Adaptive game playing using multiplicative weights*, Games and Economic Behavior **29** (1999), 79–103.

[41] Friesz T.L., Berstein D., Mehta N.J., Tobin R.L., Ganjalizadeh S., *Day-to-day dynamic network disequilibria and idealized traveler information systems*, Operations Research **42** (1994), 1120–1136.

[42] Fudenberg D., Levine D.K., *The Theory of Learning in Games*, MIT Press, Cambridge, MA, 1998.

[43] Fukushima M., *On the dual approach to the traffic assignment problem*, Transportation Research Part B **18** (1984), 235–245.

[44] Hannan J., *Approximation to Bayes risk in repeated plays*, in: Contributions to the Theory of Games, Vol. 3, edited by M. Dresher, A.W. Tucker and P. Wolfe, Princeton Univ. Press, 1957, 97–139.

[45] Hart S., *Adaptive heuristics*, Econometrica **73** (2002), 1401–1430.

[46] Hart S., Mas-Colell A., *A reinforcement procedure leading to correlated equilibrium*, in: Economics Essays; A Festschrift for Werner Hildenbrand, edited by G. Debreu, W. Neuefeind and W. Trockel, Springer, Berlin, 2001, 181–200.

[47] Hazelton M.L., *Day-to-day variation in Markovian traffic assignment models*, Transportation Research B **36** (2002), 637–648.

[48] Hofbauer J., Sandholm W.H., *On the global convergence of stochastic fictitious play*, Econometrica **70** (2002), 2265–2294.

[49] Horowitz J., *The stability of stochastic equilibrium in a two-link transportation network*, Transportation Research Part B **18** (1984), 13–28.

[50] Kushner H.J., Yin G.G., *Stochastic Approximations Algorithms and Applications*, Applications of Mathematics 35, Springer-Verlag, New York, 1997.

[51] Larsson T., Liu Z., Patriksson M., *A dual scheme for traffic assignment problems*, Optimization **42(4)** (1997), 323–358.

[52] Leurent F.M., *Contribution to Logit assignment model*, Transportation Research Record **1493** (1996), 207–212.

[53] Laslier J.-F., Topol R., Walliser B., *A behavioral learning process in games*, Games and Economic Behavior **37** (2001), 340–366.

[54] Ljung L., *Analysis of recursive stochastic algorithms*, IEEE Trans. Automatic Control AC-22(4) (1977), 551–575.

[55] Maher M., *Algorithms for Logit-based stochastic user equilibrium assignment*, Transportation Research Part B **32** (1998), 539–549.

[56] Mastroenni G., *A Markov chain model for traffic equilibrium problems*, RAIRO Operations Research **36** (2002), 209–226.

[57] McKelvey R., Palfrey T., *Quantal response equilibria for normal form games*, Games and Economic Behavior **10** (1995), 6–38.

[58] Miyagi T., *On the formulation of a stochastic user equilibrium model consistent with random utility theory*, in: Proceedings of the 4th World Conference for Transportation Research, 1985, 1619–1635.

[59] Monderer D., Shapley L., *Potential games*, Games and Economic Behavior **14** (1996), 124–143.

[60] Pemantle R., *A survey of random processes with reinforcement*, Probability Surveys **4** (2007), 1–79.

[61] Posch M., *Cycling in a stochastic learning algorithm for normal form games*, J. Evol. Econ. **7** (1997), 193–207.

[62] Powell W.B., Sheffi Y., *The convergence of equilibrium algorithms with predetermined step sizes*, Transportation Science **16** (1989), 45–55.

[63] Robinson J., *An iterative method of solving a game*, Ann. of Math. **54** (1951), 296–301.

[64] Rosenthal R.W., *A class of games possessing pure-strategy Nash equilibria*, International Journal of Game Theory **2** (1973), 65–67.

[65] San Martín J., personal communication, 2003.

[66] Sandholm W., *Potential games with continuous player sets*, J. Economic Theory **97** (2001), 81–108.

[67] Sandholm W., *Evolutionary implementation and congestion pricing*, Review of Economic Studies **69** (2002), 667–689.

[68] Selten R., Chmura T., Pitz T., Kube S., Schreckenberg M., *Commuters route choice behaviour*, Games and Economic Behavior **58** (2007), 394–406.

[69] Selten R., Güth W., *Equilibrium point selection in a class of market entry games*, in: M. Diestler, E. Fürst and G. Schwadiauer (eds.), Games Economic Dynamics and Time Series Analysis, Physica-Verlag, Wien-Würzburg, 1982, 101–116.

[70] Sheffi Y., Powell W., *An algorithm for the equilibrium assignment with random link times*, Networks **12** (1982), 191–207.

[71] Sheffi Y., Daganzo C.F., *Another paradox of traffic flow*, Transportation Research **12** (1978), 43–46.

[72] Shor N.Z., *Minimization Methods for Non-differentiable Functions*, Springer Series in Computational Mathematics, Springer-Verlag, Berlin, 1985.

[73] Smith M., *The stability of a dynamic model of traffic assignment: an application of a method of Lyapunov*, Transportation Science **18** (1984), 245–252.

[74] Trahan M., *Probabilistic assignment: an algorithm*, Transportation Science **8** (1974), 311–320.

[75] Train K., *Discrete Choice Methods with Simulation*, Cambridge University Press, 2003.

[76] Wardrop J., *Some theoretical aspects of road traffic research*, Proceedings of the Institution of Civil Engineers Vol. 1 (1952), 325–378.

[77] Watling D.P., Hazelton M.L., *The dynamics and equilibria of day-to-day assignment models*, Networks and Spatial Economics **3** (2003), 349–370.

[78] Williams H.C.W.L., *On the formulation of travel demand models and economic evaluation measures of user benefit*, Environment and Planning A **9** (1977), 285–344.

[79] Young P., *Strategic Learning and Its Limits*, Oxford University Press, 2004.

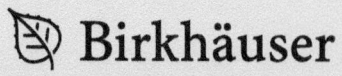

GPSR Compliance

The European Union's (EU) General Product Safety Regulation (GPSR) is a set of rules that requires consumer products to be safe and our obligations to ensure this.

If you have any concerns about our products, you can contact us on ProductSafety@springernature.com

In case Publisher is established outside the EU, the EU authorized representative is:

Springer Nature Customer Service Center GmbH
Europaplatz 3
69115 Heidelberg, Germany

Batch number: 09484943

Printed by Printforce, the Netherlands

Batch number: 09484943

Printed by Printforce, the Netherlands